Remote Sensing of the Terrestrial Hydrologic Cycle

Remote Sensing of the Terrestrial Hydrologic Cycle

Special Issue Editors

Qiuhong Tang
Youcun Qi
Zhihui Wang
Yun Pan

MDPI • Basel • Beijing • Wuhan • Barcelona • Belgrade • Manchester • Tokyo • Cluj • Tianjin

Special Issue Editors
Qiuhong Tang
Institute of Geographic Sciences and Natural Resources Research, Chinese Academy of Sciences
China

Youcun Qi
Chinese Academy of Sciences
China

Zhihui Wang
Yellow River Institute of Hydraulic Research, Yellow River Conservancy Commission
China

Yun Pan
Capital Normal University
China

Editorial Office
MDPI
St. Alban-Anlage 66
4052 Basel, Switzerland

This is a reprint of articles from the Special Issue published online in the open access journal *Remote Sensing* (ISSN 2072-4292) (available at: https://www.mdpi.com/journal/remotesensing/special_issues/hydro_cycle).

For citation purposes, cite each article independently as indicated on the article page online and as indicated below:

LastName, A.A.; LastName, B.B.; LastName, C.C. Article Title. *Journal Name* **Year**, *Article Number*, Page Range.

ISBN 978-3-03928-807-6 (Pbk)
ISBN 978-3-03928-808-3 (PDF)

Cover image courtesy of Yuanyuan Zhou.

Contents

About the Special Issue Editors

Qiuhong Tang (Dr.) received his Ph.D. degree from the University of Tokyo, Japan, in 2006. He has been a professor at the Institute of Geographic Sciences and Natural Resources Research, Chinese Academy of Sciences (CAS), since 2010. He is the Director of the Key Laboratory of Water Cycle and Related Land Surface Processes, CAS. He serves as an Associate Editor of the *Journal of Hydrometeorology, Wiley Interdisciplinary Reviews: Water*, and the *Journal of Geographical Sciences*.

Youcun Qi (Dr.) received his Ph.D. degree from the Nanjing University of Information Science and Technology, China, in 2011. He has been a professor at the Institute of Geographic Sciences and Natural Resources Research, Chinese Academy of Sciences (CAS), since 2017. He serves as an Associate Editor of the *Journal of Hydrometeorology* and Editor of *Advances in Atmospheric Sciences*.

Zhihui Wang (Dr.) is a senior engineer in the Yellow River Institute of Hydraulic Research, Yellow River Conservancy Commission, Ministry of Water Resources, Zhengzhou, China. He received his Ph.D. degree in cartography and geographical information system from the Institute of Remote Sensing and Digital Earth, Chinese Academy of Sciences (CAS), Beijing, China, in 2015. He finished post-doctoral research at the Institute of Geographic Sciences and Natural Resources Research, CAS. Dr. Wang's research interests are satellite and airborne image processing, vegetation dynamics mapping, land surface parameters inversion, hydrologic modeling, and water cycle response to a changing environment. Dr. Wang is also a recipient of the Young Talents Award from the China Association of Science and Technology.

Yun Pan (Dr.) received his Ph.D. degree from Hiroshima University, Japan, in 2009. He has been a professor at the Capital Normal University (CNU) since 2017 and serves as the Dean of the College of Resources Environment and Tourism, CNU.

remote sensing

MDPI

Editorial

Editorial for the Special Issue "Remote Sensing of the Terrestrial Hydrologic Cycle"

Qiuhong Tang [1,2,*], Youcun Qi [1], Zhihui Wang [3] and Yun Pan [4]

1 Key Laboratory of Water Cycle and Related Land Surface Processes, Institute of Geographical Sciences and Natural Resources Research, Chinese Academy of Sciences, Beijing 100101, China; youcun.qi@igsnrr.ac.cn
2 University of Chinese Academy of Sciences, Beijing 100049, China
3 Yellow River Institute of Hydraulic Research, Yellow River Conservancy Commission, Ministry of Water Resources, Zhengzhou 450003, China; wangzhihui@hky.yrcc.gov.cn
4 College of Resources Environment and Tourism, Capital Normal University, Beijing 100048, China; pan@cnu.edu.cn
* Correspondence: tangqh@igsnrr.ac.cn

Received: 16 March 2020; Accepted: 20 March 2020; Published: 23 March 2020

To address global water security issues, it is important to understand the evolving global water system and its natural and anthropogenic influencing factors [1]. The emerging remote sensing technologies enable relatively long-term consistent observations of the key variables of the terrestrial water cycle with unprecedented spatial coverage, providing precious data to broaden our understanding of the causes and consequences of change in the terrestrial water cycle [2]. The data-rich environment largely created by advances in remote sensing has boosted research in global change hydrology [3].

In this Special Issue, most studies used data acquired by various satellite and/or ground-based sensors to characterize the change in the terrestrial water cycle. The key hydrologic variables such as evapotranspiration (ET) and precipitation were derived and evaluated at varying spatial and temporal resolutions. Wang et al. [4] proposed a latent heat flux (LE) algorithm based on a novel microwave vegetation index (EDVI) instead of the optical vegetation index that was commonly used in previous studies. This algorithm was driven by multiple-sensor satellite products of vegetation water content index, solar radiation, and cloud properties, with some aid from a reanalysis dataset. The result showed that the performance of the proposed algorithm was very promising with correlation coefficients of 0.56–0.88 and a mean bias of 16% (23.0 W/m^2) for instantaneous LE estimations, and with correlations of 0.84–0.95 and a mean bias of less than 14.3% for monthly LE estimations based on in situ measurements at three Chinese Terrestrial Ecosystem Flux Research Network (ChinaFLUX) forest sites. Due to the insensitivity of microwave data to clouds, this algorithm shows great potential for estimating ET under both clear and cloudy skies on a global scale. Zhong et al. [5] estimated the ET of major exorheic catchments in China with a water balance method using Gravity Recovery and Climate Experiment (GRACE) data. Although the method has been demonstrated in previous studies [6], the authors addressed a potential approach for understanding the model performance associated with the errors in precipitation input. This study further highlights the unique ability of gravity satellites to capture the response of total water storage changes to both natural and anthropogenic causes, as well as its additional values of closing the water budget. Xu et al. [7] applied various statistical indicators to evaluate the main current satellite-based quantitative precipitation estimate (QPE) products over China. They found that the Chinese Fengyun (FY)-2G QPE and the Integrated Multi-satellitE Retrievals for Global Precipitation Measurement (IMERG) products performed significantly better than FY-2E QPE. The IMERG product agreed well with rain gauge data at the monthly scale, but it performed worse than FY-2G QPE at hourly and daily scales. The FY-2G QPE underestimated precipitation in summer, and the FY-2E QPE and IMERG QPE generally overestimated precipitation. The authors also found both FY and GPM-based products performed worse during 06:00 to 10:00 UTC than other periods. Their findings can provide valuable references for improving satellite-based QPE retrieval algorithms. Qiu et al. [8] evaluated the performance of three "radar-gauge" merging algorithms. Their results show that the radar-gauge integration method performs better than the others. quality of the blending QPE product is not only related to the blending algorithm, but also related to radar QPE and gauge observations. In order to further evaluate the merging QPE product, the authors applied the merging QPE product for flood forecasting and found that a higher quality of the merging products indicates a better agreement between the observed and the simulated runoff. Yang et al. [9] used a machine learning algorithm, K-nearest neighbor (KNN), to classify precipitation types. The authors used six Doppler radar data sources from China as training and classification samples, and used the 2A23 product of the Tropical Precipitation Measurement Mission (TRMM) to obtain the training labels and evaluate the classification performance. Three types of cases, namely the squall line, embedded convective and stratiform cases, were classified by KNN. The results show that the KNN method can accurately classify the location and area of stratiform and convective systems and suggest that the KNN method has great potential for classifying precipitation types.

Some studies used remote sensing techniques to detect anthropogenic impacts on the water cycle, providing valuable knowledge of the interrelations between humans and water. Zhang and Gao [10] combined the water surface area estimated from Moderate Resolution Imaging Spectroradiometer (MODIS) images and the reservoir Area–Elevation (A-H) relationship derived from the Digital Elevation Model (DEM) data collected by the Shuttle Radar Topography Mission (SRTM) to monitor water storage variation. The water storage variation data estimated from the proposed methodology cover reservoirs with 46.6% of the overall reservoir storage capacity in South Asia, providing valuable information for flood monitoring and water resource management in this region. Xu et al. [11] demonstrated an interesting attempt to distinguish irrigated fields from rainfed fields using higher-resolution images (30 m) from Landsat together with hydroclimatic data. It is a challenge to detect the difference in humid and sub-humid areas as these fields have similar land surface characteristics. Efforts have been made to enhance the contrast between neighboring rainfed and irrigated fields, and a machine learning method was adopted to generate maps of irrigated areas in southwestern Michigan. The success of mapping suggested that the subtle difference of land surface characteristics caused by water management practice is detectable from space. The maps also showed that the irrigated area in southwestern Michigan tripled during 2001–2016. Information on the change in irrigated areas would be highly relevant for water and food management. Hao et al. [12] developed a method to detect the irrigation signal (frequency, timing and area) based on multisource time-series data including soil moisture active passive (SMAP), MODIS-normalized difference vegetation index (NDVI) and evapotranspiration (ET), and precipitation from the meteorological stations. The detection result showed that irrigation signals can be effectively detected by removing the precipitation effect and setting the soil moisture change threshold with an overall accuracy of 77.08% in a typical crop-producing region in China. To solve the problem of the coarse resolution of SMAP pixels, a downscaling method was proposed by combing the winter wheat area extracted from MODIS NDVI, and the proposed method can indicate the true winter wheat irrigation timing, area and frequency with an 82.72% growth consistency in the surface water irrigation period. Zou et al. [13] measured the ET rate and quantified the cooling effects of urban hedges using the "three-temperature model and infrared remote sensing (3T+IR)"—a fetch-free and high-spatiotemporal-resolution method—in the urban area of Shenzhen in China. The study discovered that the "3T+IR" technique was a reasonable method for measuring the ET of urban hedges. The hedges could consume 68.44% and 60.81% of the net radiation through the latent heat of ET on a summer day, while their cooling rates for air temperature were 1.29 °C min^{-1} m^{-2} and 1.13 °C min^{-1} m^{-2}, respectively. In addition, urban hedges could also significantly cool the underlying surface, and the surface temperatures of the two hedges were 19 °C lower than that of the asphalt pavement on the summer day. The findings provide new insights in understanding the process of ET in urban hedges and the vegetation cooling effect in the urban environment.

A few studies leveraged remotely sensed hydrologic information to understand the causes and consequences of changes in the terrestrial water cycle. Liu et al. [14] investigated the transition characteristics of the dry–wet regime and vegetation dynamic responses over the Yarlung Zangbo River (YZR) Basin, using NDVI data from the Global Inventory Modeling and Mapping Studies (GIMMS)-NDVI3g dataset together with the Standardized Precipitation Evapotranspiration Index (SPEI) from the Noah land surface model simulations in the Global Land Data Assimilation System (GLDAS). The widely-used remote sensing datasets helped find that the spatiotemporal characteristics of the dry–wet regime exhibited a reversal phenomenon before and after 2000, and the soil water content was an important indicator to identify the dry–wet transition in the YZR basin. This provided another solid demonstration of the value of widely available remote-sensing datasets in helping us to better understand the hydrological responses to global changes in space and time. Yao et al. [15] investigated the impacts of land cover change on the hydrologic regime in Northwestern, Northern, and Northeastern China. The Global Land Surface Satellite (GLASS) leaf area index (LAI) data retrieved from the MODIS reflectance data (MOD-09A1) were used as a primary parameter to reflect land cover change in a hydrological model. The hydrological simulations showed that in relatively humid areas, urbanization increases runoff and the consequent flood risk, whereas in arid and semi-arid regions, the increase of greenness resulting from the Three-North Forest Shelterbelt (TNFS) ecological restoration program increased evapotranspiration and reduced runoff and soil moisture. The study suggested that land cover change would heighten the risk of dryland expansion and flooding more than climate change alone in the future, providing new insights into global change impacts on the terrestrial water cycle. Chen et al. [16] investigated the area dynamics of two closed lakes in the semi-arid areas of the Inner Mongolian Plateau by using Landsat images and found that it expanded in dry seasons and degraded in wet seasons. Such an unexpected phenomenon was believed to be related to the external groundwater recharge from the leakage of the fault zone in this area and provided valuable information for understanding the impact of human activities on lake shrinkage in this area. Although the deep groundwater recharge is being debated, this study provided an easy way of using satellite images to infer interactions between surface water and groundwater.

The articles published in this Special Issue cover a wide range of innovative remote sensing methods to meet the needs of hydrologic practices and remote sensing applications to detect the changes in the terrestrial water cycle and to understand the causes and consequences of the changes. It highlights the potential of remote sensing in addressing global water issues under a changing environment.

Acknowledgments: We would like to thank all the authors and reviewers who contributed to this Special Issue on "Remote Sensing of the Terrestrial Hydrologic Cycle". Q.T. was supported by the National Natural Science Foundation of China (41730645, 41790424), the Strategic Priority Research Program of Chinese Academy of Sciences (XDA20060402), and Newton Advanced Fellowship.

Conflicts of Interest: The authors declare no conflict of interest.

References

1. Tang, Q.; Oki, T. *Terrestrial Water Cycle and Climate Change: Natural and Human-Induced Impacts*; John Wiley & Sons: Hoboken, NJ, USA, 2016.
2. Tang, Q.; Gao, H.; Lu, H.; Lettenmaier, D.P. Remote sensing: Hydrology. *Prog. Phys. Geogr.* **2009**, *33*, 490–509. [CrossRef]
3. Tang, Q. Global change hydrology: Terrestrial water cycle and global change. *Sci. China Earth Sci.* **2020**, *63*, 459–462. [CrossRef]
4. Wang, Y.; Li, R.; Min, Q.; Zhang, M.; Yu, G.; Bergeron, Y. Estimation of vegetation latent heat flux over three forest sites in ChinaFLUX using satellite microwave vegetation water content index. *Remote Sens.* **2019**, *11*, 1359. [CrossRef]

5. Zhong, Y.; Zhong, M.; Mao, Y.; Ji, B. Evaluation of evapotranspiration for exorheic catchments of China during the GRACE era: From a water balance perspective. *Remote Sens.* **2020**, *12*, 511. [CrossRef]
6. Pan, Y.; Zhang, C.; Gong, H.; Yeh, P.J.F.; Shen, Y.; Guo, Y.; Huang, Z.; Li, X. Detection of human-induced evapotranspiration using GRACE satellite observations in the Haihe River basin of China. *Geophys. Res. Lett.* **2017**, *44*, 190–199. [CrossRef]
7. Xu, J.; Ma, Z.; Tang, G.; Ji, Q.; Min, X.; Wan, W.; Shi, Z. Quantitative evaluations and error source analysis of Fengyun-2-based and GPM-based precipitation products over mainland China in summer, 2018. *Remote Sens.* **2019**, *11*, 2992. [CrossRef]
8. Qiu, Q.; Liu, J.; Tian, J.; Jiao, Y.; Li, C.; Wang, W.; Yu, F. Evaluation of the Radar QPE and Rain Gauge Data Merging Methods in Northern China. *Remote Sens.* **2020**, *12*, 363. [CrossRef]
9. Yang, Z.; Liu, P.; Yang, Y. Convective/stratiform precipitation classification using ground-based Doppler radar data based on the K-nearest neighbor algorithm. *Remote Sens.* **2019**, *11*, 2277. [CrossRef]
10. Zhang, S.; Gao, H. Using the Digital Elevation Model (DEM) to improve the spatial coverage of the MODIS based reservoir monitoring network in South Asia. *Remote Sens.* **2020**, *12*, 745. [CrossRef]
11. Xu, T.; Deines, J.M.; Kendall, A.D.; Basso, B.; Hyndman, D.W. Addressing challenges for mapping irrigated fields in subhumid temperate regions by integrating remote sensing and hydroclimatic data. *Remote Sens.* **2019**, *11*, 370. [CrossRef]
12. Hao, Z.; Zhao, H.; Zhang, C.; Wang, H.; Jiang, Y. Detecting winter wheat irrigation signals using SMAP gridded soil moisture data. *Remote Sens.* **2019**, *11*, 2390. [CrossRef]
13. Zou, Z.; Yang, Y.; Qiu, G.Y. Quantifying the evapotranspiration rate and its cooling effects of urban hedges based on three-temperature model and infrared remote sensing. *Remote Sens.* **2019**, *11*, 202. [CrossRef]
14. Liu, L.; Niu, Q.; Heng, J.; Li, H.; Xu, Z. Transition characteristics of the dry-wet regime and vegetation dynamic responses over the Yarlung Zangbo river basin, southeast Qinghai-Tibet Plateau. *Remote Sens.* **2019**, *11*, 1254. [CrossRef]
15. Yao, Y.; Xie, X.; Meng, S.; Zhu, B.; Zhang, K.; Wang, Y. Extended dependence of the hydrological regime on the land cover change in the Three-North Region of China: An evaluation under future climate conditions. *Remote Sens.* **2019**, *11*, 81. [CrossRef]
16. Chen, J.; Lv, J.; Li, N.; Wang, Q.; Wang, J. External groundwater alleviates the degradation of closed lakes in semi-arid regions of China. *Remote Sens.* **2020**, *12*, 45. [CrossRef]

remote sensing

MDPI

Article

Estimation of Vegetation Latent Heat Flux over Three Forest Sites in ChinaFLUX using Satellite Microwave Vegetation Water Content Index

Yipu Wang [1], Rui Li [1,*], Qilong Min [2], Leiming Zhang [3], Guirui Yu [3] and Yves Bergeron [4]

1 School of Earth and Space Sciences, University of Science and Technology of China, Hefei 230026, China; wypustc@mail.ustc.edu.cn
2 Atmospheric Science Research Center, State University of New York, Albany, NY 12203, USA; qmin@albany.edu
3 Key Laboratory of Ecosystem Network Observation and Modeling, Institute of Geographic Sciences and Natural Resources Research, Chinese Academy of Sciences, Beijing 100101, China; zhanglm@igsnrr.ac.cn (L.Z.); yugr@igsnrr.ac.cn (G.Y.)
4 Institut de recherche sur les forêts, Université du Québec en Abitibi-Témiscamingue (UQAT), Rouyn-Noranda, QC J9X 5E4, Canada; yves.bergeron@uqat.ca
* Correspondence: rli7@ustc.edu.cn

Received: 18 March 2019; Accepted: 28 May 2019; Published: 6 June 2019

Abstract: Latent heat flux (LE) and the corresponding water vapor lost from the Earth's surface to the atmosphere, which is called Evapotranspiration (ET), is one of the key processes in the water cycle and energy balance of the global climate system. Satellite remote sensing is the only feasible technique to estimate LE over a large-scale region. While most of the previous satellite LE methods are based on the optical vegetation index (VI), here we propose a microwave-VI (EDVI) based LE algorithm which can work for both day and night time, and under clear or non-raining conditions. This algorithm is totally driven by multiple-sensor satellite products of vegetation water content index, solar radiation, and cloud properties, with some aid from a reanalysis dataset. The satellite inputs and the performance of this algorithm are validated with in situ measurements at three ChinaFLUX forest sites. Our results show that the selected satellite observations can indeed serve as the inputs for the purpose of estimating ET. The instantaneous estimations of LE (LE_{cal}) from this algorithm show strong positive temporal correlations with the in situ measured LE (LE_{obs}) with the correlation coefficients (R) of 0.56–0.88 in the study years. The mean bias is kept within 16.0% (23.0 W/m^2) across the three sites. At the monthly scale, the correlations between the retrieval and the in situ measurements are further improved to an R of 0.84–0.95 and the bias is less than 14.3%. The validation results also indicate that EDVI-based LE method can produce stable LE_{cal} under different cloudy skies with good accuracy. Being independent of any in situ measurements as inputs, this algorithm shows great potential for estimating ET under both clear and cloudy skies on a global scale for climate study.

Keywords: Microwave emissivity difference vegetation index (EDVI); evapotranspiration (ET); satellite remote sensing; cloudy sky; clouds and earth's radiation energy system (CERES); ChinaFLUX

1. Introduction

The process of latent heat flux (LE) and evapotranspiration (ET) is associated with the exchange of water and energy between the land surface and the atmosphere [1,2]. As a major component of the land surface energy budget, LE consumes about three-fifths of net radiation income, ranging from 48% to 88% based on different models [3,4]. Over 80% of terrestrial ET is from plant transpiration [5]. Globally, the ET from forests account for about 45% of the total ET [6]. Therefore, the accurate and real-time estimation of forest LE is critical to improve our understanding of water resource management, the hydrological cycle and the associated energy balance between the terrestrial ecosystem and the atmosphere.

The conventional ground-based flux tower methods (Bowen ratio and eddy covariance) can provide relatively accurate estimates of LE within a small footprint area (200–500 m^2 radius depending on the height of tower) around flux towers at local scales with an uncertainty of about 10%–30% [5]. However, the spatial representativeness of point LE estimates is limited to the regional scale. Large uncertainties will appear if point measurements are extended to regional scales because of the heterogeneity of land surfaces.

Satellite remote sensing has become the only feasible technique for scaling the point LE estimates from flux towers to mean LE over a large-scale area. Most of the current remote sensing approaches to estimate LE are based on optical vegetation indices (VIs) such as the normalized difference vegetation index (NDVI) and the enhanced vegetation index (EVI) [7]. Researchers utilize these optical VIs in order to estimate foliage density and to predict ET over a vegetated surface [8]. A number of satellite optical VI methods have been successfully proposed to estimate ET for multiple landscape types from the site to regional scale [9–16]. Among these methods, the resistance-based Penman-Monteith (PM) model is considered as the preferred method due to its precise physical mechanism. VIs are typically used for the estimation of canopy resistance (or conductance) in the PM model [8]. Previous studies have shown that optical VI-based resistance and LE estimation show good performances at the 8-day or 16-day time scale and under clear or less cloudy sky conditions [11,17–19]. However, since the optical VIs are retrieved based on the visible, near-infrared and shortwave-infrared bands [19] which are very sensitive to clouds and aerosols, their applicability under cloudy and overcast skies is limited.

When compared with land surface reflection at visible and infrared channels, the microwave radiative properties of land surface are less affected by atmospheric conditions such as cloud and aerosol, and the passive microwave signal is independent of solar radiation and is thus available at both the daytime and the nighttime. Microwave-based VIs thus have the advantage of being used under all sky conditions. Several microwave VIs have been proposed for monitoring vegetation, such as the frequency index (FI) [20], microwave polarization difference index (MPDI) [21] and microwave vegetation index (MVI) [22]. Most of those microwave vegetation indexes are derived under clear sky only and are designed for short vegetation. It is currently still a challenge to use these microwave VIs to quantify LE with a physical scheme due to their complexities.

Barraza et al. [23,24] used the combined empirical method of microwave and optical VIs to estimate conductance and LE over forest and savanna ecosystems. They illustrated the superiority of such combinations from multiple sensors for LE estimation. However, such an empirical method between canopy conductance and VIs [17–19,23,24] is insufficient to reflect the fast dynamics of conductance and LE affected by local environments during short-term periods (e.g., diurnal and daily scale). Furthermore, most of the validations of estimated LE are conducted at the 8-day (or 16-day) time scale and under clear or less cloudy sky since optical VIs are easily contaminated by cloud cover. In practice, the LE under cloudy sky differs significantly from that under clear sky. Thus, LE estimation under cloudy and overcast sky is equally important, particularly for the study of short-term interaction of vegetation-cloud. However, few studies have conducted ET or LE estimations under such sky conditions [25].

Min and Lin [26,27] proposed a new satellite remote-sensed microwave emissivity difference vegetation index (EDVI) at a mid-latitude forest (Harvard forest). The EDVI is defined as $(MLSE_{19}-MLSE_{37})/(MLSE_{19}+MLSE_{37})$, where $MLSE_{19}$ and $MLSE_{37}$ are the microwave land surface emissivities at 19 and 37 GHz. EDVI was found closely related to vegetation water content based on in-situ measured leave amount and their analysis on microwave radiative transfer [26,27]. In addition, a stronger relationship between the EDVI and evaporative fraction (EF) than that between NDVI and EF was found [26,27]. EDVI retrievals became available at the regional scale over a long time period [28]. Based on the above work, a quantitative algorithm was developed on the basis of satellite EDVI and local measurements in order to estimate EF and ET fluxes at the Harvard forest [29].

By using observations from multiple satellites, they even observed the diurnal variations of ET at the site as well. However, in their study, several in situ measurements were required as inputs, namely: net radiation, air temperature, photosynthetically active radiation, etc [29]. Consequently, the method of Li et al. [29] was limited to sites where the above inputs are measured.

For most applications in modeling and climate study, ET and LE estimation at large scale and over a long term is required [30–32]. A satellite-based ET retrieval algorithm independent of in situ measurements is thus required.

With the rapid progression of remote sensing technology, an increasing number of geophysical parameters are available from satellites. It is promising to develop an LE algorithm independent of in situ measurements that is driven only by satellite observations and reanalysis datasets. In this study, we tested this hypothesis and applicability of the LE method over dense vegetation in combination with the EDVI from advanced microwave scanning radiometer for EOS (AMSR-E), the net radiation flux from the clouds and earth's radiation energy system (CERES), the vegetation fraction information from moderate-resolution imaging spectroradiometer (MODIS) and the associated meteorology parameters from the reanalysis dataset of European Centre for Medium-Range Weather Forecasts (ECMWF). The results of this method were validated against the in situ measurements at selected ChinaFLUX forests sites.

In ChinaFLUX, part of a "network of regional networks" (FLUXNET), the eddy covariance (EC) technique was used to measure the H_2O, CO_2, and heat fluxes between the atmosphere and the ecosystems in China [33]. Because the concept of EDVI was originally developed in forest, we selected three typical forest ecosystems for the validation study. These sites are the Dinghushan (DHS) subtropical evergreen broad-leaved forest, Qianyanzhou (QYZ) subtropical plantation forest, and Changbaishan (CBS) temperate deciduous mixed forest. More information is provided in Section 2.1.

2. Data and Method

2.1. Site Descriptions

The three selected forest towers in this study are shown in Figure 1. Three years of in-situ measurements (2003–2005) at these sites are available in this study. And at each site, we split the measurements into different time periods for calibration and validation study. These sites are in different latitude zones and cover a wide range of temperature and precipitation. The annual mean precipitation is 1956 mm, 1485 mm and 695 mm, while the annual mean temperature is 21 °C, 17.9 °C and 3.6 °C at DHS, QYZ and CBS, respectively. The terrain is also complex among the forest sites: DHS is on a steep 30° slope, QYZ is on slightly choppy terrain, and CBS is on flat terrain [34]. Previous studies have indicated that forest sites of ChinaFLUX network achieve 57%–73% energy balance closure during the daytime [33,35]. More information about these sites are provided in Table 1. ChinaFLUX data in this study is available from the website: http://159.226.110.139/pingtai/LoginRe/opendata.jsp.

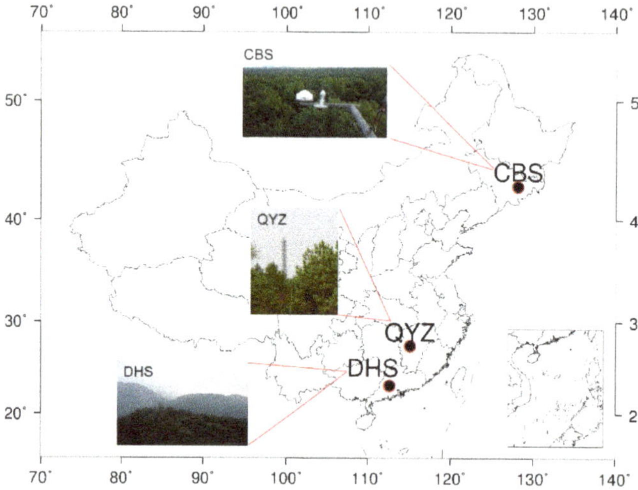

Figure 1. Locations of three ChinaFLUX forests sites. More information regarding these pictures is available at: http://www.chinaflux.org/.

Table 1. The background of three ChinaFLUX forest sites [33]. Precipitation and temperature are annual mean values (http://www.chinaflux.org/).

Sites and Period	Locations and Altitude	Precipitation and Temperature	Vegetation	Canopy Height	Measuring Height	LAI
Dinghushan (DHS) 2003–2005	112°34′E,23°10′N; 300 m	1956 mm, 21 °C	subtropical evergreen broad-leaved forest	20 m	27 m	4
Qianyanzhou (QYZ) 2003–2005	115°03′E,26°44′N; 102 m	1485 mm, 17.9 °C	subtropical monsoon plantation forest	12 m	39.6 m	3.5
Changbaishan (CBS) 2003–2005	128°05′E,42°24′N; 738 m	695 mm, 3.6 °C	Temperate deciduous broad-leaved mixed forest	26 m	40 m	6.1

Carbon and water fluxes of the three forests are measured using open-path eddy covariance (EC) system. The system consists of an open-path infrared gas analyser (Li-7500, Licor Inc., Lincoln, NB, USA) and a 3-D sonic anemometer (CSAT3, Campbell Scientific Ltd., USA). The signals of fluxes with 10Hz sampling frequency are recorded by system and 30-min averaged fluxes are derived from black averaging within 30-min step. Other meteorological variables are measured simultaneously with 30-min temporal resolution. Solar radiation is measured using radiometers (Model CM11 and Model CNR-1, Kipp & Zonnen, Delft, Netherlands). Photosynthetic active radiation (PAR) is measured using a quantum sensor (Model LI190SB, LICOR Inc.), air temperature (Ta) is measured using shielded and aspirated probes (Model HMP45C, Campbell Scientific Inc.). Precipitation is recorded using a rain gauge above the canopy (Model 52203, Rm Young, Traverse City, MI, USA). More detailed information can be found in previous studies [33,34,36] and the references therein.

The 30-min in situ measurements of LE (LE_{obs}), solar radiation, photosynthetic active radiation (PAR), air temperature (Ta), and precipitation from 2003 to 2005 are used for validation in this study. To compare with satellite observations and retrievals obtained around 13:30 local time, the averaged in-situ measurements from 12:30–14:30 are used. Temperature is a key factor affecting vegetation activity at higher latitudes. Low Ta will induce weak metabolism and suppress the LE of vegetation.

Therefore, the days with midday Ta lower than 0 degrees are excluded from the validation study at CBS. This makes no impact on the samples in DHS and QYZ.

2.2. Satellite Inputs of EDVI-based LE Method

Table 2 provides the basic information of the employed satellite products and reanalysis datasets and variables in this study. EDVI plays a fundamental role in the current LE algorithm as it determines the spatial and temporal variations of vegetation resistance to evapotranspiration process. The satellite EDVI retrievals during 2003-2005 over China were derived from a similar method to Min et al. [28] using multiple channel microwave measurements from the AMSR-E on the Aqua satellite with a spatial resolution of ~20 km at local time 13:30 each day. It should note that the instantaneous retrievals of EDVI are conducted under no rain conditions which are identified by AMSR-E rain rate/type product [28]. Since snow on the ground could significantly affect the value of EDVI, at the CBS site we only retrieve LE during the growing season from April to October, based on the study of Li et al. [37].

Table 2. Multiple data sources used in EDVI-based LE method.

Variables	Units	For	Datasets	Type	Resolution
EDVI	-	f3(VPD)f4(Ψ)f5(CO2), Minimal canopy resistance (rc_{min})	EDVI	Satellite	Daily, ~20 km
Downward shortwave radiation (dSW)	W/m^2	Photosynthetically active radiation (PAR), f2(PAR)	CERES SSF	Satellite	Daily, ~20 km
Net shortwave radiation (nSW)	W/m^2	Net radiation (Rn)	CERES SSF	Satellite	Daily, ~20 km
Net longwave radiation (nLW)	W/m^2	Net radiation (Rn)	CERES SSF	Satellite	Daily, ~20 km
NDVI	-	Vegetation fractional coverage (VFC), Ground heat flux (G),	MYD13C1	Satellite	16 day, 0.05°
2 m temperature (t2m)	K	Slope of saturated vapor pressure (Δ), f1(Ta)	ERA-20C	Reanalysis	Daily, 0.125°
Wind speed at 10 and 100 m (U10, U100)	m/s	Aerodynamic resistance (ra)	ERA-20C	Reanalysis	Daily, 0.125°

The product of single scanner footprint toa/surface fluxes and clouds (SSF) of CERES provides the estimates of instantaneous daytime net and downward shortwave (nSW and dSW) and longwave (nLW and dLW) radiation fluxes at surface at ~20 km resolution (Table 2) [38]. Those fluxes are estimated based on the NASA Langley parameterized shortwave/longwave algorithm (LPSA/LPLA) methods [39,40]. The CERES SSF datasets are available at https://eosweb.larc.nasa.gov/project/ceres/ssf_aqua-fm3_ed4a_table.

NDVI, derived from the observations of MODIS on the Aqua satellite, is also an optical VI which is highly correlated to the green foliage of vegetation. We use NDVI to estimate the vegetation fractional coverage. To do this, 16-day MODIS NDVI product (MYD13C1) with 0.05° resolution was used. To estimate the temporal variation of vegetation fraction, the 16-day NDVI was further linearly interpolated to achieve daily NDVI. MYD13C1 is available from the Land Processes Distributed Active Archive Center (LPDAAC; https://lpdaac.usgs.gov/).

ERA-20C reanalysis dataset of ECMWF provides estimations of air temperature at 2 m (t2m) and wind at 10 and 100 m (U_{10m}, U_{10m}) required by the ET algorithm. ERA-20C has a temporal resolution of 3 h and 0.125° × 0.125° spatial resolution (Table 2), and the values at 6:00 UTC (14:00 local time in China) were used in this study. The dataset is available on the ECMWF public datasets website (http://apps.ecmwf.int/datasets/).

2.3. Description of EDVI-based LE Algorithm

The primary driving force of LE is the available energy at the surface: the received net radiation Rn minus the energy transport to ground (G). An evaporative fraction (EF) is used to define the ratio of ET to the total available energy (Rn-G). This idea is often used to estimate instantaneous ET at the time of a satellite overpass [41–45].

In forest ecosystems, transpiration generally dominates evapotranspiration and accounts for 80-90% of the ET amount [46,47]. We therefore scale the available energy in given satellite footprint to vegetation part (with the subscript veg) by the vegetation fraction coverage (VFC) and only take into account the ET from vegetation using the following formula.

$$ET_{veg} = EF_{veg} \cdot (Rn - G)_{veg} = EF_{veg} \cdot (Rn - G) \cdot VFC \quad (1)$$

2.3.1. Estimation of Available Energy for Vegetation

Radiation fluxes under all sky conditions are required in our algorithm. CERES SSF directly provides the all-sky surface net shortwave radiation (nSW) and surface long-wave radiation (nLW) (Table 2). Therefore, we use the sum of nSW and nLW as the input of Rn in our ET algorithm. Validations of CERES surface radiation flux with in-situ measurements are conducted in results.

Surface ground heat flux can be estimated in combination with vegetation cover and net radiation. We use the model of Su [48] to calculate G, such that:

$$G = Rn\left[f_{veg} + (1 - VFC)\left(f_{soil} - f_{veg}\right)\right] \quad (2)$$

where f_{veg} (set as 0.05) and f_{soil} (set as 0.315) are the ratios of G to Rn over full vegetation and bare soil, respectively [48]. VFC is the vegetation fractional coverage which can be obtained based on NDVI according to the method of Gutman and Ignatov [49]:

$$VFC = \frac{NDVI - NDVI_0}{NDVI_\infty - NDVI_0} \quad (3)$$

where $NDVI_\infty$ and $NDVI_0$ are the NDVI of full vegetation (i.e., when VFC = 1) and bare soil (i.e., when VFC = 0), respectively. The values of $NDVI_\infty$ depend on different vegetation types, but they are relatively stable over forest types [50]. Values of $NDVI_0$ have very small variations when VFC = 0 [51]. Following Li and Zhang [50] and Zeng et al. [51], we adopt 0.90 and 0.1 as the values of $NDVI_\infty$ and $NDVI_0$, respectively. When the calculated VFC is less than 0, VFC is set to 0, and when the calculated VFC is larger than 1, VFC is set to 1. In this equation, the interpolated daily NDVI from 16-day values is used to calculate daily VFC under the assumption that vegetation foliage or greenness changes slowly and linearly within each 16-day period. We provide the variations of 16-day NDVI in Figure 2 in Section 3.1.

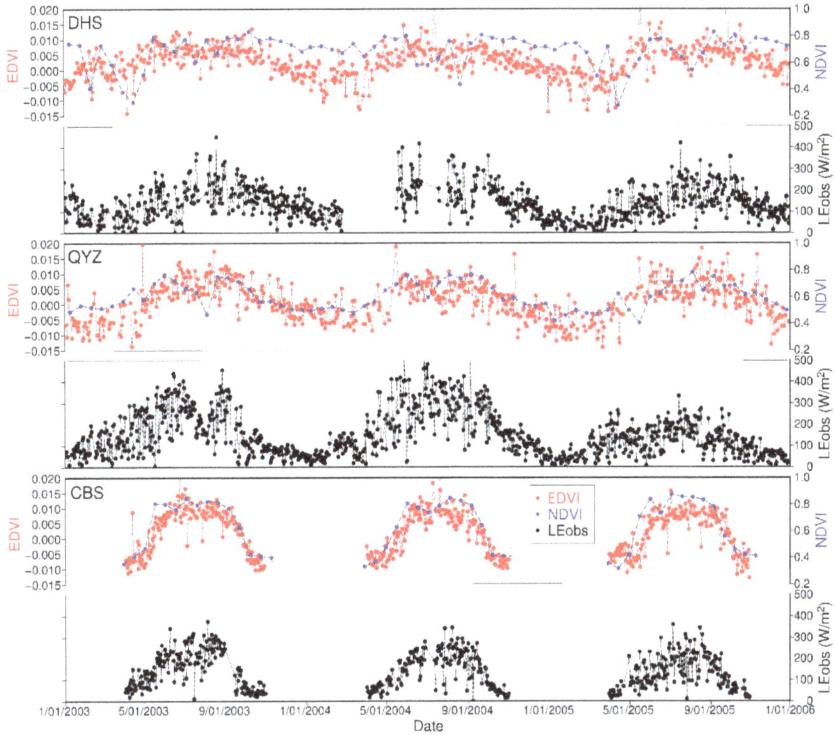

Figure 2. Time series of EDVI, in-situ measured LE (averaged over 2 h around the satellite overpass), and 16-day NDVI from 2003 to 2005 at three forest sites of ChinaFLUX. QYZ forest site had a serious imbalance of energy in 2005.

2.3.2. Estimation of Evaporative Fraction of Forest

The estimation of EF_{veg} is from the same method of Nishida et al. [52] and Li et al. [29]:

$$EF_{veg} = \frac{\alpha\Delta}{\Delta + \gamma(1 + r_c/2r_a)} \tag{4}$$

where α is the Priestley-Taylor's parameter (1.26), Δ is the slope of saturated vapor pressure at air temperature (hPa/K), γ is the psychometric constant (Pa/K) (66.5 Pa/K), r_a is the aerodynamic resistance (s/m), and r_c is the canopy resistance (s/m) which is highly related to EDVI in this study.

We calculate Δ based on the formula in study of Murray [53]:

$$\Delta = \frac{26{,}297.76}{(Ta - 29.65)^2}\exp\left(\frac{17.67(Ta - 273.15)}{Ta - 29.65}\right) \tag{5}$$

where Ta is the air temperature at 2 m (t2m) from ERA-20C.

The aerodynamic resistance r_a is determined by wind. Kondo [54,55] proposed two empirical formulas for forest canopy and grasslands. The formulas have been used for the estimation of ET in other studies [52,56]. We thus follow their study and use equation (6) to calculate the r_a of forests:

$$r_a = \begin{cases} \frac{1}{0.008U_{50m}} & \text{forest types} \\ \frac{1}{0.003U_{1m}} & \text{grass and crops} \end{cases} \tag{6}$$

where U_{50m} is the wind speed at 50 m (ms^{-1}) which is calculated by averaging the mean values of 10-m wind (U_{10m}) and 100-m wind (U_{100m}) from ERA-20C.

The canopy resistance r_c can be parameterized by the use of Jarvis-type equation.

$$r_c = \left[\frac{f_1(Ta)f_2(PAR)f_3(VPD)f_4(\Psi)f_5(CO_2)}{r_{cmin}} + \frac{1}{r_{cuticle}}\right]^{-1} \quad (7)$$

where r_c has two components: the stomatal resistance ($r_{stomatal}$) and the cuticle resistance ($r_{cuticle}$). $r_{cuticle}$ was set to be 10^5 ms^{-1} based on other studies [11].

The stomatal resistance $r_{stomatal}$ consists of five stress functions and the minimum canopy resistance (r_{cmin}). The five stress functions are used to quantify the impacts on r_c imposed by environmental conditions. The functions of air temperature ($f_1(Ta)$) and photosynthetically active radiation ($f_2(PAR)$) were adopted from Jarvis [57] and Kosugi [58], respectively. The day-to-day variation of EDVI (dEDVI) represents the response of canopy leaves to environmental factors such as water vapor pressure deficit (VPD), water potential (Ψ) and ambient carbon dioxide concentration (CO_2). In this study, we parameterize their stress functions (i.e., $f_3f_4f_5$) as a whole by using their fair correlation with dEDVI based on Li et al. [29]:

$$f_3(VPD)f_4(\Psi)f_5(CO_2) = [a - b \times dEDVI]^{-1} \quad (8)$$

where a and b are the coefficients. It is worth noting that the accurate determination of a and b require the field measurements of stomatal and canopy resistance. Li et al. [29] used the field measurements of resistances and developed the relationship at forest. Since such field measurements are unavailable in this study, we thus follow their study and adopt 1.186 and 105.755 as the values of a and b for these forests.

To describe the mean seasonal variation of EDVI associated with the different stages in the growing season of the forest, we define the normalized EDVI (^{N}EDVI) as (EDVI-$EDVI_{min}$)/($EDVI_{max}$-$EDVI_{min}$), where $EDVI_{min}$ and $EDVI_{max}$ are respectively the minimum and maximum EDVI values during growing seasons at each site [27,29]. The seasonal variation of EDVI (i.e., ^{N}EDVI) represents the slow change of VWC which is highly correlated to the vegetation resistance of LE. ^{N}EDVI is thus used to determine r_{cmin} which is the key parameter for the estimation of $r_{stomatal}$. Li et al. [29] found that there was a general anti-correlation between ^{N}EDVI and canopy resistance. The minimum canopy resistance can be parameterized as:

$$r_{cmin} = r_{cmin0}/{}^{N}EDVI \quad (9)$$

The canopy resistance will decrease when the plant leaves develop. In the case of Li et al. [29], the measurements of canopy resistance at Harvard forest were available to determine the r_{cmin0}. Based on the field studies of Kelliher et al. [59], the maximum vegetation conductance ranges from 13.0 mm/s (tropical rainforest) to 32.5 mm/s (cereals), while the corresponding r_{cmin} (G_{smax}^{-1}) ranges from 76.9 s/m to 30.8 s/m. On the basis of this, r_{cmin0} was set to be a reference value, 50 s/m, for all three forest types in our study.

3. Results

Since our objective of this study is to improve the EDVI-based LE method driven by satellite and reanalysis data under all sky conditions, the inputs of the method are crucial and were first investigated in this section. Section 3.1 shows the consistency between LE_{obs} and two vegetation indices. Section 3.2 provides the validation of accuracy of satellite and reanalysis data. The estimated instantaneous LE (LE_{cal}) under all skies are validate in Section 3.3. Then the accuracy of LE_{cal} under different cloudy sky conditions are further investigated in Section 3.4.

3.1. Time Series of EDVI, NDVI and In-Situ Measured LE

The time series of midday in situ measured LE_{obs}, satellite-based EDVI and NDVI at the three forest sites in this study are shown in Figure 2. The EDVI and NDVI show the consistent seasonal cycles with the LE_{obs} at all forests (maximum in summer and minimum in winter). EDVI vary from -0.015 to 0.02 during growing seasons when NDVI vary from 0.3 to 0.9. The LE_{obs} of forest ecosystems can have significant variations during the short-term period due to environmental and weather effects (e.g., precipitation and solar radiation). Daily EDVI can indicate such fast changes of vegetation status and show the similar day-to-day variations to the LE_{obs} at all sites (Figure 2), which suggests that EDVI can be potentially used to retrieve LE over the three forests. As an indicator of vegetation foliage, 16-day NDVI are found to have the small and relatively slow changes within each 16-day interval, suggesting that the use of interpolated daily NDVI from 16-day NDVI for the calculation of VFC may introduce small errors.

3.2. Validation of Satellite Radiation Flux and Air Temperature Inputs

Studies have been conducted to validate the CERES surface fluxes over different surface types (i.e., island, coastal, polar, continental, and desert) and have found good accuracies of the data [38,60]. In this study, since there is no in-situ measurements of net radiation (Rn) available, we therefore conduct the validation of downward shortwave radiation fluxes (dSW) between satellite observations and in-situ measurements instead.

Figure 3 shows the comparison of all-sky CERES dSW and in situ measured downward solar radiation flux at midday from 2003 to 2005. Over all forest sites, CERES dSW agrees well with the in-situ measured solar radiation, with high positive correlation coefficients (R) of 0.81, 0.95, and 0.90 at DHS, QYZ, and CBS, respectively. The relative biases of all sites are less than 9%. Because dSW is the dominant term of Rn, we therefore conclude that CERES SSF estimated Rn is suitable for estimating the real net radiation for calculating ET.

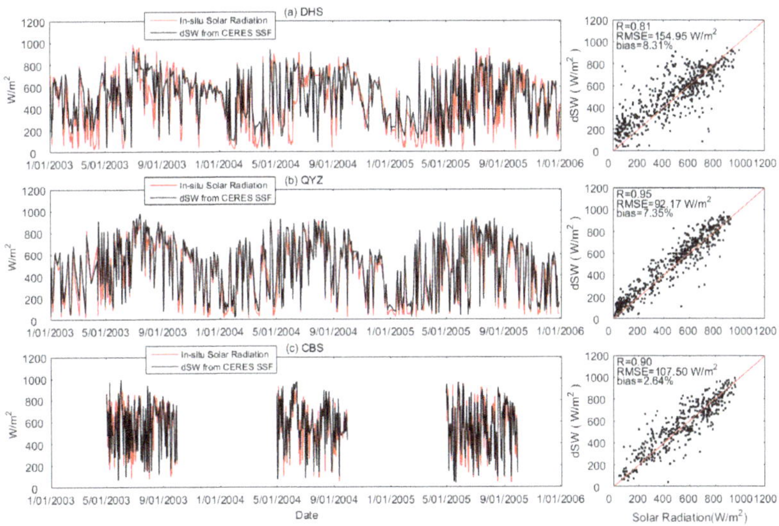

Figure 3. Time series and scatter plots of in-situ measured solar radiation (averaged over 2 h around the satellite overpass) and matched surface downward shortwave radiative fluxes (dSW) from CERES SSF under all-sky conditions.

Photosynthetically active radiation (PAR) is required in the estimation of EDVI-based canopy resistance. Studies have reported that PAR is linearly related to incident solar radiation [61–63]. In our study, all-sky CERES dSW is used to estimate PAR by multiplying it by a constant of 1.70 μmol/W. The results in Figure 4 indicate these estimations agree well with in-situ observations across three sites. Correlation coefficients strongly vary from 0.81 to 0.95 with a small bias of no more than 11%.

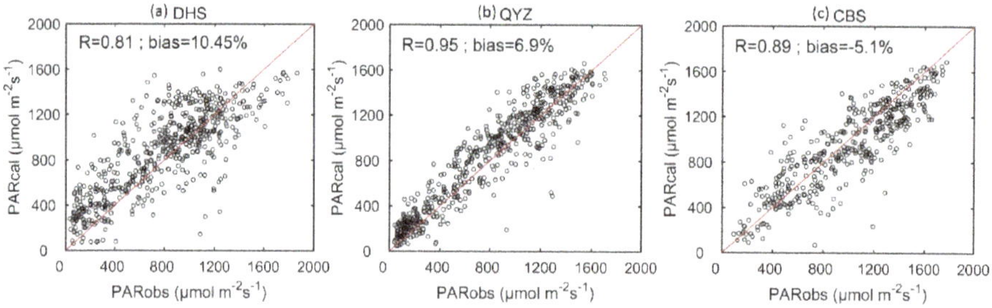

Figure 4. Comparisons between in-situ measured PAR and estimated PAR (dSW*1.70). dSW is surface downward shortwave radiative fluxes from CERES SSF in all sky conditions.

Air temperature is an important parameter for estimating forest ET. We compare the 2-m air temperature (t2m) from ERA-20C (t2m-ECMWF) and from NCEP FNL (t2m-NCEP) with in-situ measurements. As shown in Figure 5, both t2m-ECMWF and t2m-NCEP have a high correlation with in-situ measured temperature. However, t2m-NCEP estimations are systematically lower by 3–5 K than in-situ measurements. In contrast, t2m-ECMWF are in better agreement with in situ measurements with a smaller bias (0.1–2.7 K). The overall biases of t2m-ECMWF for all sites are within 3K. Therefore, we determine to use ERA-20C rather than NCEP FNL to provide complementary parameters to the satellite observations.

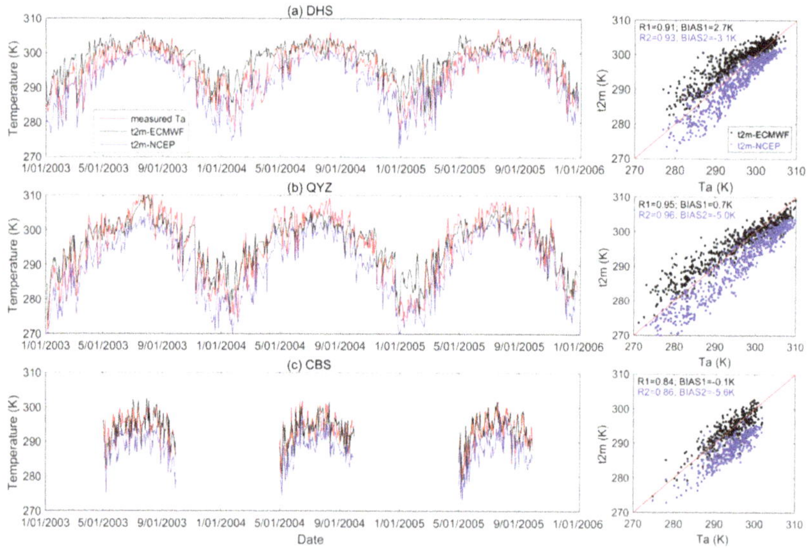

Figure 5. Time series and scatter plots of in-situ measured air temperature (Ta), 2-meter temperature from ECMWF reanalysis (t2m-ECMWF) and 2-meter temperature from NCEP reanalysis (t2m-NCEP). In-situ measurements at DHS, QYZ, and CBS are at 20 m, 23 m, and 26 m, respectively. In scatterplots, t2m-ECMWF and t2m-NCEP are marked by black and blue color, respectively. R1 represents the correlation coefficient (R) of in-situ Ta and t2m-ECMWF. R2 represents the R of in-situ Ta and t2m-NCEP. BIAS1 represents the difference between mean t2m-ECMWF and in-situ Ta. BIAS2 represents the difference between mean t2m-NCEP and in-situ Ta.

3.3. The EDVI-Based LE Estimation

The time series of EDVI-based LE_{cal}, in situ LE_{obs}, and their differences at the three forest sites are shown in Figure 6. Statistic results are shown in Table 3. Generally, LE_{cal} has the capability to capture the seasonality of forest LE_{obs} from 2003 to 2005 correctly. Both of them reach maximums in the mature stage of growing seasons in mid-summer due to the ample water and solar radiation for evapotranspiration (Figure 6). There is a strong correlation between LE_{cal} and LE_{obs} with the overall R being 0.56–0.88. In terms of magnitude, $LE_{cal,}$ ranging from 0 to 500 Wm^{-2} at DHS and QYZ in southern China, and from 0 to 400 Wm^{-2} at CBS in northeastern China, matches LE_{obs} well. Large discrepancies occur at QYZ in 2005 due to the serious imbalance of energy, which leads to a significant underestimation of in situ LE_{obs} [64,65] and should be responsible for the large bias in this year (Table 3). Because of this, we exclude the result of 2005 in QYZ in the following discussion.

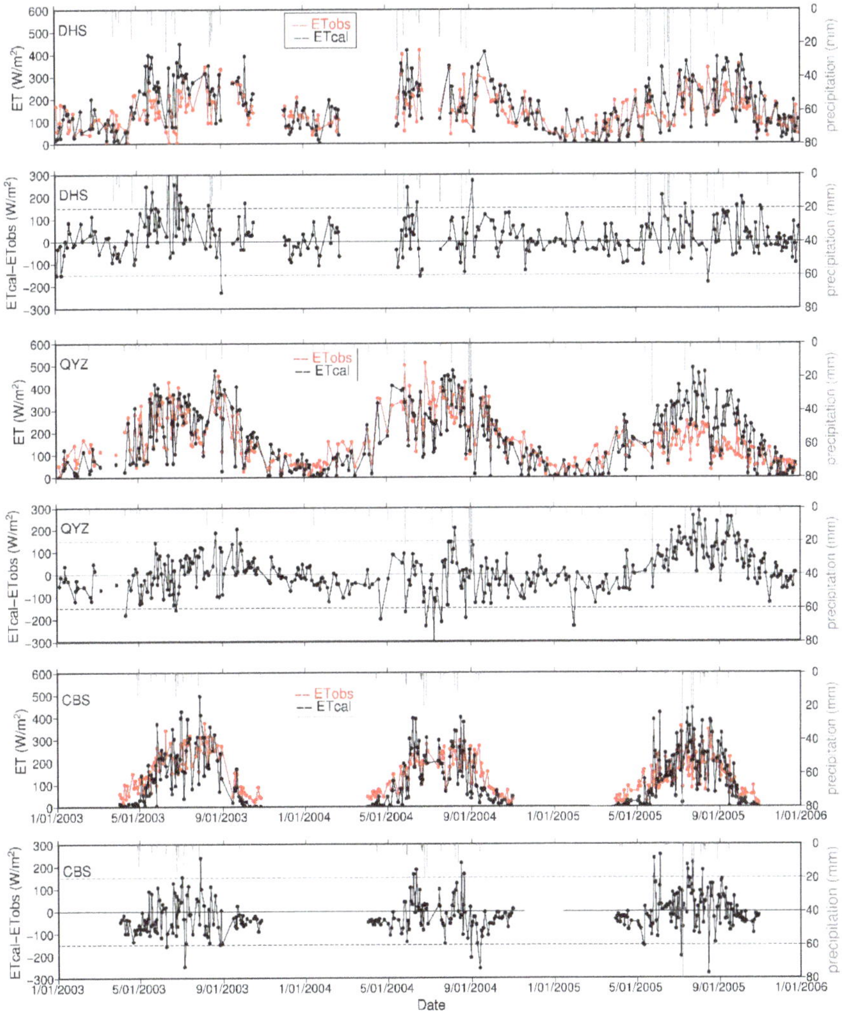

Figure 6. Time series of in-situ LE (LE_{obs}), estimated LE (LE_{cal}), and their differences at three forest sites from 2003 to 2005. LE_{obs} was averaged over 2 h around the satellite overpass. The dashed lines are ± 150 Wm^{-2}.

Table 3. Summary of comprehensive metrics for the results of three forest sites. $\overline{LE_{cal}}$ is the mean value of all estimations. $\overline{LE_{obs}}$ is the mean value of all in-situ measurements. BIAS $= \overline{LE_{cal}} - \overline{LE_{obs}}$.

Sites	Year	LEobs = k * LEcal + b		$\overline{LE_{obs}}$	$\overline{LE_{cal}}$	RMSE	R	BIAS	Bias/$\overline{LE_{obs}}$
		k	b	W/m2	W/m2	W/m2		W/m2	(%)
	2003	0.97	26.90	143.60	176.38	90.90	0.58	32.78	22.83
DHS	2004	0.70	63.60	157.97	177.04	79.10	0.56	19.06	12.07
	2005	1.01	6.09	132.65	150.06	73.10	0.73	17.40	13.12
	mean	0.89	32.20	144.74	167.83	81.03	0.62	23.08	16.00
	2003	0.99	4.58	179.28	182.32	74.20	0.82	3.04	1.70
QYZ	2004	1.00	−21.90	200.35	176.48	74.30	0.85	−23.87	−11.92
	2005	1.68	−28.90	119.39	172.30	83.50	0.79	52.90	44.31
	mean	1.22	−15.41	166.34	177.03	77.33	0.82	10.69	11.36
	2003	1.06	−46.80	165.59	134.85	74.01	0.81	−30.74	−18.56
CBS	2004	1.10	−42.04	146.64	123.95	56.00	0.88	−22.69	−15.48
	2005	1.29	−31.20	130.54	137.26	73.70	0.80	6.72	5.15
	mean	1.15	−40.01	147.59	132.02	67.90	0.83	−15.57	-9.63

The samples with the differences between LE_{cal} and LE_{obs} less than 150 Wm^{-2} account for 94%, 93%, and 92% of total samples at DHS, QYZ, and CBS, respectively (Figure 6). Our algorithm tends to underestimate the LE during transient periods with about 30–100 Wm^{-2} (particularly in non-snowy wintertime and early spring). In spite of this, results in Table 3 show that the method can produce the small bias varying from −30.7 to 32.8 Wm^{-2} with the RMSE from 56.0 to 90.9 Wm^{-2}, respectively. The relative bias at three forests was kept within 23% for most of the years and range from −18.6% to 22.8%. The regression lines between LE_{cal} and LE_{obs} are well close to a 1:1 line at all sites with slopes of 0.70–1.29 (Figure 7, Table 3).

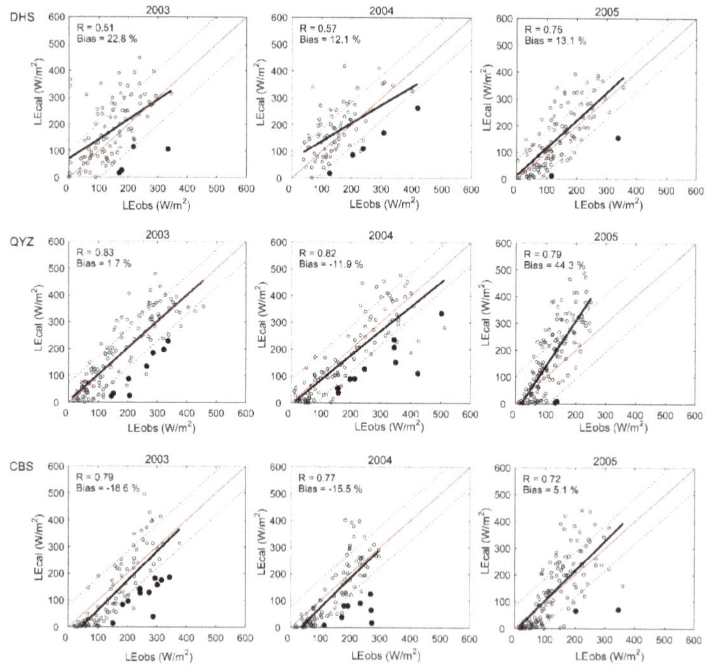

Figure 7. Comparisons between daily LE_{obs} and LE_{cal} at three sites from 2003 to 2005. The two dashed lines are the 1:1 line±RMSE. Solid circles are samples severely contaminated by precipitation.

The algorithm performs better at QYZ and CBS in terms of R (0.80–0.88) and RMSE (56–83.5 Wm^{-2}). CBS has the smallest mean RMSE (67.9 Wm^{-2}), highest R (0.83) and lowest relative bias (−9.6%) for all study years (Table 3). The best LE estimations at DHS, QYZ, and CBS occur in 2005 (R = 0.73, bias = 17.4 Wm^{-2}, RMSE = 73.1 Wm^{-2}), 2003 (R = 0.82, bias = 3.0 Wm^{-2}, RMSE = 74.2 Wm^{-2}), and 2004 (R = 0.88, bias = −22.7 Wm^{-2}, RMSE = 56.0 Wm^{-2}), respectively. However, the estimation performed relatively poorly at DHS in 2003 (R = 0.58, bias = 32.8 Wm^{-2}, RMSE = 90.9 Wm^{-2}) (Figure 7, Table 3).

Our algorithm could be affected by heavily rainy events occurring before or after when Aqua satellite overpasses. For these rainy days, our retrieval algorithm is expected to underestimate the LE because of the reduced microwave EDVI over wet surfaces and the omission of interception evaporation. Some rainy samples severely contaminated by such precipitation are marked by solid circles in Figure 7. If we excluded these samples, the performances would be generally improved at all sites, particularly for R and RMSE. This comparison results are shown in Figure 8. The R would be improved to 0.66–0.91 for all years and the RMSE would be reduced to 48.2–84.5 Wm^{-2}.

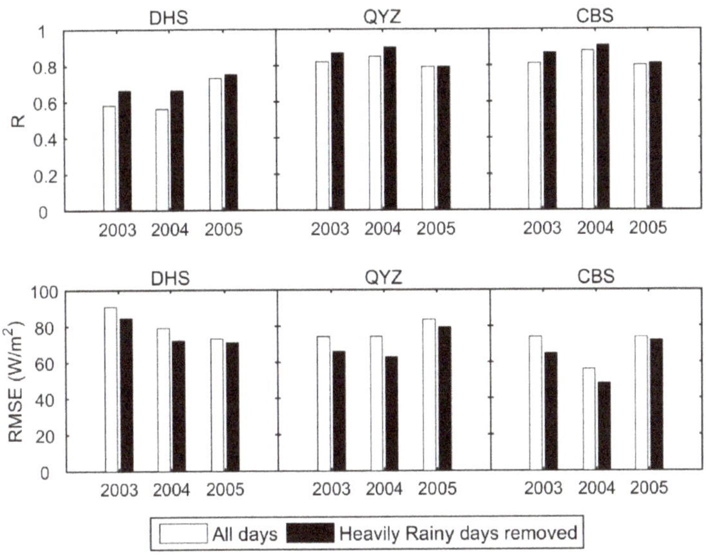

Figure 8. Comparison results of all LE estimations (white bars) and the LE estimations after removing the heavily rain-contaminated days (black bars).

Additionally, a few samples with much larger LE_{cal} than LE_{obs} occurred during growing seasons, such as 2003 at DHS (mainly in summer with less precipitation). A possible explanation is that under high temperature and less precipitation conditions, the plant water deficit will induce leaf stomatal closure to prevent excessive water deficits in the plants during summer time [66]. As a result, the real plant physiological activities, such as leaf transpiration, carbon gain, and growth, are remarkably suppressed [66].

The comparison results between LE_{cal} and LE_{obs} at monthly scale show the better performance (Figure 9). R can reach 0.84, 0.88, and 0.95 for DHS, QYZ, and CBS with the bias of 14.3%, 0.3%, and −12.9%, respectively. The standard deviations of monthly mean LE_{cal} and LE_{obs} are comparable. As discussed, the results can be improved after removing the heavily rain-contaminated days (Figure 9).

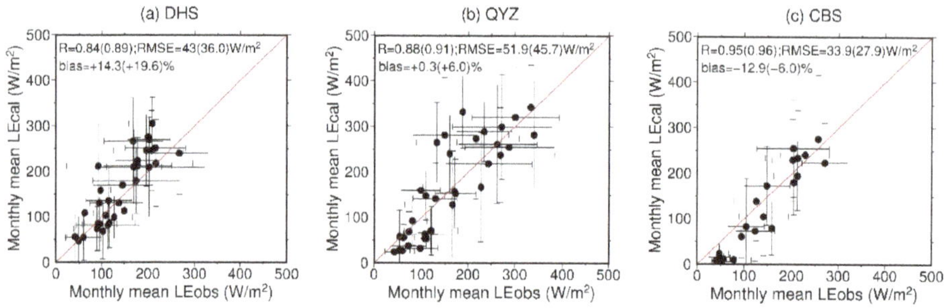

Figure 9. Comparisons between monthly mean LE_{obs} and LE_{cal} at DHS, QYZ, and CBS. Horizontal and vertical error bars stand for the standard deviations of LE_{obs} and LE_{cal}. Numbers in the brackets are the statistic results after removing the heavily rain-contaminated samples.

3.4. Validation of EDVI-based LE under Different Cloudy Sky

Since EDVI is able to indicate vegetation hydrological states under both clear and cloudy sky [26,27], EDVI-based LE method combined with all-sky satellite-retrieved radiation also has the capability of estimating LE under different cloud covers (Frc). Figure 10 shows the comparison of LE_{cal} and LE_{obs} under a partly cloudy sky (Frc < 40%), cloudy sky (Frc >= 40%) and all cloudy sky (0% <= Frc <= 100%). Their corresponding statistical metrics are shown in Figure 11. In general, our method has good performance at three sites. The slopes of fit lines are all close to 1.0 with the range of 0.88 to 1.16 (Figures 10 and 11), suggesting that the method can produce small systematic bias in the estimation of instantaneous forest LE under different cloud covers. The capability is also illustrated in the results of bias and relative bias in Figure 11. For all three sites, the method produces bias less than 35 W/m^2 (26%) at instantaneous scale when compared with in-situ measurements. These biases vary from −34.2 W/m^2 to 25.1 W/m^2 with the relative values from −25.9% to 16.7% (Figure 11), respectively. The mean bias (relative bias) under Frc < 40%, Frc ≥ 40% and all cloudy sky are well kept within 11 W/m^2 with the values of 10.6 W/m^2 (6.3%), 0.32 W/m^2 (0.8%) and 7.7 W/m^2 (5%), respectively. In addition, a good correlation between LE_{cal} and LE_{obs} under different cloudy skies are also found at three sites with the R of 0.62–0.80, which suggests that the seasonal dynamics of forest LE under cloud cover can be well recaptured by the EDVI-based method. Because of the relatively coarse spatial resolution (see the related discussion in Section 4), the method could produce relatively large RMSE (59 to 90 W/m^2) illustrated by the scattering of samples in Figure 10.

Most importantly, it can be found that the EDVI-based method is able to produce stable statistic metrics under different cloud cover conditions for three typical forests. These results indicate that the developed EDVI-based method in this study, completely driven by satellite and reanalysis datasets, can be used to estimate forest LE effectively from clear sky to cloudy sky.

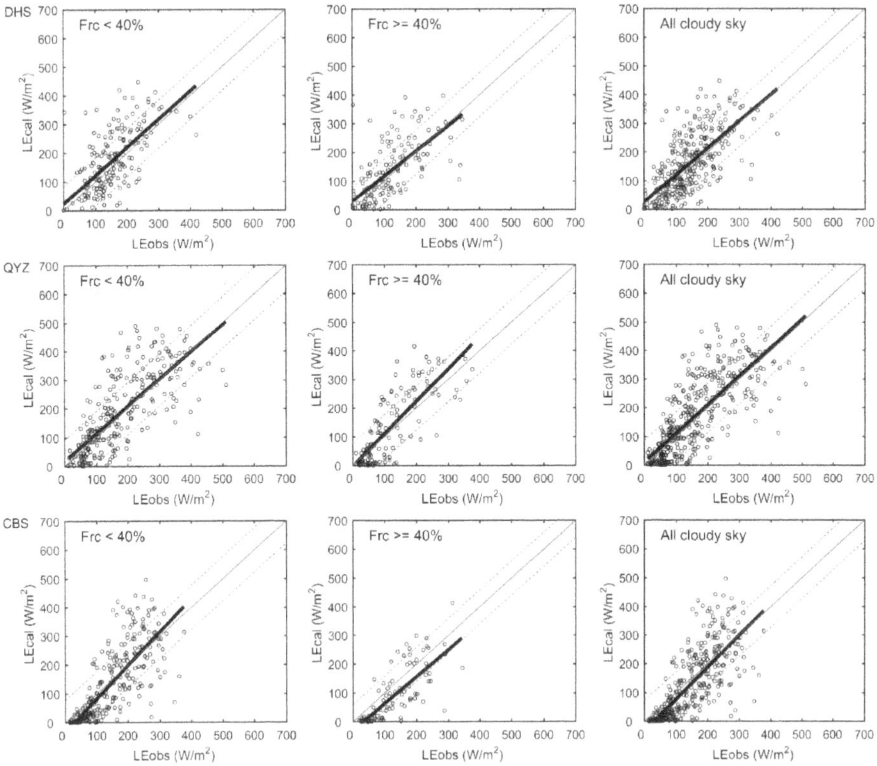

Figure 10. Validation of EDVI-based LE_{cal} under different cloud cover (Frc). Partly cloudy sky (Frc < 40%), Cloudy sky (Frc >= 40%) and all cloudy sky (0% <= Frc <= 100%).

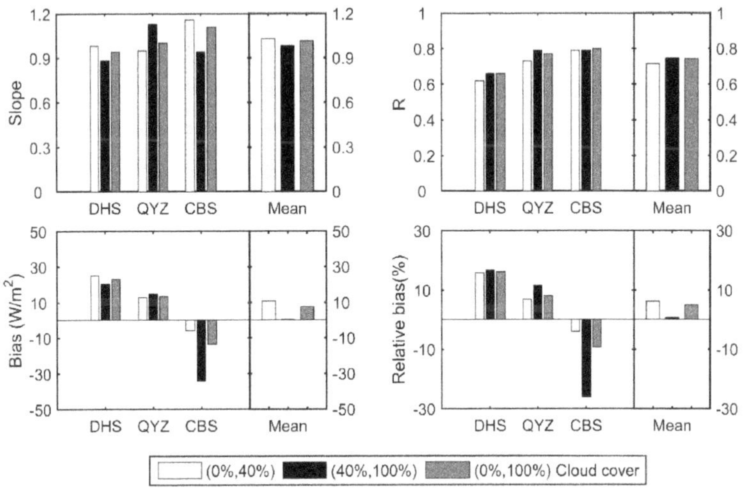

Figure 11. Statistical metrics at three sites under different cloud cover. Slope is the slope of fit line between LE_{cal} and LE_{obs}.

4. Discussion

4.1. Summary of the Method

The above results suggest that the new EDVI-based LE algorithm can effectively estimate instantaneous LE during midday at three different forests. Compared with the early EDVI method of Li et al. [29] based on in situ measurements as inputs, the developed method in this study was completely driven by all-sky satellite-retrieved radiation fluxes and reanalysis-based meteorological data. Therefore, this method has the potential capability to estimate LE in unfrequented places, such as area in the deep forests and high mountains.

Which is central to our method is using microwave EDVI to quantify the stomatal and canopy resistance under both clear sky and cloudy sky. Most of previous resistance-based LE methods are based on optic VIs [17–19] which cannot be effectively implemented under cloudy sky. Their validations are thus limited to under clear or partly cloudy sky. Barraza et al. [23,24] compared the estimated surface conductance based on the different regression models of microwave and optical VIs in forest and savanna ecosystems. Although they concluded that the combined microwave-optical VIs method produced the best conductance and LE estimation at the 8-day mean scale, the performances under different sky conditions were not investigated at finer temporal resolution. EDVI-based LE method is designed for the instantaneous all-sky LE. The validations under different cloudy skies show that our method can produce stable forest LE from clear to cloudy sky with good accuracy, which could be important for the study of EF and LE under cloudy sky.

4.2. Uncertainties in the Results

In spite of these, some uncertainties in our EDVI-based LE method should be noted.

The retrieved EDVI ($\overline{EDVI}$) over a landscape is the integration of EDVI from vegetation (EDVIveg) and bare soil (EDVIsoil). During growing seasons, $\overline{EDVI}$ is dominated by EDVIveg and can be used as a good indicator of vegetation states. Thus, the LE estimated from $\overline{EDVI}$ approximates the in-situ measured LE. However, for some vegetation regions such as deciduous forests, $\overline{EDVI}$ can be affected by soil signals to different extent during transient periods.

Since the EDVI-based LE method is implemented over forests in this study, the evaporation of intercepted water and bare soil are simply omitted in the forest LE estimation. This simplification could result in some underestimation, particularly during the non-growing seasons of deciduous forests when soil evaporation may dominate the total LE. Further, as we discussed in Section 3.3, the algorithm could be affected by heavily rainy events occurring near the time when Aqua satellite overpasses. The wet canopy surface can reduce EDVI and thus result in the underestimation of LE in our method. In spite of these, results in this study (Table 3) indicate that our current method for instantaneous LE estimation at forests with the overall mean bias of 16% (DHS), 11.4% (QYZ) and −9.6% (CBS) (Table 3) is well within the typical error range of 30% for satellite VI-based LE method [8].

As the surrogates for in situ measurements, the quality of the satellite remotely sensed and reanalyzed data, which are the inputs of our algorithm, particularly net radiation, which is the direct driving force for evapotranspiration, would highly affect the accuracy of the LE estimate, and the effect of this would vary at different sites (Figure 3). Yan et al. [60] compared CERES surface radiation fluxes with in situ measurements over Loess Plateau and found CERES surface downward radiation fluxes have higher accuracies in clear sky than those in cloudy sky. The standard deviations of the dSW differences rise from ~30 W m^{-2} to ~130 W m^{-2} when cloud coverage increases from 5% to 80%. The errors of CERES estimated net radiation will be directly transferred to the results of EDVI-based $LE_{cal.}$

Due to the inhomogeneity of vegetation coverage on the ground and the inconsistency of spatial resolution among the input datasets (Table 2), the matching of these data over the selected area (0.15° × 0.15°) around each forest flux tower may introduce errors. In addition, spatial mismatch of the geolocation between satellite field of view to the forest sites also can result in some poor performances in validations. For example, the flux tower at DHS is on a steep 30° slope and is close to a city, and there is a river located in the southeast. However, in our algorithm, the input data averaged over the 0.15° × 0.15° area surrounding the flux tower are used to estimate the regional mean LE. This certainly introduced additional discrepancies between the satellite EDVI-based LE_{cal} and the in situ measured LE_{obs}.

In addition, it should be noted that the validation results are also affected by the accuracies of the in situ measurements. The EC method may produce potential errors or uncertainty of 10%–30% [5]. Also, the EC flux towers frequently suffer energy closure errors. In this study, the three ChinaFLUX forest sites achieve approximately only 70% energy balance closure during daytime, with about 30% of variation unexplained [35].

Although affected by the above uncertainties, the accuracies of instantaneous LE estimations in this study are comparable with those in Li et al. [29] which utilized both local in-situ measurements and satellite retrieved EDVI as data input. The results of this study demonstrate that it is feasible to operate the EDVI-based LE algorithm under both clear and cloudy skies with all satellite observations as data input.

4.3. Pros and Cons of this method

In comparison to Li et al. [29], the biggest improvement of this method is that its inputs do not depend on any in-situ measurements. This feature makes it applicable to any places with available satellite observations. Further, with proper calibration and validation, it can be developed to be a global algorithm in the future.

Compared to other optical VIs-based methods, the most significant advantage of this method is its capability for estimating ET related LE under cloudy sky. As shown in Section 3.4, no matter in what cloud fraction condition, the retrieved LE keeps small bias and good consistency with the in situ measurements. The most serious disadvantage of this method could be its poor spatial resolution, i.e., ~20 km. It is certain that the vegetation states and the evapotranspiration rate can vary significantly over this scale since the heterogeneous surfaces. This shortcoming may offset its merit of being useful under all weather conditions. A study of downscaling this method to the finer spatial resolution of optical VI is undergoing in our lab.

It should be noted that some parameterization schemes in our EDVI-based ET method need to be further improved in the future, especially for those resistance estimations based on EDVI and dEDVI. Since the selected three forests are similar to the Harvard forest where the initial method was developed [29], EDVI-based resistance schemes (e.g., a and b in Equation (8)) are thus assumed to be the same in this study. In addition, this method is recommended in flat or moderate terrain due to the associated errors with large scale satellite imagery of steep slope and LE estimation.

The real LE in the forest can be significantly affected by heavy rainfall due to the enhanced evaporation from interception water on the leaves. However, such an effect cannot be captured by the current method because it only takes into account one evapotranspiration source from inner vegetation water. In contrast, the plant water deficit will induce leaf stomatal closure to prevent excessive water deficits. However, this response is not described in the current model. A further study to consider more ET sources and the phenology response of vegetation to drought should be conducted to improve the retrieving performance.

Besides EDVI, there are several other vegetation water content related indexes that have been developed. For example, the normalized difference water index (NDWI) [67] and land surface water index (LSWI) [68]. Due to the differences in their physical connections to vegetation water and their distinct spatial and temporal resolution, they may be used to estimate ET and LE with particular advantages and disadvantages, respectively. A comprehensive comparison study among them will be valuable for improving the satellite based global estimation of evapotranspiration and latent heat flux.

5. Conclusions

The microwave-based vegetation water content index, Emissivity Difference Vegetation Index (EDVI) has a close connection to the evapotranspiration process in forests [26–29] and responses to precipitation process dynamically [69]. In this study, we designed an algorithm for forest LE estimation driven by EDVI from an advanced microwave scanning radiometer for EOS (AMSR-E), vegetation fraction information derived from a moderate-resolution imaging spectroradiometer (MODIS), net radiation flux derived from clouds and earth's radiation energy system (CERES), which are all on the Aqua satellite, and the associated parameters of atmospheric states from the European Centre for Medium-Range Weather Forecasts (ECMWF).

The satellite inputs and the results of this algorithm are validated against the in situ measurements at three ChinaFLUX sites located at the Dinghushan (DHS) covered by subtropical evergreen broad-leaved forest, Qianyanzhou (QYZ) covered by subtropical plantation forest, and Changbaishan (CBS) covered by temperate deciduous mixed forest from 2003 to 2005. Validation results show that the mean correlation coefficients (R) between instantaneous LE_{cal} and LE_{obs} in the study years of DHS, QYZ and CBS are 0.62, 0.82 and 0.83 with small bias errors of +23.08 Wm^{-2}, +10.69 Wm^{-2} and −15.57 Wm^{-2}, respectively. These biases were well kept within 16% of the in situ measurements for three sites. At a monthly scale, the R between LE_{cal} and LE_{obs} can reach 0.84, 0.88, and 0.95 at DHS, QYZ, and CBS, with bias of +14.3%, −0.3%, and −12.9%, respectively. Validation results can be further improved after removing the samples in severely rainy days.

Our method is also validated under different cloudy sky conditions. The results indicate that EDVI-based LE_{cal} have stable performances with good accuracy under cloudy sky for all three forests. Slopes of fit lines are close to 1.0 and the bias are less than 35 Wm^{-2} (26%) for different cloud cover. A good temporal correlation between LE_{cal} and LE_{obs} under clear and cloudy skies is indicated by the R of 0.62–0.80. These results indicate that this EDVI-based LE algorithm, using only satellite and reanalysis datasets as inputs, has great potential for estimating LE at large scale in forest areas under cloudy sky in China.

The extensive application and improvement of our algorithm is warranted in more different biome types. Potential improvements can be achieved by (i) taking into account the evaporation components from bare soil and canopy-intercepted water, (ii) considering the inhomogeneity within the satellite field of view at microwave regions and (iii) extending the current LE estimation from one point to the regional scale.

Author Contributions: Conceptualization, Y.W. and R.L.; Methodology, Y.W. and R.L.; Validation, Y.W.; Writing—original draft preparation, Y.W.; Writing—review and editing, Y.W. and R.L.; Supervision, R.L., Q.M., L.Z., G.Y. and Y.B. All authors discussed the results and revised the manuscript.

Funding: This work was funded by the National Natural Science Foundation of China NSFC (Grant No. 41675022, 41375148, 41830104), National Key R&D Program on Monitoring, Early Warning and Prevention of Major Natural Disasters under grant (Grant No. 2017YFC1501402), Belmont Forum and JPI-Climate Collaborative Research Action with NSFC (Grant No. 41661144007), the "Hundred Talents Program" of the Chinese Academy of Sciences, the "Hundred Talents Program" of Anhui Province, and the Jiangsu Provincial 2011 Program (Collaborative Innovation Center of Climate Change).

Acknowledgments: We thank the anonymous reviewers and members of the editorial team for their helpful comments and valuable suggestions which have helped us greatly improve the manuscript. We thank Zongting Gao, in Institute of Meteorological Sciences of Jilin Province, Jilin Provincial Key Laboratory of Changbai Mountain Meteorology & Climate Change, Laboratory of Research for Middle-High Latitude Circulation and East Asian Monsoon in Changchun, for his significant help in data processing and data analysis in this work.

Conflicts of Interest: The authors declare no conflict of interest.

References

1. Monteith, J.; Unsworth, M. *Principles of Environmental Physics: Plants, Animals, and the Atmosphere*, 3rd ed.; Academic Press: London, UK, 2008.
2. Priestley, C.H.B.; Taylor, R. On the assessment of surface heat flux and evaporation using large-scale parameters. *Mon. Weather Rev.* **1972**, *100*, 81–92. [CrossRef]
3. Trenberth, K.E.; Fasullo, J.T.; Kiehl, J. Earth's global energy budget. *Bull. Am. Meteorol. Soc.* **2009**, *90*, 311–324. [CrossRef]
4. Wang, K.; Dickinson, R.E. A review of global terrestrial evapotranspiration: Observation, modeling, climatology, and climatic variability. *Rev. Geophys.* **2012**, *50*, 1–54. [CrossRef]
5. Glenn, E.P.; Huete, A.R.; Nagler, P.L.; Hirschboeck, K.K.; Brown, P. Integrating remote sensing and ground methods to estimate evapotranspiration. *Crit. Rev. Plant Sci.* **2007**, *26*, 139–168. [CrossRef]
6. Oki, T.; Kanae, S. Global hydrological cycles and world water resources. *Science* **2006**, *313*, 1068–1072. [CrossRef] [PubMed]
7. Huete, A.; Didan, K.; Miura, T.; Rodriguez, E.P.; Gao, X.; Ferreira, L.G. Overview of the radiometric and biophysical performance of the MODIS vegetation indices. *Remote Sens. Environ.* **2002**, *83*, 195–213. [CrossRef]
8. Glenn, E.P.; Nagler, P.L.; Huete, A.R. Vegetation index methods for estimating evapotranspiration by remote sensing. *Surv. Geophys.* **2010**, *31*, 531–555. [CrossRef]
9. Carlson, T. An overview of the "triangle method" for estimating surface evapotranspiration and soil moisture from satellite imagery. *Sensors* **2007**, *7*, 1612–1629. [CrossRef]
10. Jiang, L.; Islam, S. Estimation of surface evaporation map over southern Great Plains using remote sensing data. *Water Resour. Res.* **2001**, *37*, 329–340. [CrossRef]
11. Mu, Q.; Zhao, M.; Running, S.W. Improvements to a MODIS global terrestrial evapotranspiration algorithm. *Remote Sens. Environ.* **2011**, *115*, 1781–1800. [CrossRef]
12. Murray, R.S.; Nagler, P.L.; Morino, K.; Glenn, E.P. An empirical algorithm for estimating agricultural and riparian evapotranspiration using MODIS enhanced vegetation index and ground measurements of ET. II. application to the lower Colorado river, U.S. *Remote Sens.* **2009**, *1*, 1125–1138. [CrossRef]
13. Nagler, P.; Scott, R.; Westenburg, C.; Cleverly, J.; Glenn, E.; Huete, A. Evapotranspiration on western U.S. rivers estimated using the Enhanced Vegetation Index from MODIS and data from eddy covariance and Bowen ratio flux towers. *Remote Sens. Environ.* **2005**, *97*, 337–351. [CrossRef]
14. Wang, K.; Li, Z.; Cribb, M. Estimation of evaporative fraction from a combination of day and night land surface temperatures and NDVI: A new method to determine the Priestley–Taylor parameter. *Remote Sens. Environ.* **2006**, *102*, 293–305. [CrossRef]
15. Wang, K.; Liang, S. An improved method for estimating global evapotranspiration based on satellite determination of surface net radiation, vegetation index, temperature, and soil moisture. *J. Hydrometeorol.* **2008**, *9*, 712–727. [CrossRef]
16. Wang, K.; Wang, P.; Li, Z.; Cribb, M.; Sparrow, M. A simple method to estimate actual evapotranspiration from a combination of net radiation, vegetation index, and temperature. *J. Geophys. Res. Atmos.* **2007**, *112*, D15107. [CrossRef]
17. Bai, Y.; Zhang, J.; Zhang, S.; Koju, U.A.; Yao, F.; Igbawua, T. Using precipitation, vertical root distribution, and satellite-retrieved vegetation information to parameterize water stress in a Penman-Monteith approach to evapotranspiration modeling under Mediterranean climate. *J. Adv. Modeling Earth Syst.* **2017**, *9*, 168–192. [CrossRef]
18. Hu, X.; Shi, L.; Lin, L.; Zhang, B.; Zha, Y. Optical-based and thermal-based surface conductance and actual evapotranspiration estimation, an evaluation study in the North China Plain. *Agric. For. Meteorol.* **2018**, *263*, 449–464. [CrossRef]
19. Yebra, M.; Van Dijk, A.; Leuning, R.; Huete, A.; Guerschman, J.P. Evaluation of optical remote sensing to estimate actual evapotranspiration and canopy conductance. *Remote Sens. Environ.* **2013**, *129*, 250–261. [CrossRef]
20. Ferrazzoli, P.; Guerriero, L. Passive microwave remote sensing of forests: A model investigation. *IEEE Trans. Geosci. Remote Sens.* **1996**, *34*, 433–443. [CrossRef]

21. Becker, F.; Choudhury, B.J. Relative sensitivity of normalized difference vegetation index (NDVI) and microwave polarization difference index (MPDI) for vegetation and desertification monitoring. *Remote Sens. Environ.* **1988**, *24*, 297–311. [CrossRef]
22. Shi, J.; Jackson, T.; Tao, J.; Du, J.; Bindlish, R.; Lu, L.; Chen, K. Microwave vegetation indices for short vegetation covers from satellite passive microwave sensor AMSR-E. *Remote Sens. Environ.* **2008**, *112*, 4285–4300. [CrossRef]
23. Barraza, V.; Restrepo-Coupe, N.; Huete, A.; Grings, F.; Van Gorsel, E. Passive microwave and optical index approaches for estimating surface conductance and evapotranspiration in forest ecosystems. *Agric. For. Meteorol.* **2015**, *213*, 126–137. [CrossRef]
24. Barraza, V.; Restrepo-Coupe, N.; Huete, A.; Grings, F.; Beringer, J.; Cleverly, J.; Eamus, D. Estimation of latent heat flux over savannah vegetation across the North Australian Tropical Transect from multiple sensors and global meteorological data. *Agric. For. Meteorol.* **2017**, *232*, 689–703. [CrossRef]
25. Pan, X.; Liu, Y.; Gan, G.; Fan, X.; Yang, Y. Estimation of Evapotranspiration Using a Nonparametric Approach Under All Sky: Accuracy Evaluation and Error Analysis. *IEEE J. Sel. Top. Appl. Earth Obs. Remote Sens.* **2017**, *10*, 2528–2539. [CrossRef]
26. Min, Q.; Lin, B. Determination of spring onset and growing season leaf development using satellite measurements. *Remote Sens. Environ.* **2006**, *104*, 96–102. [CrossRef]
27. Min, Q.; Lin, B. Remote sensing of evapotranspiration and carbon uptake at Harvard Forest. *Remote Sens. Environ.* **2006**, *100*, 379–387. [CrossRef]
28. Min, Q.; Lin, B.; Li, R. Remote sensing vegetation hydrological states using passive microwave measurements. *IEEE J. Sel. Top. Appl. Earth Obs. Remote Sens.* **2010**, *3*, 124–131. [CrossRef]
29. Li, R.; Min, Q.; Lin, B. Estimation of evapotranspiration in a mid-latitude forest using the Microwave Emissivity Difference Vegetation Index (EDVI). *Remote Sens. Environ.* **2009**, *113*, 2011–2018. [CrossRef]
30. Dickinson, R.E. Modeling evapotranspiration for three-dimensional global climate models. *Clim. Process. Clim. Sensit.* **1984**, *29*, 58–72.
31. Shukla, J.; Mintz, Y. Influence of land-surface evapotranspiration on the earth's climate. *Science* **1982**, *215*, 1498–1501. [CrossRef]
32. Vinukollu, R.K.; Wood, E.F.; Ferguson, C.R.; Fisher, J.B. Global estimates of evapotranspiration for climate studies using multi-sensor remote sensing data: Evaluation of three process-based approaches. *Remote Sens. Environ.* **2011**, *115*, 801–823. [CrossRef]
33. Yu, G.-R.; Wen, X.-F.; Sun, X.-M.; Tanner, B.D.; Lee, X.; Chen, J.-Y. Overview of ChinaFLUX and evaluation of its eddy covariance measurement. *Agric. For. Meteorol.* **2006**, *137*, 125–137. [CrossRef]
34. Yu, G.; Song, X.; Wang, Q.; Liu, Y.; Guan, D.; Yan, J.; Sun, X.; Zhang, L.; Wen, X. Water-use efficiency of forest ecosystems in eastern China and its relations to climatic variables. *New Phytol.* **2008**, *177*, 927–937. [CrossRef] [PubMed]
35. Li, Z.; Yu, G.; Wen, X.; Zhang, L.; Ren, C.; Fu, Y. Energy balance closure at ChinaFLUX sites. *Sci. China Ser. D* **2005**, *48*, 51–62.
36. Zhu, X.-J.; Yu, G.-R.; Hu, Z.-M.; Wang, Q.-F.; He, H.-L.; Yan, J.-H.; Wang, H.-M.; Zhang, J.-H. Spatiotemporal variations of T/ET (the ratio of transpiration to evapotranspiration) in three forests of Eastern China. *Ecol. Indic.* **2015**, *52*, 411–421. [CrossRef]
37. Li, M.; Wu, Z.-F.; Du, H.-B.; Zong, S.; Meng, X.; Zhang, L. Growing-season trends determined from SPOT NDVI in Changbai Mountains, China, 1999–2008. *Sci. Geogr. Sin.* **2011**, *31*, 1242–1248.
38. Kratz, D.P.; Gupta, S.K.; Wilber, A.C.; Sothcott, V.E. Validation of the CERES edition 2B surface-only flux algorithms. *J. Appl. Meteorol. Climatol.* **2010**, *49*, 164–180. [CrossRef]
39. Gupta, S.K.; Darnell, W.L.; Wilber, A.C. A parameterization for longwave surface radiation from satellite data: Recent improvements. *J. Appl. Meteorol.* **1992**, *31*, 1361–1367. [CrossRef]
40. Gupta, S.K.; Kratz, D.P.; Stackhouse, P.W., Jr.; Wilber, A.C. *The Langley Parameterized Shortwave Algorithm (LPSA) for Surface Radiation Budget Studies. 1.0*; Technical Report; NASA Langley Research Center: Hampton, VA, USA, 2001.
41. Brutsaert, W.; Sugita, M. Application of self-preservation in the diurnal evolution of the surface energy budget to determine daily evaporation. *J. Geophys. Res. Atmos.* **1992**, *97*, 18377–18382. [CrossRef]
42. Crago, R.D. Conservation and variability of the evaporative fraction during the daytime. *J. Hydrol.* **1996**, *180*, 173–194. [CrossRef]

43. Gurney, R.; Hsu, A. Relating evaporative fraction to remotely sensed data at the FIFE site. In Proceedings of the Symposium on FIFE: Fist ISLSCP Field Experiment, American Meteorological Society, Boston, MA, USA, 7–9 February 1990; pp. 112–116.
44. Jackson, R.D.; Hatfield, J.L.; Reginato, R.; Idso, S.; Pinter Jr, P. Estimation of daily evapotranspiration from one time-of-day measurements. *Agric. Water Manag.* **1983**, *7*, 351–362. [CrossRef]
45. Shuttleworth, W.J.; Gurney, R.J.; Hsu, A.Y.; Ormsby, J.P. FIFE: The variation in energy partitioning at surface flux sites. In *Remote Sensing and Large-Scale Global Processes*; International Association of Hydrologic Science: Wallingford, Oxfordshire, UK, 1989; pp. 67–74.
46. Kurpius, M.; Panek, J.; Nikolov, N.; McKay, M.; Goldstein, A.H. Partitioning of water flux in a Sierra Nevada ponderosa pine plantation. *Agric. For. Meteorol.* **2003**, *117*, 173–192. [CrossRef]
47. Wilson, K.B.; Hanson, P.J.; Mulholland, P.J.; Baldocchi, D.D.; Wullschleger, S.D. A comparison of methods for determining forest evapotranspiration and its components: Sap-flow, soil water budget, eddy covariance and catchment water balance. *Agric. For. Meteorol.* **2001**, *106*, 153–168. [CrossRef]
48. Su, Z. The Surface Energy Balance System (SEBS) for estimation of turbulent heat fluxes. *Hydrol. Earth Syst. Sci.* **2002**, *6*, 85–100. [CrossRef]
49. Gutman, G.; Ignatov, A. The derivation of the green vegetation fraction from NOAA/AVHRR data for use in numerical weather prediction models. *Int. J. Remote Sens.* **1998**, *19*, 1533–1543. [CrossRef]
50. Li, X.; Zhang, J. Derivation of the Green Vegetation Fraction of the Whole China from 2000 to 2010 from MODIS Data. *Earth Interact.* **2016**, *20*, 1–16. [CrossRef]
51. Zeng, X.; Dickinson, R.E.; Walker, A.; Shaikh, M.; DeFries, R.S.; Qi, J. Derivation and evaluation of global 1-km fractional vegetation cover data for land modeling. *J. Appl. Meteorol.* **2000**, *39*, 826–839. [CrossRef]
52. Nishida, K.; Nemani, R.R.; Running, S.W.; Glassy, J.M. An operational remote sensing algorithm of land surface evaporation. *J. Geophys. Res. Atmos.* **2003**, *108*. [CrossRef]
53. Murray, F.W. On the computation of saturation vapor pressure. *J. Appl. Meteorol. Climatol.* **1967**, *6*, 203–204. [CrossRef]
54. Kondo, J. *Meteorology in Aquatic Environments*; Asakura Publishing Ltd.: Tokyo, Japan, 1994; p. 350.
55. Kondo, J. *Atmospheric science near the ground surface*; University of Tokyo Press: Tokyo, Japan, 2000; p. 90.
56. Tang, Q.; Peterson, S.; Cuenca, R.H.; Hagimoto, Y.; Lettenmaier, D.P. Satellite-based near-real-time estimation of irrigated crop water consumption. *J. Geophys. Res. Atmos.* **2009**, *114*. [CrossRef]
57. Jarvis, P. The interpretation of the variations in leaf water potential and stomatal conductance found in canopies in the field. *Phil. Trans. R. Soc. Lond. B* **1976**, *273*, 593–610. [CrossRef]
58. Kosugi, Y. Leaf-Scale Analysis of the CO_2 and H_2O Exchange Processes between Trees and the Atmos Phere. Ph.D. Thesis, Kyoto University, Kyoto, Japan, 1996.
59. Kelliher, F.; Leuning, R.; Raupach, M.; Schulze, E.-D. Maximum conductances for evaporation from global vegetation types. *Agric. For. Meteorol.* **1995**, *73*, 1–16. [CrossRef]
60. Yan, H.; Huang, J.; Minnis, P.; Wang, T.; Bi, J. Comparison of CERES surface radiation fluxes with surface observations over Loess Plateau. *Remote Sens. Environ.* **2011**, *115*, 1489–1500. [CrossRef]
61. Frouin, R.; Pinker, R.T. Estimating photosynthetically active radiation (PAR) at the earth's surface from satellite observations. *Remote Sens. Environ.* **1995**, *51*, 98–107. [CrossRef]
62. Jacovides, C.P.; Tymvios, F.S.; Asimakopoulos, D.N.; Theofilou, K.M.; Pashiardes, S. Global photosynthetically active radiation and its relationship with global solar radiation in the Eastern Mediterranean basin. *Theor. Appl. Climatol.* **2003**, *74*, 227–233. [CrossRef]
63. Zhang, X.; Zhang, Y.; Zhoub, Y. Measuring and modelling photosynthetically active radiation in Tibet Plateau during April–October. *Agric. For. Meteorol.* **2000**, *102*, 207–212. [CrossRef]
64. Huanqi, W.; Honglin, H.; Min, L.; Li, Z.; Gui-Rui, Y.; Cheng-cheng, M.; Hui-Min, W.; Ying, L. Modeling evapotranspiration and its components in Qianyanzhou plantation based on remote sensing data. *J. Nat. Resour.* **2012**, *27*, 778–789.
65. Qian-Qian, L.U.; Hong-Lin, H.E.; Zhu, X.J.; Gui-Rui, Y.U.; Wang, H.M.; Zhang, J.H.; Yan, J.H. Study on the Variations of Forest Evapotranspiration and Its Components in Eastern China. *J. Nat. Resour.* **2015**, *30*, 1436–1448.
66. Cochard, H.; Coll, L.; Le Roux, X.; Ameglio, T. Unraveling the Effects of Plant Hydraulics on Stomatal Closure during Water Stress in Walnut. *Plant Physiol.* **2002**, *128*, 282–290. [CrossRef]

67. Gao, B.-C. NDWI—A normalized difference water index for remote sensing of vegetation liquid water from space. *Remote Sens. Environ.* **1996**, *58*, 257–266. [CrossRef]
68. Chandrasekar, K.; Sesha Sai, M.; Roy, P.; Dwevedi, R. Land Surface Water Index (LSWI) response to rainfall and NDVI using the MODIS Vegetation Index product. *Int. J. Remote Sens.* **2010**, *31*, 3987–4005. [CrossRef]
69. Li, R.; Min, Q. Dynamic response of microwave land surface properties to precipitation in Amazon rainforest. *Remote Sens. Environ.* **2013**, *133*, 183–192. [CrossRef]

remote sensing

MDPI

Article

Evaluation of Evapotranspiration for Exorheic Catchments of China during the GRACE Era: From a Water Balance Perspective

Yulong Zhong [1,2,3], Min Zhong [3,4], Yuna Mao [5] and Bing Ji [6,*]

1 School of Geography and Information Engineering, China University of Geosciences (Wuhan), Wuhan 430078, China; zhongyl@cug.edu.cn
2 Artificial Intelligence School, Wuchang University of Technology, Wuhan 430223, China
3 State Key Laboratory of Geodesy and Earth's Dynamics, Institute of Geodesy and Geophysics, Chinese Academy of Sciences, Wuhan 430077, China; zmzm@asch.whigg.ac.cn
4 College of Earth and Planetary Sciences, University of Chinese Academy of Sciences, Beijing 100049, China
5 School of Environment, Beijing Normal University, Beijing 100875, China; myn@bnu.edu.cn
6 Department of Navigation Engineering, Naval University of Engineering, Wuhan 430033, China
* Correspondence: jibing1978@126.com; Tel.: +86-276-5461-026

Received: 18 December 2019; Accepted: 4 February 2020; Published: 5 February 2020

Abstract: Evapotranspiration (ET) is usually difficult to estimate at the regional scale due to scarce direct measurements. This study uses the water balance equation to calculate the regional ET with observations of precipitation, runoff, and terrestrial water storage changes (TWSC) in nine exorheic catchments of China. We compared the regional ET estimates from a water balance perspective with and without considering TWSC (ET_{WB}: ET estimates with considering TWSC, and ET_{PQ}: ET estimates from precipitation minus runoff without considering TWSC). Results show that the regional annual ET ranges from 417.7 mm/yr to 831.5 mm/yr in the nine exorheic catchments based on the water balance equation. The impact of ignoring TWSC on calculating ET is notable, as the root mean square errors (RMSEs) of annual ET between ET_{WB} and ET_{PQ} range from 12.0–105.8 mm/yr (2.6–12.7% in corresponding annual ET) among the exorheic catchments. We also compared the estimated regional ET with other ET products. Different precipitation products are assessed to explain the inconsistency between different ET products and regional ET from a water balance perspective. The RMSEs between ET estimates from Gravity Recovery and Climate Experiment (GRACE) and ET from land surface models can be reduced if the deviation of precipitation forcing data is considered. ET estimates from Global Land Evaporation Amsterdam Model (GLEAM) can be improved by reducing the uncertainty of precipitation forcing data in three semiarid catchments. This study emphasizes the importance of considering TWSC when calculating the regional ET using a water balance equation and provides more accurate ET estimates to help improve modeled ET results.

Keywords: evapotranspiration; China; exorheic catchments; water balance; GRACE; terrestrial water storage changes

1. Introduction

Evapotranspiration (ET) is one of the most important components of the climate system connecting the water, energy, and carbon cycle [1,2]. ET changes can be used as an indicator of climate change, especially in areas where the water cycle is accelerated [3,4]. However, regional ET is often difficult to estimate. The flux tower observing station network can provide accurate ET observations at each site [5], but it often has too sparse sites for basin scale study. Remote sensing provides an opportunity to monitor spatial-temporal changes in ET [6,7], but regional calibration and uncertainty from vegetation cover data will also lead to large uncertainty in ET [8]. Land surface models (LSMs) can also provide grid-to-regional scale ET estimates, such as the multiple LSMs simulations using the global land data assimilation system (GLDAS) issued by NASA [9]. Regional ET can be derived from the terrestrial water budget, namely the residual between precipitation (*P*) and the sum of runoff (*Q*) and terrestrial water storage change (*ds/dt*), which have been regarded as benchmark estimates for validating ET products or estimates on a regional scale [10,11].

The Gravity Recovery and Climate Experiment (GRACE) satellites mission launched in March 2002 has provided a unique way to monitor terrestrial water storage (TWS) changes on the monthly scale with a ~300 km footprint [12]. As for a given basin, the time series of TWS changes (TWSC) in the basin can be obtained from the differential of TWS anomalies (TWSA) observed by GRACE [1,13,14]. The regional ET can be estimated from TWSC, regional precipitation, and runoff data based on the water balance equation. Rodell et al. [4] discussed the method of calculating ET from GRACE TWSA and suggested that the ET based on the GRACE water balance method can be used to evaluate the ET of the model simulations. Ramillien et al. [15] estimated the ET of 16 globally distributed basins based on the GRACE water balance method, and they are compared to outputs of four global LSMs, which shows good overall agreement. A few studies have applied this method to estimate the regional ET in several global basins, e.g., the Lake Chad basin, Africa [16], continental USA [17], and Amazon Basin [18]. However, the differences among different ET estimates are usually ignored or ascribed to the uncertainty of estimates during data processing [13,15]. Castle et al. [19] and Pan et al. [13] estimated the human-induced ET in the Colorado River Basin (USA) and Haihe River Basin (China) and attributed the differences between GRACE water balance ET and GLDAS ET to the influence of human activities. However, the quality of input precipitation also has a great influence on the ET outputs [10,20]. Badgley et al. [21] emphasized the significant uncertainty of the regional ET estimate from the choice of input forcing dataset. Liu et al. [11] used a bias-corrected water balance method to calculate annual reference ET from 1983–2006 and evaluated nine ET products in 35 global river basins on the interannual and long-term scale. They determined that different performances among the ET products may result from different forcing datasets. Given the uncertainty of ET products caused by the precipitation forcing data, in this study, we seek to explain the difference among different ET estimates by considering uncertainty from the different precipitation forcing data and modeled runoff from a water balance perspective.

The water balance equation is the classic method to calculate the ET on a regional scale. In China, Mao et al. [22] emphasized the significant impact of water storage due to reservoir construction on calculating ET trends. However, they did not consider other factors that cause the water storage changes, e.g., water withdrawal, lakes change, and glaciers melting [22,23]. Jiang et al. [24] took basin water balance as a benchmark to evaluate the MODIS MOD16 ET products in several exorheic basins. However, the assessment on the uncertainty of GRACE-derived TWSC was limited and was restricted to the Yangtze River Basin (YRB), Yellow River Basin (YeRB), and Songhuajiang River Basin (SRB) on monthly scale. Li et al. [25] used the revised Remote Sensing–Penman Monteith (RS-PM) model [26,27] to produce an ET map in China and derived an estimate of mean annual land–surface ET to 500 mm/yr. The revised RS-PM model predictions did not show a significant systematic error, but they only explained 61% of the ET variations at all the validation sites, which showed the uncertainty of the ET model in the regional estimation. Bai and Liu [10] used water balance-based ET estimates to evaluate the Global Land Evaporation Amsterdam Model (GLEAM), GLDAS and MODIS MOD16 ET products for 22 river basins in China, but the selected basins are restricted in wet basins, most of which are located in the YRB. ET calculated from the water balance equation for some exorheic basins of China are estimated by the above studies. However, little attention is paid to uncertainties from TWSC and precipitation forcing data. Therefore, we conduct a systematic assessment for the ET of exorheic basins from a water balance perspective.

Several studies assume TWSC to be zero on the annual scale due to the lack of data [22], and ET is obtained by precipitation minus runoff directly, as in the studies by Zhang et al. [28], Senay et al. [29] and Xue et al. [30]. However, TWSA can have large variability on seasonal and interannual scales due to human water consumption [31,32] and the building of reservoirs [22,23]. Zeng et al. [33] also acknowledged that ignoring TWSC would bring much bias into ET estimation, especially in regions with low ET. Wang [34] points out the importance of considering interannual TWSC in the estimation of ET. Hence, we compare the difference of ET estimates by considering and not considering TWSC in the water balance equation to explore the impact of TWSC on the ET estimate on interannual and monthly scales.

This study aims to (1) estimate the regional ET of nine exorheic catchments in China using the water balance equation considering TWSC; (2) analyze the impact of not considering TWSC and different TWSC products on the ET estimates; (3) explain the inconsistency between different ET products and regional ET from a water balance perspective. The flowchart of this study is shown in Figure 1.

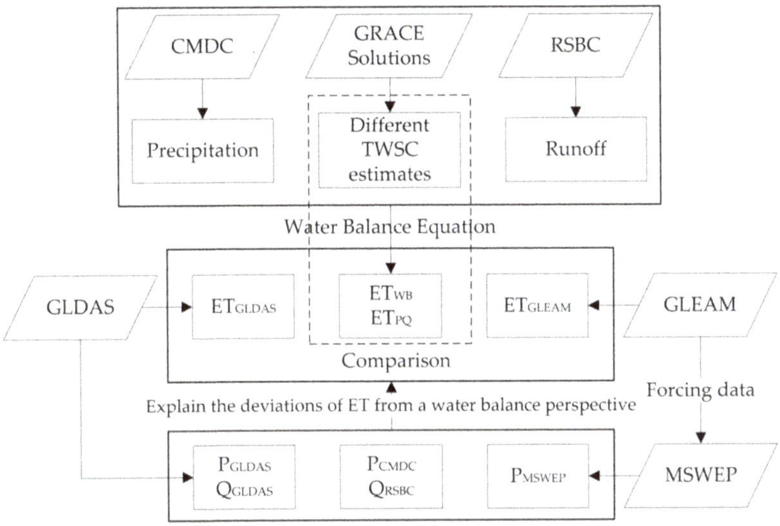

Figure 1. Flowchart of this study. It includes the used data and process of calculation and analysis in this study.

2. Materials and Methods

2.1. Study Area

Nine exorheic catchments were divided from eleven hydrological gauge stations from River Sediment Bulletin of China (RSBC) [35]. The catchment boundaries are derived from the location of gauge stations, Shuttle Radar Topography Mission (SRTM) elevation data (https://www2.jpl.nasa.gov/srtm/), and processed in ArcGIS with Soil and Water Assessment Tool (ArcSWAT) plugin (https://swat.tamu.edu/software/arcswat/). The ArcSWAT is an ArcGIS-ArcView extension and a graphical user input interface for the SWAT (Soil and Water Assessment Tool) model. The Gaoyao, Shijiao, and Boluo hydrological gauge stations are all in the Pearl River Basin (PRB), we merge them into one catchment (Figure 2). The YRB is divided into two sub-basins, based on its two hydrological gauge stations: Yichang and Datong. The information of the nine catchments is listed in Table 1, and the catchments are also presented in Figure 2.

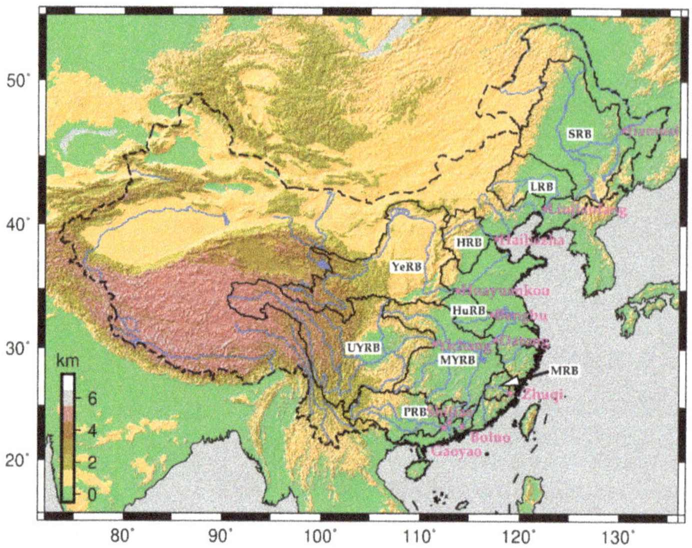

Figure 2. Map of exorheic catchments of China. The black curves show the boundaries of the nine exorheic catchments, the black texts with a white background are the abbreviation of catchments names. The magenta stars show the locations of eleven hydrological gauge stations, and the magenta texts present the name of hydrological gauge stations. Their descriptions can be further found in Table 1. The blue curves represent the main rivers and tributaries in China. The country boundary is shown with the dash line.

Table 1. Descriptions for the nine exorheic catchments of China (sorted by areas and location). The climate categories are based on annual precipitation and dryness [36].

Catchments ID	Name	Hydrological Gauge Stations	Area ($\times 10^4$ km^2)	Climate Categories
1	Upper Yangtze River Basin (UYRB)	Yichang	100.55	Humid
2	Middle Yangtze River Basin (MYRB)	Datong-Yichang	69.99	Humid
3	Yellow River Basin (YeRB)	Huayuankou	73.00	Semiarid
4	Songhuajiang River Basin (SRB)	Jiamusi	52.83	Semihumid
5	Pearl River Basin (PRB)	Gaoyao + Shijiao + Boluo	41.52	Humid
6	Liaohe River Basin (LRB)	Liujianfang	18.58	Semiarid
7	Haihe River Basin (HRB)	Haihezha	14.28	Semiarid
8	Huaihe River Basin (HuRB)	Bengbu	12.13	Humid
9	Minjiang River Basin (MRB)	Zhuqi	5.45	Humid

2.2. Water Balance Equation

The ET can be estimated from surface water balance on the basin or continental scales, which usually serves as a benchmark for other products. The equation is as follows:

$$\mathrm{ET}_{\mathrm{WB}} = P - Q - ds/dt \tag{1}$$

where $\mathrm{ET}_{\mathrm{WB}}$ is calculated ET, P is precipitation, Q is river discharge, and ds/dt is the change in terrestrial water storage for a specific time period [4,11,15]. TWSC is estimated as the temporal derivative of TWSA from the GRACE products [37,38]. ET, P, and Q are the cumulated amount in a full month [4,39]. Then ds/dt (TWSC) is the differential of two consecutive months of TWSA at the beginning of a month. To obtain the time point of every beginning of a month of TWSA time series, we interpolate TWSA by an interpolation method. A similar process of calculation can be found in Li et al. [40]. Firstly, seasonal and trend signals are estimated using unweighted least squares and then interpolated for every beginning of a month. Secondly, the residuals removed by TWSA time series subtracting seasonal and trend signals are then interpolated by linear interpolation. Finally, the sum of interpolated residuals, seasonal, and trend signals are the interpolated TWSA time series.

Root mean square error (RMSE) is used to evaluate the deviation between ET_{WB} and other types of ETs. The equation is as follows:

$$RMSE = \sqrt{\frac{1}{N}\sum_{i=1}^{N}(X_i - Y_i)^2}. \tag{2}$$

where N is the data length (time series), X_i is the ith estimated ET results from other methods, Y_i is the ith ET_{WB} result.

2.3. Data

An overview of the datasets can be found in Table 2. This table lists relevant information about the datasets used in this study, such as the TWSC, precipitation, runoff, ET, spatial resolution, and corresponding links of data access. A detailed description of these data is provided below.

Table 2. Summary of the datasets used in this study.

Type	Data Name	Data Version/Source	Spatial Resolution	Data Access
Satellite	TWSC	CSR-M	0.5° × 0.5°	http://www2.csr.utexas.edu/grace
		JPL-M	0.5° × 0.5°	https://grace.jpl.nasa.gov/data/get-data/
		CSRT-GSH	1° × 1°	https://grace.jpl.nasa.gov/data/get-data/
		CSR-DDK4	1° × 1°	http://icgem.gfz-potsdam.de/home
In situ	Precipitation	CMDC v2.0	0.5° × 0.5°	http://data.cma.cn/
	Runoff	RSBC	Station	http://www.mwr.gov.cn/sj/tjgb/zghlnsgb/
Model	ET	GLDAS-1 Noah	1° × 1°	https://ldas.gsfc.nasa.gov/gldas/
		GLDAS-2.1 Noah	1° × 1°	
		GLEAM v3.2a	0.25° × 0.25°	https://www.gleam.eu/
	Runoff	GLDAS-1 Noah	1° × 1°	https://ldas.gsfc.nasa.gov/gldas/
		GLDAS-2.1 Noah	1° × 1°	
Forcing data	Precipitation	GLDAS-1 Noah	1° × 1°	
		GLDAS-2.1 Noah	1° × 1°	
		MSWEP	0.5° × 0.5°	http://www.gloh2o.org/

2.3.1. GRACE–Derived TWSC Data

We used the Center for Space Research (CSR) GRACE RL05 Mascons data to estimate the TWSC. The mascon solutions are global and can be better applied to hydrology, oceanography, and the cryosphere without any post–processing and without applying any empirical scaling factors [41]. The data can be downloaded from http://www2.csr.utexas.edu/grace. The Mascons data is represented at a 0.5-degree lon-lat grid and is estimated with the same standards as the CSR RL05 spherical harmonics solutions using GRACE Level-1 observations. C_{20} coefficients were replaced, degree-1 coefficients (Geocenter) and glacial isostatic adjustment (GIA) corrections were applied. More details about the CSR GRACE RL05 Mascons (CSR-M) can be found in Save et al. [41]. With the development of post-processing GRACE satellite data, several GRACE solutions can be used for hydrology applications. However, different solutions would lead to different TWSC estimates.

To evaluate the impact of TWSC from different GRACE solutions on the estimate of ET, we also take JPL Mascons [42], CSR GRCTellus Land data [43], and CSR RL05 spherical harmonics solutions with the DDK4 filter applied (CSR-DDK4) [44] as a comparison. All of these above solutions are processed with the same C_{20} coefficients replaced, the same degree-1 coefficients, and GIA corrections.

The processing of JPL Mascons is based on external information provided by near-global geophysical models to constrain the solution. JPL Mascons use the coarse 3-degree spherical cap Mascons, and they are downscaled to 0.5° × 0.5° using downscaling factors (dsf) calculated from Community Land Model (CLM ver. 4.0) [42], the grid values of JPL Mascons are multiplied by downscaling factors (JPL-M.dsf). The CSR and JPL mascon solutions can be used directly without leakage corrections. CSR GRCTellus Land data is developed by Landerer and Swenson [43] from CSR data, and scaling factors are provided to account for the signal loss during processing related to truncation to degree and order 60 and application of a 300 km Gaussian smoothing filter. The grid values of CSR GRCTellus Land are multiplied by scaling factors (CSRT-GSH.sf). The DDK filter is proposed by Kusche et al. [44], and the DDK4 filter shows a good performance in the application of the upper Yellow River [45]. The process of CSR-DDK4 is similar to CSR GRCTellus Land data while replacing 300 km Gaussian smoothing filter and destriping filter with the DDK4 filter. There is no leakage correction applied in CSR-DDK4 in this study, as with in Yi et al. [45], and in fact, the results for CSR-DDK4 are at the same level with the other solutions.

2.3.2. In Situ Precipitation and Runoff Data

Gridded precipitation data was obtained from the China Meteorological Data Service Center (CMDC, hereafter, P_{CMDC}). The gridded precipitation data was generated by a thin plate spline spatial interpolation of precipitation observations from 2472 weather stations. It has a monthly temporal resolution and $0.5° \times 0.5°$ spatial resolution over all of China [46]. This data is validated by cross-validation and error analysis with gauge-based precipitation, indicating good quality. This precipitation data has been used in several studies [13,22,47]. The monthly runoff datasets are from eleven gauge stations recorded in RSBC (hereafter, Q_{RSBC}) (http://www.mwr.gov.cn/sj/tjgb/zghlnsgb/) [35], which integrates all runoff considering the upstream of the corresponding catchment. The runoff measurements are all from well-gauged rivers in China.

2.3.3. Land Evapotranspiration Products

We use two kinds of land evapotranspiration products for comparison, which included GLDAS ET and GLEAM ET (Table 2). Two versions of GLDAS LSM data are used for inter-comparison in this paper, i.e., GLDAS version 1 (GLDAS-1) and GLDAS version 2.1 (GLDAS-2.1) [9]. ET outputs from both GLDAS versions are driven by Noah LSM [48]. GLDAS-1 datasets cover the time period from 1979 to the present. GLDAS-2.1 datasets cover the period from 2000 to the present. Their temporal resolutions used here are monthly. More information and details about the GLDAS-1 and some improvements and changes about the GLDAS-2.1 are available at https://ldas.gsfc.nasa.gov/gldas/. The ET outputs from GLDAS-1 and GLDAS-2.1 are expressed as $ET_{GLDAS-1}$ and $ET_{GLDAS-2.1}$.

We use GLEAM v3.2a ET products (hereafter, ET_{GLEAM}), which were published jointly by Vrije Universiteit Amsterdam, Netherlands and Ghent University, Belgium [6]. The data has a spatial resolution of $0.25° \times 0.25°$ and daily temporal resolution. We sum them to the monthly results in this study. GLEAM uses a set of algorithms to separately estimate the different components (transpiration, bare-soil evaporation, interception loss, open-water evaporation, and sublimation) of land ET. The Priestley and Taylor equation was used in GLEAM to calculate potential evaporation based on observations of surface net radiation and near-surface air temperature. The rationale of GLEAM is to maximize the recovery of information on evaporation contained in current satellite observations of climatic and environmental variables [6].

2.3.4. Precipitation Forcing Data and Modeled Runoff Data

The precipitation forcing data from the GLDAS-1, GLDAS-2.1 (hereafter, $P_{GLDAS-1}$ and $P_{GLDAS-2.1}$), and the Multi-Source Weighted-Ensemble Precipitation (MSWEP, precipitation forcing data of GLEAM, hereafter, P_{MSWEP}) datasets [49] are used to explain the difference of ET results. They are also computed as regional averages. As runoff is another critical variable in the water balance equation, we also calculate the mean runoff outputs of the nine exorheic catchments from GLDAS-1 and GLDAS-2.1 Noah LSM (hereafter, $Q_{GLDAS-1}$ and $Q_{GLDAS-2.1}$) and compare the results with those for in situ runoff.

2.4. Methods

In the above-mentioned data sets, which are provided with different spatial resolutions, are used with regional averages results, the different spatial resolutions have little impact on ET estimates. The grids in the catchments are used to extract regional averages estimates. Their results are computed on a monthly scale. The ET results are shown at monthly mean and annual scales, as the amount of annual ET, interannual changes, and mean annual cycles of ET are the main characteristics of ET. Besides, the difference between the ET estimates can be clearer at monthly mean and annual scales.

We use the TWSC from GRACE (CSR-M), P_{CMDC}, and Q_{RSBC} to derive ET_{WB}. To explore the impact of TWSC on ET estimate, we also estimate the ET from precipitation minus runoff directly (expressed as ET_{PQ}) without considering TWSC. The results are shown in Section 3.1.1.

Here we compute TWSC results from different GRACE solutions to evaluate their impact on ET while keeping all other inputs (*P* and *Q*) unchanged (Section 3.1.2). The GRACE products include CSR-M, JPL-M.dsf, CSRT-GSH.sf, and CSR-DDK4. The TWSC used for ET estimates from JPL-M.dsf, CSRT-GSH.sf, and CSR-DDK4 are expressed as $ET_{JPL-M.dsf}$, $ET_{CSRT-GSH.sf}$, and $ET_{CSR-DDK4}$, respectively. They are further discussed in Section 4.1.

We then compare the ET estimates from GLDAS and GLEAM with ET_{WB} (Section 3.2). The discussion of RMSEs between ET_{WB} and other ET estimates are shown in Section 4.2. We attempt to analyze the deviation of precipitation and runoff to the regional ET estimate from a water balance perspective (see Sections 3.3 and 4.3), quantifying the deviations between P_{CMDC} and precipitation forcing data, Q_{RSBC}, and Q_{GLDAS} estimates. Previous studies demonstrate that the TWSA (or TWSC) from GRACE and GLDAS are comparable [50–52]. Hence, we do not compare the TWSC component in the water balance equation.

We compute the deviations between water balance ET and other ET results, precipitation results, runoff results, and results of precipitation minus runoff (Section 3.3). We assess the impact of deviation of precipitation and runoff on the estimate of ET based on RMSE and the change of RMSE (Section 4.3). The RMSEs are calculated between annual ET_{WB} minus annual P_{CMDC} and other ET estimates minus their precipitation forcing data (expressed as RMSE (ET-P)) from a water balance perspective. Similarly, the RMSE (ET-Q) represents the calculated RMSEs between annual ET_{WB} minus Q_{RSBC} and ET_{GLDAS} minus Q_{GLDAS}, and the RMSE (ET-(P-Q)) represents the calculated RMSEs between annual ET_{WB} subtracting the result of P_{CMDC} minus Q_{RSBC} and annual ET_{GLDAS} subtracting the result of P_{GLDAS} minus Q_{GLDAS} from 2003–2015. The proportions of RMSEs changed of RMSE (ET-P), RMSE (ET-Q) and RMSE (ET-(P-Q)) relative to RMSE (ET) are further computed.

2.5. Uncertainty Estimation

The TWSC estimates used in the estimate of ET_{WB} are from CSR-M. Hence, we only estimate the uncertainty of TWSC based on CSR-M. The uncertainty estimate followed the method used in Landerer and Swenson [43] and Scanlon et al. [53]. Details about the method can be found in the supporting information of Scanlon et al. [53]. As the TWSC is the differential of two consecutive months, the uncertainty of TWSC is $\sqrt{2}$ of the uncertainty of TWSA. The uncertainties of monthly precipitation and runoff data collected by gauge are estimated to 10% and 5%, respectively [2,4,13]. The uncertainty of monthly ET_{WB} is estimated by uncertainties of TWSC, precipitation, runoff based on error propagation law.

The monthly mean ET is the mean values of 13 months for the study period of 2003–2015, from the error propagation law, the uncertainty of monthly mean ET can be calculated as the uncertainty in monthly ET_{WB} divided by $\sqrt{13}$. The uncertainty of annual ET_{WB} is estimated from the uncertainty of annual *P*, *Q*, and TWSC based on error propagation law. Since the annual TWSC is estimated from the difference of the TWSA at the beginning month in one year and the next year, we estimate the uncertainty of annual TWSC equal to monthly TWSC.

3. Results

3.1. Impact of TWSC on ET Estimate

3.1.1. ET Estimated by Ignoring TWSC

Figure 3 shows the monthly mean of ET_{WB} following Equation 1, ET_{PQ}, *P*, *Q*, and TWSC for 2003–2015 in the nine catchments. The negative ET_{WB} values in January and February in the SRB and December in the Haihe River Basin (HRB) may result from the uncertainties of in situ precipitation and TWSC [13,40]. The TWSC has a significant impact on ET estimates in most catchments. The deviation of the monthly mean of ET_{WB} and ET_{PQ} reaches 34.2 mm/month in June (accounting for 52.9% of ET_{WB}) in the Upper Yangtze River Basin (UYRB). The deviations between ET_{WB} and ET_{PQ} range from 6.7 to 37.2 mm/month for twelve months in the Middle Yangtze River Basin (MYRB), and its RMSE accounts for 37.1% of variations of ET_{WB}. The RMSEs are computed following Equation 2. In the YeRB and SRB, the deviations between ET_{WB} and ET_{PQ} are small, with their RMSEs between ET_{WB} and ET_{PQ} reaching only 6.4 and 8.7 mm/month, respectively. In the MRB, the RMSE between ET_{WB} and ET_{PQ} shows the maximum value, i.e., 27.2 mm/month, accounts for 33.5% of variations of monthly mean ET_{WB}.

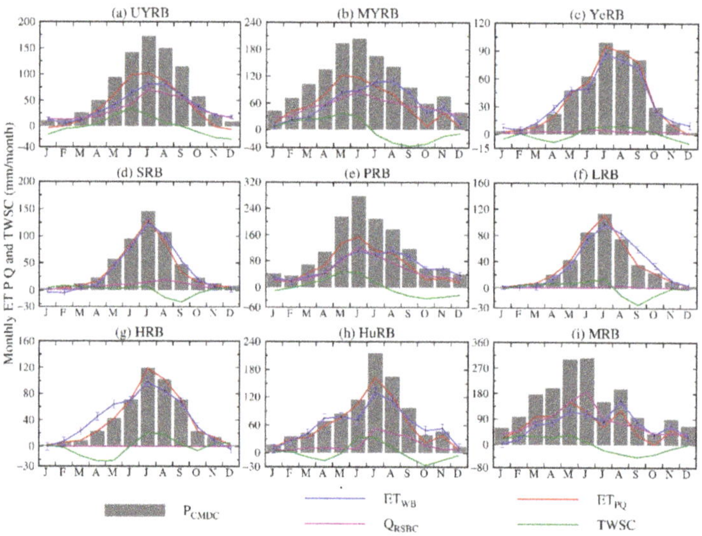

Figure 3. Monthly mean of ET_{WB} (blue curves), ET_{PQ} (red lines), P, Q (purple lines), and TWSC (green lines) for 2003–2015 in the nine exorheic catchments. The histograms represent the monthly precipitation (P). The error bars show the uncertainties of monthly mean ET_{WB}. (**a**)–(**i**) corresponding to the nine exorheic catchments in China. (**a**): UYRB; (**b**): MYRB; (**c**) YeRB; (**d**): SRB; (**e**): PRB; (**f**): LRB; (**g**) HRB; (**h**): HuRB; (**i**): MRB.

The annual ET_{WB} and ET_{PQ} estimates are shown in Figure 4; the mean annual P_{CMDC}, ET_{WB}, and ET_{PQ} results are shown in Table 3. In the UYRB, the largest deviation of annual ET between ET_{WB} and ET_{PQ} is only 27.5 mm/yr in 2014, and the RMSE between ET_{WB} and ET_{PQ} only makes up 2.6% of the mean annual ET. In the SRB, the TWSC has a large impact on annual ET, large deviations between ET_{WB} and ET_{PQ} occur almost all the years, and the proportion of the RMSE accounting for the mean annual ET_{WB} reaches 11.5%. In the Minjiang River Basin (MRB), the RMSE represents 12.7% of the mean annual ET, with the largest deviation (291.8 mm/yr, 39.1% of total ET_{WB} in this year) occurring in 2003.

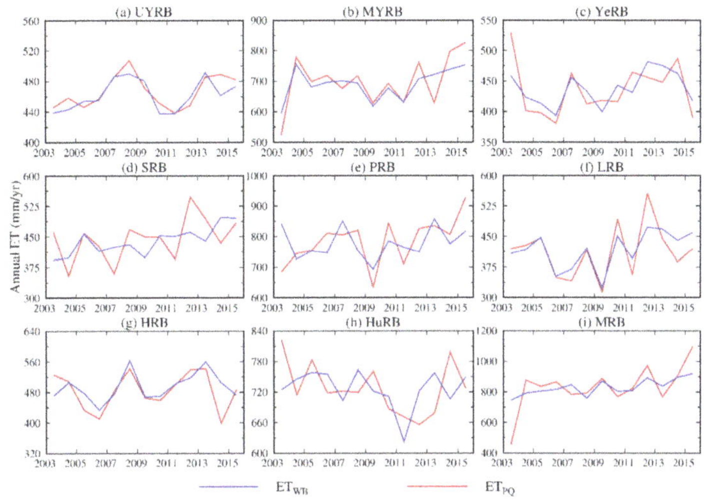

Figure 4. Annual ET_{WB} and ET_{PQ} from 2003–2015. (**a**)–(**i**) corresponding to the nine exorheic catchments in China.

Table 3. The mean annual precipitation from CMDC, mean annual ET_{WB} and ET_{PQ} estimation from GRACE in the exorheic catchments of China from 2003–2015, with one standard deviation (unit: mm/yr).

Catchments ID	Name	P_{CMDC}	ET_{WB}	ET_{PQ}
1	UYRB	864.7 ± 56.2	461.9 ± 20.3	466.4 ± 21.4
2	MYRB	1331.4 ± 149.7	689.7 ± 50.3	698.0 ± 83.2
3	YeRB	473.0 ± 45.2	437.5 ± 28.1	435.6 ± 43.4
4	SRB	545.3 ± 80.0	439.6 ± 34.3	445.0 ± 53.4
5	PRB	1411.6 ± 183.1	778.3 ± 50.4	784.9 ± 77.6
6	LRB	426.5 ± 72.4	417.7 ± 46.5	413.1 ± 66.1
7	HRB	486.9 ± 49.3	494.2 ± 37.2	484.2 ± 48.9
8	HuRB	944.1 ± 168.5	726.4 ± 37.8	727.6 ± 50.8
9	MRB	1786.3 ± 391.7	831.5 ± 53.3	833.9 ± 146.6

3.1.2. Impact of Different GRACE Solutions on ET Estimate

The monthly mean TWSC from different GRACE solutions is shown in Figure A1, where their mean TWSC is the arithmetical mean from all TWSC estimates for the corresponding calendar month. Note that the TWSC derived from CSR-M compare favorably with the mean TWSC (Figure A1), thus the CSR-M TWSC is used for the estimate of ET_{WB} in this study. The TWSC from CSRT-GSH.sf show significant differences among the four TWSC results, as they exaggerate the monthly mean TWSC in the UYRB, MYRB, YeRB, SRB, and PRB, and the differences may result from the scaling factor derived from CLM4.5 [43]. The spatial distribution of scaling factors is checked in our study (not shown), and the spatial variability of scaling factors varies greatly in the basins, indicating exaggerated TWSC. The maximum deviations are calculated between each two monthly mean TWSC estimates, which range from 10.7 to 35.6 mm/month, the deviations occur in the YeRB (10.7 mm/month) and MRB (35.6 mm/month) (Figure A1c,i), respectively. As the area of MRB is the smallest, and with the most abundant precipitation, it is understandable that the MRB shows the largest deviation of TWSC. The large deviation of TWSC between JPL-M.dsf and other GRACE solutions in the HRB and MRB (Figure A1g,i) may result from the processing strategy and coarse resolution in the spatial of JPL-M since the areas of the two basins are small [54].

The annual ET estimates based on different GRACE solutions are shown in Figure A2. The RMSEs among ET_{CSR-M} (=ET_{WB}), $ET_{JPL-M.dsf}$, $ET_{CSRT-GSH.sf}$, and $ET_{CSR-DDK4}$ are understandably less than those RMSEs between ET_{WB} and ET_{PQ}, and their interannual fluctuations are more consistent than that of ET_{PQ}. We compute the standard deviations (STDs) between the four ET estimates from different GRACE solutions for every single year, and the results show that the max STD is only 51.2 mm/yr, occurring in the MRB. The mean STD for the years from 2003–2015 in the corresponding catchment is also computed, ranging from 9.7 to 27.1 mm/yr (accounting for 1.8–3.9% of annual ET_{WB}), with the least occurring in the YeRB and the largest occurring in the MRB. In three catchments, the max STDs occur in 2003 in the PRB, the Huaihe River Basin (HuRB), and the MRB, which are located in Southeast China. In the other three catchments, the max STDs appear in 2011, which are the YeRB, SRB, and Liaohe River Basin (LRB), in North China.

3.2. Comparison of Different ET Products

Figure 5 shows the monthly mean of ET estimates from different ET products. Their mean annual cycles are similar among all the catchments. In the humid catchments: in the UYRB, the other three ET products overestimate the ET compared with ET_{WB} for all months except December, the maximum deviation exists in July, which has the most precipitation (Figure 5a). In the MYRB, other ET estimates are bigger than ET_{WB} estimates for all months except November when the ET_{WB} increases to respond to increased precipitation. In the PRB, the mean of ET_{WB} in July is less than that in June and August, and the mean of ET_{WB} in October is also less than September and November (Figure 5e). In the HuRB, ET_{WB} shows a rapid increase response for sharply increased precipitation in July (Figure 5h), while the three other ET results do not catch it. The ET_{WB} also can capture the irregular monthly mean precipitation changes from June to December in the MRB (Figure 5i). In the semihumid and semiarid catchments: the two versions of GLDAS both show the maximum deviation in September with ET_{WB} in the YeRB. Most months of $ET_{GLDAS\text{-}2.1}$ are more than other ET estimates in the LRB. During the intense irrigation period of April and May, the ET_{WB} is significantly greater than other ET estimates in the HRB. From the above, these ET results all show similar annual cycles, while ET_{WB} can capture some irregular variations in monthly precipitation.

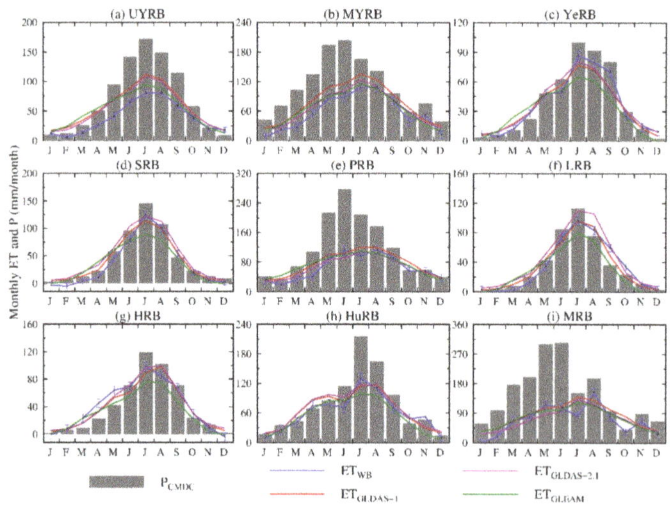

Figure 5. Monthly mean of ET estimates in the exorheic catchments of China. The error bars show the uncertainties of monthly mean ET_{WB}. (**a**)–(**i**) corresponding to the nine exorheic catchments in China.

The maximum RMSEs between the monthly mean of ET_{WB} and other ET results are in the MRB, which are 27.5 (vs $ET_{GLDAS\text{-}1}$), 26.3 (vs $ET_{GLDAS\text{-}2.1}$), and 24.7 (vs ET_{GLEAM}) mm/month (Table A1). In the YeRB, the RMSEs are the least, which are 7.2, 7.5, and 13.0 mm/month. We also compute the proportion of the RMSEs accounting for an average of the monthly mean of ET_{WB}. The proportions in the UYRB show the maximum values, which are 50.5% (vs $ET_{GLDAS-1}$), 43.3% (vs $ET_{GLDAS-2.1}$), and 43.1% (vs ET_{GLEAM}). The HuRB experienced the minimum proportions, which are 20.4%, 22.3%, and 25.9%, respectively.

The annual ET from different sources is illustrated in Figure 6, which shows huge gaps among different ET estimates. In terms of the humid catchments: In the UYRB and MYRB, it is obvious that other annual ET estimates are all larger than ET_{WB}. Their mean deviations between ET_{WB} and ET_{GLEAM} reaching 144.7 mm/yr (31.3% in mean annual ET_{WB}) and 88.0 mm/yr (12.8% in mean annual ET_{WB}) in the two catchments (Figure 6a,b). In the PRB, $ET_{GLDAS-1}$ and ET_{GLEAM} both overestimate the annual ET, and $ET_{GLDAS-2.1}$ shows different interannual variations with respect to ET_{WB} (Figure 6e). In the HuRB, all the ET estimates capture the drop of ET in 2011 due to reduced precipitation (Figure A3h), but there is some discrepancy among mean annual ET. In the MRB, the ET results show large differences in the interannual variations. The $ET_{GLDAS-1}$ even verges on 1100 mm/yr after 2012. Concerning the semihumid and semiarid catchments: In the YeRB, the ET_{WB} is consistent with $ET_{GLDAS-2.1}$ except for the years 2006 and 2009, while ET_{GLEAM} underestimates the annual ET for all the years. In the SRB, the $ET_{GLDAS-2.1}$ is significantly greater than the other annual ET. Nevertheless, the ET_{GLEAM} is close to ET_{WB}. In the LRB, the $ET_{GLDAS-1}$ is close to ET_{WB} in mean annual ET, and their interannual variations are similar. In the HRB, $ET_{GLDAS-2.1}$ is closest to ET_{WB}. ET_{GLEAM} somewhat underestimates the annual ET for the other three results. For the two catchments in Northeast China (i.e., SRB and LRB, Figure 6d,f), both $ET_{GLDAS-2.1}$ results overestimate the annual ET. Additionally, the four ET results show consistent interannual changes in most catchments.

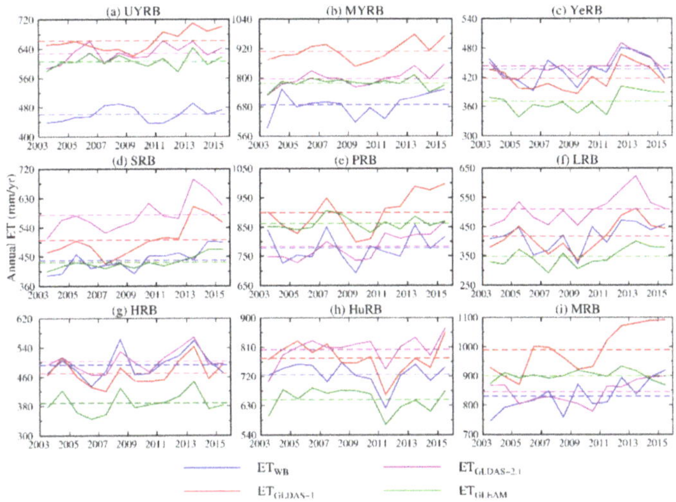

Figure 6. Annual ET from different products in the exorheic catchments of China. The dotted lines with different colors corresponding to their mean annual ET. (**a**)–(**i**) corresponding to the nine exorheic catchments in China.

3.3. Comparison of Different Precipitation and Runoff Inputs for ET Estimation

In all the catchments, the interannual fluctuations of precipitation from different sources show similar patterns (Figure A3), while the mean annual precipitation shows some differences (Figure 7). In the MYRB and YeRB, the P_{CMDC} is higher than $P_{GLDAS-1}$, which is similar to the comparison from Lv et al. [55] in corresponding regions. In the SRB and LRB (Northeast China), $P_{GLDAS-2.1}$ is prominently larger than the other three precipitation sources (Figure A3d,f, and Figure 7b). Caution should be taken when using the $P_{GLDAS-2.1}$ in the two catchments. It should be noted that the annual

P_{MSWEP} is the least for all the catchments (Figure A3), and mean $P_{GLDAS\text{-}1}$ are all less than those of P_{CMDC} (Figure 7a).

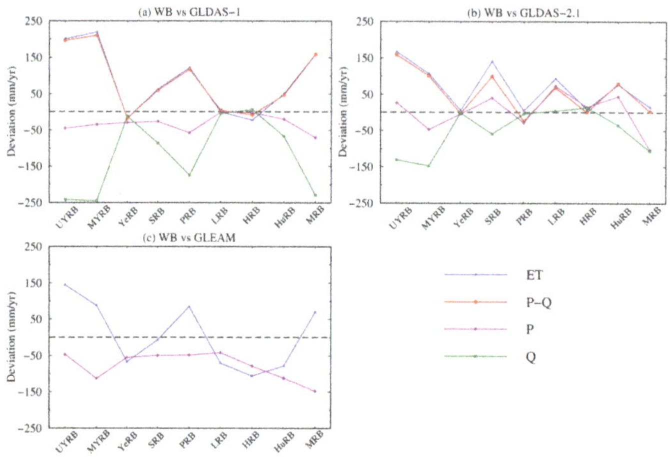

Figure 7. Deviations of annual ET, precipitation, and runoff between (**a**): WB and GLDAS–1; (**b**): WB and GLDAS–2.1; (**c**): WB and GLEAM. The blue curves with triangles represent the deviations between annual ET and other ET estimates. The red curves with dots represent the deviations between P_{CMDC} minus Q_{RSBC} and P_{GLDAS} minus the Q_{GLDAS}. The purple lines with rhombus points represent the deviations between P_{CMDC} and other precipitation forcing data. The green lines with square points represent the deviations of runoff between Q_{RSBC} and two versions of Q_{GLDAS}.

Q_{RSBC}, $Q_{GLDAS\text{-}1}$, and $Q_{GLDAS\text{-}2.1}$ show larger discrepancies than that for precipitation. The comparison of annual runoff from 2003–2015 can be found in Figure A4. Since the runoff is modeled results, it faces more uncertainties than precipitation. As the similar results showed in YRB (UYRB and MYRB) and YeRB in Lv et al. [55], the in situ runoff was significantly larger than that from $Q_{GLDAS\text{-}1}$. The interannual variations of runoff from different sources show similar patterns in most catchments except in the YeRB and HRB, which experienced small amounts of runoff. In all the catchments, $Q_{GLDAS\text{-}2.1}$ is larger than that from $Q_{GLDAS\text{-}1}$, and they are closer to Q_{RSBC} in most catchments, presumably due to some modification for GLDAS-2.1 [56].

To explore the impact of precipitation and modeled runoff (GLDAS Noah LSM outputs) on ET estimates, we analyze the difference in both sides of the water balance equation. We first compute the deviation between mean annual ET, precipitation, runoff, and precipitation minus runoff for 2003–2015 (Figure 7). In the UYRB, the mean annual deviation between ET_{WB} and $ET_{GLDAS\text{-}1}$ reaches 200.9 mm/yr, the mean annual deviation between P_{CMDC} minus Q_{RSBC} (expressed as P-Q) and $P_{GLDAS\text{-}1}$ minus $Q_{GLDAS\text{-}1}$ are close to the deviation of ET (Figure 7a). This deviation is mostly contributed by the deviation of runoff (−241.6 mm/yr). As Figure 7a shows, the deviations of P-Q are close to the deviations of ET in all the catchments. For the water balance (WB) with GLDAS-2.1 (Figure 7b), the deviations of P-Q are close to the deviations of ET in all the catchments except PRB. In the PRB, the deviation of mean annual ET is only 5.3 mm/yr, while the deviation of mean annual precipitation reaches −27.9 mm/yr, and the deviation of mean annual runoff is −4.34 mm/yr (Figure 7b). Based on the water balance method, we only assess the precipitation forcing data variable for GLEAM ET. As Figure 7c shows, in the YeRB, LRB, and HRB, the deviation of precipitation may explain the most differences between annual ET_{WB} and ET_{GLEAM}. In other catchments, it is somewhat opposite between the deviation of annual precipitation and ET.

3.4. Uncertainty Estimation Results

Uncertainties of TWSC, P_{CMDC}, Q_{RSBC}, and ET_{WB} are shown in Table 4. Large uncertainties of TWSC appear in the PRB, HuRB, and MRB, which are more than 30 mm/month. The large uncertainties of TWSC may result from the small study area (HuRB and MRB) and large variations of TWSC caused by abundant precipitation (PRB and MRB). In three catchments (MYRB, PRB, and MRB), their annual precipitation is more than 1300 mm/yr, and their uncertainties of precipitation are also more than 11 mm/month. The uncertainties of runoff are similar to those of precipitation except the YeRB, LRB, and HRB, where the water use is intense and the runoff is little.

Table 4. The uncertainties of monthly TWSC, precipitation, runoff, ET, monthly mean ET and annual ET from 2003–2015 (unit: mm/month and mm/yr).

Name	TWSC	P_{CMDC}	Q_{RSBC}	ET_{WB}	Monthly Mean ET_{WB}	Annual ET_{WB}
UYRB	± 14.1	± 7.2	± 1.7	± 15.9	± 4.4	± 89.9
MYRB	± 26.0	± 11.1	± 5.0	± 28.8	± 8.0	± 149.8
YeRB	± 14.1	± 3.9	± 0.2	± 14.7	± 4.1	± 49.4
SRB	± 18.8	± 4.5	± 0.4	± 19.3	± 5.4	± 57.9
PRB	± 33.8	± 11.8	± 2.6	± 35.9	± 10.0	± 148.5
LRB	± 17.4	± 3.6	± 0.1	± 17.8	± 4.9	± 46.1
HRB	± 22.0	± 4.1	± 0.0	± 22.4	± 6.2	± 53.4
HuRB	± 33.8	± 7.9	± 0.9	± 34.7	± 9.6	± 100.8
MRB	± 35.3	± 14.9	± 4.0	± 38.5	± 10.7	± 188.2

From Table 4, we can conclude that the uncertainties of monthly ET are mainly from TWSC, which is similar to the conclusions in Long et al. [1] and Pan et al. [13]. Almost in all the catchments, the uncertainties of TWSC are two times or even three times larger than the uncertainties of P_{CMDC}, and they are also much larger than the uncertainties of Q_{RSBC}. The uncertainties of annual ET are all larger than 45 mm/yr, while the uncertainties are mainly from the uncertainties of annual precipitation.

4. Discussions

4.1. Impact of TWSC on ET Estimates in Local Catchments

As we can see, in the UYRB, MYRB, PRB, LRB, and HuRB, the ET_{WB} is typically smaller than ET_{PQ} in the wet season from May to July (Figure 3). Meanwhile, from September to December, the ET_{WB} is larger than ET_{PQ}. From the water balance equation, it is because, during the wet season, TWSA usually increases, and TWSC (*ds/dt*) is greater than 0 (Figure 3). In contrast, in the dry season with less precipitation, TWSA generally decreases, and TWSC (*ds/dt*) is smaller than 0, then ET_{WB} is larger than ET_{PQ} (Figure 3).

It should be noted that the impacts of TWSC on ET estimates are region–specific. On the monthly scale, ET_{WB} is obviously larger than ET_{PQ} from March to May in the HRB (Figure 3g), which is caused by the spring irrigation of wheat [13]. The ET_{WB} is larger than ET_{PQ} from March to May, which is 14.4, 22.1, and 21.5 (sum: 57.9) mm/month, respectively. This result is similar to the human-induced ET (60.0 ± 24.2 mm) estimated by Pan et al. [13] for the same months. While for the SRB and LRB, ET_{WB} is obviously larger than ET_{PQ} from August to October. Since the main crop is corn in this region, the water consumption of the growth period is ongoing in the corresponding period. Meanwhile, there is a significant reduction in P_{CMDC} (−40.3 mm/month) in September relative to August (Figure 3f), which is different from the HRB. As the different water consumption in agriculture, the monthly TWSC are region-specific, then the deviations between ET_{WB} and ET_{PQ} are region-specific.

On the other hand, in the MYRB, PRB, and MRB, the RMSEs between the monthly mean of ET_{WB} and ET_{PQ} is significantly larger than those for other catchments. As Figure A2 shows, the amplitudes of monthly mean TWSC are stronger than other catchments. These catchments are all located in South China, with abundant precipitation [57]. During the rainy season, as the water is stored and TWSC is positive, the monthly ET_{WB} is smaller than ET_{PQ} in all the catchments. As the corresponding months of the rainy season are different with respect to different catchments, these catchments also show regional heterogeneities. However, the region-specific impacts of TWSC on ET deserves more research.

On the annual scale, the sizeable variations between annual ET_{WB} and ET_{PQ} can mostly be explained by large precipitation anomalies (Figure A3), such as the first year in the Figure 4c (YeRB) and 4i (MRB) and year 2012 in the Figure 4f (LRB). Their variations correspond to the lowest or highest precipitation during the whole study period in the corresponding catchment.

It is interesting that all the STDs of mean annual ET_{WB} are less than those for ET_{PQ} (see Table 3), indicating smaller interannual ET_{WB} fluctuations. In the year with more precipitation, such as the years 2008 and 2014 in the UYRB, 2012 and 2013 in the SRB, based on the water balance equation, as the TWSA increases, TWSC is greater than 0, then ET_{WB} should be less than ET_{PQ} in the given year. On the contrary, in the years with precipitation deficit, as the TWSA usually decreases, the ET_{WB} would be higher than ET_{PQ}. We would deem that the TWSA plays a role as a reservoir in the terrestrial water cycle, impounding water and reducing the amount of water that returns to the atmosphere through evapotranspiration or other forms in the wet years, but discharging water in the dry years. We can conclude that estimating annual ET simply by subtracting runoff from precipitation would overestimate the interannual fluctuations of ET.

The difference between mean annual ET_{WB} and ET_{PQ} reflects the long-term rate of TWSA in a catchment. The ET_{WB} is significantly higher than ET_{PQ} in the HRB, where the water depletion (mainly from groundwater) is fast [58]. In the LRB and YeRB, the mean annual ET_{WB} is also larger than ET_{PQ}, which also indicates the water depletion there [58]. Conversely, TWSA increases in the UYRB, MYRB, SRB, and PRB, and therefore the mean annual ET_{WB} is typically less than that for ET_{PQ}.

4.2. The Differences between ET_{WB} and other ET Estimates

On the monthly scale, the RMSEs between ET_{WB} and ET from different GRACE solutions are smaller, and the $ET_{CSR\text{–}DDK4}$ is closest to ET_{WB} among the three GRACE solutions (Table A1). In the YeRB, SRB, and LRB, the maximum RMSEs are between ET_{WB} and ET_{GLEAM}, and all of these catchments are located in North China and are semiarid catchments. In the UYRB and MRB, the RMSEs between ET_{WB} and $ET_{GLDAS\text{–}1}$ show the maximum values. In the MYRB, PRB, HRB, and HuRB, the RMSEs between ET_{WB} and ET_{PQ} show the maximum values, which indicates that impacts of ignoring TWSC on the ET estimate is the most, and it should be noted that all of these catchments are humid catchments except HRB with intense water consumption.

On the annual scale, the RMSEs between ET_{WB} and ET from the three GRACE solutions show small values, while $ET_{CSRT\text{–}GSH.sf}$ is closest to ET_{WB} among the three solutions (Table A2). It indicates some differences in the TWSC estimate on the monthly and annual scales. The RMSEs between ET_{WB} and ET from other products markedly exceed those between ET_{WB} and ET from other GRACE solutions. In the UYRB, MYRB, PRB, and MRB, the RMSEs between ET_{WB} and $ET_{GLDAS\text{–}1}$ show the maximum values, which are all in humid regions. It should be noted that in the UYRB, the RMSE between ET_{WB} and ET_{PQ} is even less than the RMSEs between ET_{WB} and ET from other GRACE solutions, which indicates that the interannual variations of TWSC are very small in this catchment. ET estimates from different GRACE solutions generally show relatively small deviations in all the catchments, and ET estimates from different products are generally relatively large deviations in the humid catchments.

4.3. Impact of Precipitation and Modeled Runoff from a Water Balance Perspective

The RMSEs between annual ET_{WB} and other ET results are further analyzed. Their results are shown in Table A3 (WB – GLDAS-1), Table A4 (WB – GLDAS-2.1), and Table A5 (WB – GLEAM). For Table A3, in the UYRB, MYRB, SRB, PRB, and MRB, the RMSEs between ET_{WB} and $ET_{GLDAS\text{–}1}$ can be markedly reduced if the deviation of $P_{GLDAS\text{–}1}$ and $Q_{GLDAS\text{–}1}$ can both be taken into consideration. Generally speaking, though GLDAS ET outputs are not computed based on the water balance method [9,59]. If the accuracy of $P_{GLDAS\text{–}1}$ and $Q_{GLDAS\text{–}1}$ can be improved in China, e.g., modeled

$Q_{GLDAS-1}$ verified by in situ runoff. Then the ET estimate would also benefit from improved runoff outputs based on the water balance equation during the simulation process. In the YeRB, LRB, HRB, and HuRB, the RMSEs are also reduced, with smaller proportions reduced than above catchments. In the YeRB and LRB, if we only consider the difference of runoff, the RMSEs would even increase, and in the HRB, the RMSE is also slightly reduced. Since the outflows are much smaller in these three catchments than other catchments, the deviation of runoff is small itself (Figure 7a). Unlike the other humid catchments, in the three semiarid catchments (YeRB, LRB, and HRB), the proportions of the RMSEs of ET–P as opposed to ET are reduced, which indicates that deviations of precipitation forcing data indeed contribute to deviations of ET.

In Table A4, the RMSEs between ET_{WB} and $ET_{GLDAS-2.1}$ reduced in all the catchments except HRB when the deviations of precipitation and runoff can be considered. In the YRB (UYRB and MYRB), the RMSEs between Q_{RSBC} and $Q_{GLDAS-2.1}$ account for most of the deviations. In the HRB and MRB, the deviations between ET_{WB} and $ET_{GLDAS-2.1}$ do not result from the precipitation difference. It should be noted that in the LRB, if the precipitation inconsistency is considered (Figure A3f), the RMSE between ET_{WB} and $ET_{GLDAS-2.1}$ is dramatically reduced, which can explain the cause of overestimation of the annual ET for $ET_{GLDAS-2.1}$. In the HRB, with the deviation of precipitation and modeled runoff considered, the proportion of the RMSE increased (Table A4). Since the HRB is heavily influenced by human activities [13,31], the RMSE between mean annual ET_{WB} and GLDAS ET outputs is mainly contributed by anthropogenic activities [13].

As for the RMSEs between ET_{WB} and ET_{GLEAM}, we only compute their precipitation difference (Table A5). In the YeRB, LRB, HRB, and HuRB, if the precipitation difference can be taken into consideration, the RMSEs between ET_{WB} and ET_{GLEAM} would be reduced, the YeRB, LRB, and HRB are semiarid catchments. Figure A3 also shows a large deviation between P_{CMDC} and P_{MSWEP}. The proportion of the RMSE reduced in the YeRB reaches 72.8%. In the UYRB, MYRB, SRB, PRB, and MRB, the RMSEs would even increase, which indicates that the deviations of precipitation do not contribute or contribute little to the deviation between ET_{GLEAM} and ET_{WB}. The RMSE between ET and ET–P rapidly increases from 86.9 to 228.7 mm/yr in the MRB, there is a small difference between their annual precipitation actually (Figure A3i).

Here we try to explore the deviation between ET_{WB} and GLDAS or GLEAM ET based on the water balance equation. The RMSE (ET) would decrease if the deviation of P_{GLDAS} and modeled Q_{GLDAS} in the GLDAS LSM can be taken into consideration. In four catchments (YeRB, LRB, HRB, and HuRB), precipitation differences contribute to the deviation between ET_{WB} and ET_{GLEAM}. However, the increased RMSE (ET-P), RMSE (ET-Q) and RMSE (ET-(P-Q)) relative to RMSE (ET) should be further explored. We do not investigate other forcing variables except precipitation to derive ET, e.g., radiation, air temperature, and snow water equivalent [6,9,60]. Therefore, a future intercomparison can be performed to identify the impact of these variables on ET estimates.

4.4. *Impact of Groundwater Baseflow and Water Diversion on ET Estimates*

Based on the water balance equation, the groundwater inflow and outflow across the basin boundary would also affect the estimate of ET. As an example, in the LRB, according to the estimate of groundwater outflow from Zhang and Li [61], the outflow is 0.61×10^8 m^3/yr, and its impact on the annual ET is only ~0.3 mm/yr. Therefore, it can be negligible relative to the annual ET (417.7 ± 46.5 mm/yr).

Water diversion in the basin inside and outside is also a part of basin water balance. In China, there is South-to-North water diversion, which includes the east route, the middle route, and the west route projects (http://nsbd.mwr.gov.cn/). The west route project has not been built yet. The starting point of the east route is in the mainstream of Lower Yangtze River, transporting water to Shandong Province, which is not in our study area. The middle route transports water from the MYRB to the HRB, is going through the HuRB and the YeRB. It transported water to the North in October 2014 for the first time, with a water volume of 21.67×10^8 m^3 in the first year. The impact on the ET estimate is

3.1 mm/yr for the MYRB, which is relatively small compared to annual ET (689.7 ± 50.3 mm/yr). If the water is totally supplied to the HRB, the impact on the ET estimate will reach 15.18 mm/yr, exerting a certain influence on the ET estimate in the HRB (494.2 ± 37.2 mm/yr). If we estimate the ET after 2015 in this region, it is necessary to account for the water diversion.

4.5. Impact of Spatial Scale on ET Estimate

The area of the MRB is only 5.45×10^4 km^2, which is less than the typical GRACE footprint (20×10^4 km^2). However, some studies have demonstrated that GRACE is capable of detecting TWSA in local regions with an area smaller than GRACE resolution if the signal amplitude is large enough [44,62,63]. As the MYRB receives the most abundant precipitation among these catchments (Table 3), TWSA should have higher SNR (Signal to Noise Ratio), and TWSC tends to have higher reliability. On the other hand, the maximum uncertainty of monthly ET estimate is indeed in the MRB, where the uncertainties of monthly TWSC, precipitation, and runoff are also large (Table 3). Thus, we recommend that caution should be exercised when using TWSA estimates in regions with a small area.

5. Conclusions

In this study, the ET was calculated based on the water balance equation in nine exorheic catchments of China. The impacts of ignoring terrestrial water storage changes and different terrestrial water storage changes from GRACE solutions on ET estimates were analyzed. The intercomparison between ET_{WB} and ET estimates from GLEAM, and GLDAS land surface models was also conducted. The comparison was carried out on the monthly and annual scales.

We found that the impact of ignoring terrestrial water storage changes on the estimate of ET is noteworthy. The RMSEs of between monthly mean ET_{WB} and ET_{PQ} range from 6.4–27.2 mm/yr (17.5–45.2% in corresponding mean monthly ET). The annual RMSEs between ET_{WB} and ET_{PQ} in the estimate of ET range from 12.0–105.8 mm/yr (2.6–12.7% in corresponding annual ET) among these catchments. The STDs of annual ET_{WB} for study periods are all less than those from ET_{PQ}, which simply estimate the annual ET by subtracting runoff from precipitation would overestimate the interannual variations of ET. Thus, TWSC should not be ignored in the estimate of ET.

The ET estimates from different GRACE solutions show relatively small deviations. The RMSEs among different GRACE solutions in most catchments are less than 10 mm/month on the monthly scale and 30 mm/yr on the annual scale. In all the catchments except the HRB and MRB, CSR-GSH.sf solutions exaggerate the monthly mean TWSC, and caution should be taken when applying this solution to derive TWSC.

Different precipitation products are assessed to explain the inconsistency between different ET products and regional ET from a water balance perspective. The difference between ET_{WB} and ET from GLDAS land surface model results can be partly explained from deviation from precipitation forcing data in several catchments, especially in the LRB. Furthermore, the ET estimates would also benefit from improved runoff outputs during the simulation process. In the three semiarid catchments and the HuRB, the RMSEs between ET_{WB} and ET_{GLEAM} can be reduced, provided that the difference of precipitation can be taken into consideration. However, the increased RMSEs with deviations of precipitation forcing data and modeled runoff considering in the estimate of ET deserves further exploration.

The ET estimates show some arresting interannual fluctuations, which warrants further study. In the SRB and MRB, there may exist some positive trends, which are likely resulting from increased precipitation or other effects. The trends are also worthy of further research. In summary, our study emphasizes the capability of GRACE in estimating the ET on the basin scale. The ET estimate based on water balance can be a benchmark to other ET products, which would benefit the GLDAS LSMs and remote sensing ET estimates.

Author Contributions: Conceptualization, M.Z. and Y.Z.; Data curation, Y.Z.; Funding acquisition, Y.Z., M.Z. and B.J.; Investigation, Y.Z.; Methodology, Y.Z. and Y.M.; Supervision, M.Z. and B.J.; Writing—original draft, Y.Z.;

Writing—review and editing, M.Z., Y.M. and B.J. All authors have read and agreed to the published version of the manuscript.

Funding: The research is funded by the National Natural Science Foundation of China (41874091, 41774094, and 41474061); Fundamental Research Funds for the Central Universities, China University of Geosciences (Wuhan) (G1323519314); and Open Research Fund Program of State Key Laboratory of Geodesy and Earth's Dynamics (SKLGED2019-2-5-E, SKLGED2019-3-2-E).

Acknowledgments: The authors thank Fei Li, Wei Feng, and Haoming Yan for their insightful suggestion and discussions and thank Fan Xie for her help. We thank four anonymous reviewers for their comments, which help to improve this paper.

Conflicts of Interest: The authors declare no conflict of interest.

Appendix A

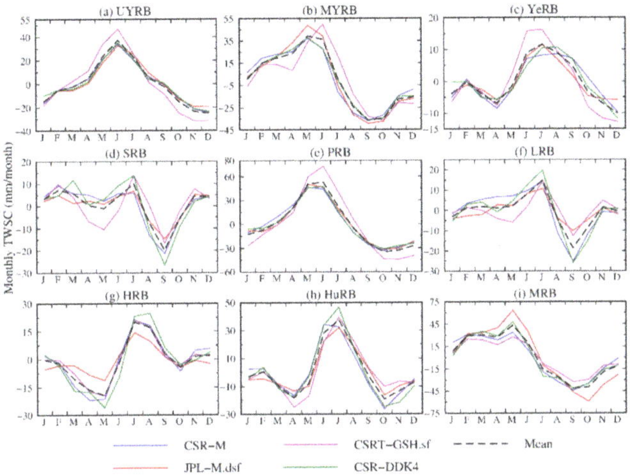

Figure A1. Monthly mean TWSC from different GRACE solutions and products in the exorheic catchments of China from 2003–2015. (**a**)–(**i**) corresponding to the nine exorheic catchments in China.

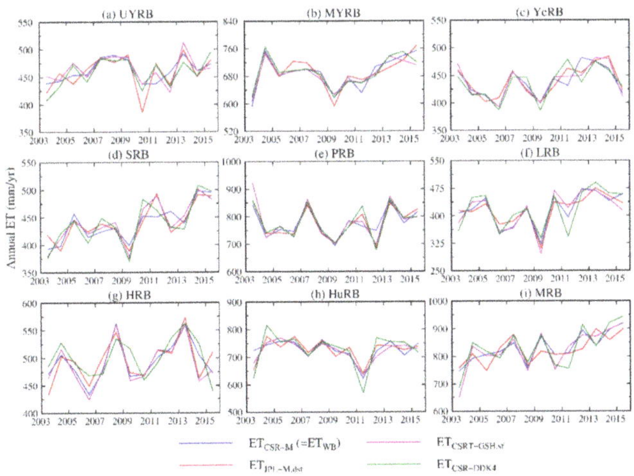

Figure A2. Annual ET based on different GRACE solutions and products in the exorheic catchments of China from 2003 to 2015. (**a**)–(**i**) corresponding to the nine exorheic catchments in China.

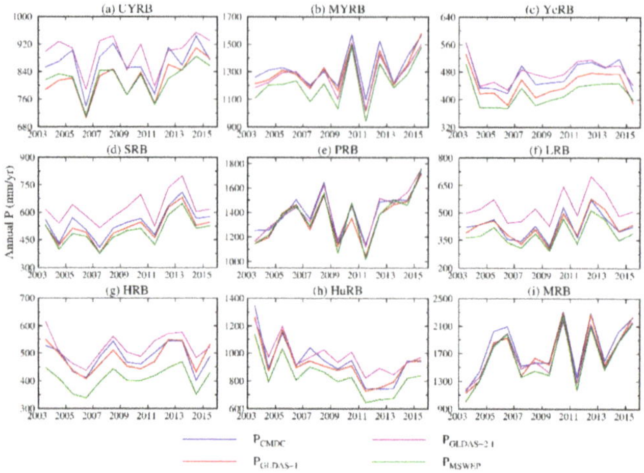

Figure A3. Annual precipitation from CMDC, and the forcing data of two versions of GLDAS and GLEAM from 2003–2015. (**a**)–(**i**) corresponding to the nine exorheic catchments in China.

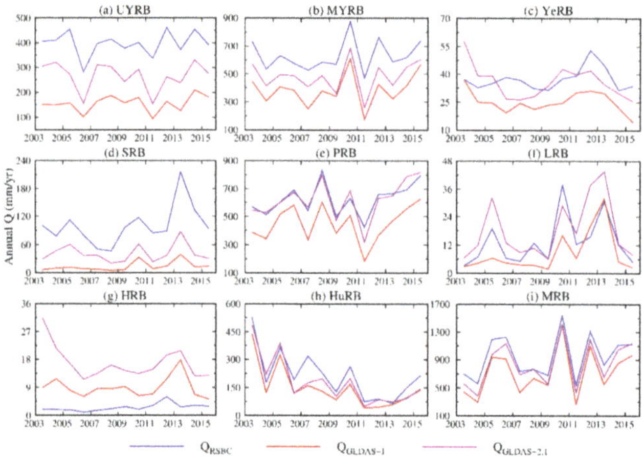

Figure A4. Annual runoff from RSBC, GLDAS-1 Noah LSM and GLDAS-2.1 Noah LSM for 2003–2015. (**a**)–(**i**) corresponding to the nine exorheic catchments in China.

Table A1. The root mean square errors (RMSEs) between monthly mean ET_{WB} and other ET estimations from 2003–2015 (unit: mm/month).

Name	RMSEs between ET_{WB} and other ET Results						
	ET_{PQ}	$ET_{JPL\text{–}M.dsf}$	$ET_{CSRT\text{–}GSH.sf}$	$ET_{CSR\text{–}DDK4}$	$ET_{GLDAS\text{–}1}$	$ET_{GLDAS\text{–}2.1}$	ET_{GLEAM}
UYRB	17.4	3.1	8.3	2.3	19.5	16.7	16.6
MYRB	24.9	6.9	14.8	4.8	21.4	13.2	13.8
YeRB	6.4	3.0	5.2	2.3	7.2	7.5	13.0
SRB	8.7	3.5	7.5	4.1	9.0	14	16.5
PRB	26.4	4.7	16.3	3.3	15.3	11.5	17.5
LRB	10.6	6.4	8.2	3.9	7.2	11.1	14.1
HRB	13.0	7.6	3.2	4.6	8.0	7.7	12.1
HuRB	17.8	7.2	9.9	5.4	12.3	13.5	15.7
MRB	27.2	17.5	8.8	8.1	27.5	26.3	24.7

Table A2. The RMSEs between annual ET_{WB} and other ET estimations from 2003–2015 (unit: mm/yr).

Name	RMSEs between ET_{WB} and other ET Results						
	ET_{PQ}	$ET_{JPL-M.dsf}$	$ET_{CSRT-GSH.sf}$	$ET_{CSR-DDK4}$	$ET_{GLDAS-1}$	$ET_{GLDAS-2.1}$	ET_{GLEAM}
UYRB	12.0	21.2	15.0	18.5	203.5	169.3	145.9
MYRB	46.5	19.5	18.5	18.2	222.8	113.3	100.0
YeRB	29.6	15	12.5	19.7	23.9	17.2	69.8
SRB	50.5	19.1	17.7	18.9	73.4	146	19.6
PRB	71.2	23.9	29.4	31.9	131.2	49.1	95.2
LRB	35.4	18.9	20.1	25.6	31.7	99.8	76.1
HRB	37.7	22.5	15.0	24.0	35.4	20.5	106.9
HuRB	54.7	26.9	20.1	43.2	60.3	86.3	82.7
MRB	105.8	37.2	35.5	33.7	168.4	52.9	86.9

Table A3. The RMSEs between mean annual ET_{WB} and $ET_{GLDAS-1}$ (expressed as RMSE(ET)). The RMSEs between annual ET_{WB} minus P_{CMDC} and annual $ET_{GLDAS-1}$ minus $P_{GLDAS-1}$ (expressed as RMSE(ET–P)). The RMSEs between annual ET_{WB} minus Q_{RSBC} and annual $ET_{GLDAS-1}$ minus $Q_{GLDAS-1}$ (expressed as RMSE(ET–Q)). The RMSEs between annual ET_{WB} subtracting the result of P_{CMDC} minus Q_{RSBC} and annual $ET_{GLDAS-1}$ subtracting the result of $P_{GLDAS-1}$ minus $Q_{GLDAS-1}$ from 2003–2015 (expressed as RMSE(ET–(P–Q))). The proportion of RMSE changed of RMSE(ET–P) as opposed to RMSE (ET), RMSE(ET–Q) as opposed to RMSE (ET) and RMSE(ET–(P–Q)) as opposed to RMSE (ET).

Name	RMSE (mm/yr)				Proportion of RMSE Changed (%)		
	ET	ET-P	ET-Q	ET-(P-Q)	ET-P	ET-Q	ET-(P-Q)
UYRB	203.5	247.3	55.6	22.0	21.5	−72.7	−89.2
MYRB	222.8	258.4	51.6	31.7	16	−76.8	−85.8
YeRB	23.9	17.0	34.2	13.1	−29	43	−45
SRB	73.4	98.1	28.5	13.1	33.6	−61.2	−82.2
PRB	131.2	188.7	79.0	33.8	43.9	−39.8	−74.3
LRB	31.7	18.2	36.3	22.2	−42.6	14.6	−30.1
HRB	35.4	30.8	32.0	27.9	−12.9	−9.5	−21.2
HuRB	60.3	91.3	37.9	32.8	51.4	−37.2	−45.6
MRB	168.4	244.3	100.2	35.1	45.1	−40.5	−79.1

Table A4. The RMSEs between mean annual ET_{WB} and $ET_{GLDAS-2.1}$ (expressed as RMSE(ET)). The RMSEs between annual ET_{WB} minus P_{CMDC} and annual $ET_{GLDAS-2.1}$ minus $P_{GLDAS-2.1}$ (expressed as RMSE(ET–P)). The RMSEs between annual ET_{WB} minus Q_{RSBC} and annual $ET_{GLDAS-2.1}$ minus $Q_{GLDAS-2.1}$ (expressed as RMSE(ET–Q)). The RMSEs between annual ET_{WB} subtracting the result of P_{CMDC} minus Q_{RSBC} and annual $ET_{GLDAS-2.1}$ subtracting the result of $P_{GLDAS-2.1}$ minus $Q_{GLDAS-2.1}$ from 2003–2015 (expressed as RMSE(ET–(P–Q))). The proportion of RMSE changed of RMSE(ET–P) as opposed to RMSE (ET), RMSE(ET–Q) as opposed to RMSE (ET) and RMSE(ET–(P–Q)) as opposed to RMSE (ET).

Name	RMSE (mm/yr)				Proportion of RMSE Changed (%)		
	ET	ET-P	ET-Q	ET-(P-Q)	ET-P	ET-Q	ET-(P-Q)
UYRB	169.3	139.8	45.9	15.1	−17.5	−72.9	−91.1
MYRB	113.3	165.8	59.4	32.7	46.4	−47.6	−71.1
YeRB	17.2	14.9	16.4	14.4	−13.6	−4.4	−16.3
SRB	146.0	76.3	86.3	22.9	−47.7	−40.9	−84.3
PRB	49.1	37.7	61.9	35.2	−23.4	25.9	−28.3
LRB	99.8	28.0	105.6	28.9	−71.9	5.8	−71
HRB	20.5	38.0	30.8	30.1	84.9	50.2	46.4
HuRB	86.3	75.3	64.3	33.3	−12.7	−25.5	−61.4
MRB	52.9	120.7	117.5	36.9	128.1	122.1	−30.2

Table A5. The RMSEs between mean annual ET^{WB} and ET^{GLEAM} (expressed as RMSE(ET)), RMSEs between annual ET^{WB} minus P^{CMDC} and annual ET^{GLEAM} minus P^{MSWEP} from 2003–2015 (expressed as RMSE(ET–P)).

Name	RMSE (mm/yr)		Proportion of RMSE Changed (%)
	ET	ET-P	ET-P
UYRB	145.9	192.8	32.1
MYRB	100.0	212.6	112.6
YeRB	69.8	19.0	−72.8
SRB	19.6	45.2	130.5
PRB	95.2	152.5	60.3
LRB	76.1	35.2	−53.8
HRB	106.9	34.3	−67.9
HuRB	82.7	51.3	−37.9
MRB	86.9	228.7	163.0

References

1. Long, D.; Longuevergne, L.; Scanlon, B.R. Uncertainty in evapotranspiration from land surface modeling, remote sensing, and GRACE satellites. *Water Resour. Res.* **2014**, *50*, 1131–1151. [CrossRef]
2. Wang, K.C.; Dickinson, R.E. A review of global terrestrial evapotranspiration: observation, modeling, climatology, and climatic variability. *Rev. Geophys.* **2012**, *50*. [CrossRef]
3. Miralles, D.G.; van den Berg, M.J.; Gash, J.H.; Parinussa, R.M.; de Jeu, R.A.M.; Beck, H.E.; Holmes, T.R.H.; Jiménez, C.; Verhoest, N.E.C.; Dorigo, W.A.; et al. El Niño–La Niña cycle and recent trends in continental evaporation. *Nat. Clim. Change* **2013**, *4*, 122–126. [CrossRef]
4. Rodell, M.; Famiglietti, J.S.; Chen, J.; Seneviratne, S.I.; Viterbo, P.; Holl, S.; Wilson, C.R. Basin scale estimates of evapotranspiration using GRACE and other observations. *Geophys. Res. Lett.* **2004**, *31*, 1–4. [CrossRef]
5. Yu, G.R.; Wen, X.F.; Sun, X.M.; Tanner, B.D.; Lee, X.H.; Chen, J.Y. Overview of ChinaFLUX and evaluation of its eddy covariance measurement. *Agric. For. Meteorol.* **2006**, *137*, 125–137. [CrossRef]
6. Martens, B.; Miralles, D.G.; Lievens, H.; van der Schalie, R.; de Jeu, R.A.M.; Fernandez–Prieto, D.; Beck, H.E.; Dorigo, W.A.; Verhoest, N.E.C. GLEAM v3: satellite–based land evaporation and root–zone soil moisture. *Geosci. Model Dev.* **2017**, *10*, 1903–1925. [CrossRef]
7. Mu, Q.Z.; Zhao, M.S.; Running, S.W. Improvements to a MODIS global terrestrial evapotranspiration algorithm. *Remote Sens. Environ.* **2011**, *115*, 1781–1800. [CrossRef]
8. Yang, Y.T.; Long, D.; Shang, S.H. Remote estimation of terrestrial evapotranspiration without using meteorological data. *Geophys. Res. Lett.* **2013**, *40*, 3026–3030. [CrossRef]
9. Rodell, M.; Houser, P.R.; Jambor, U.; Gottschalck, J.; Mitchell, K.; Meng, C.J.; Arsenault, K.; Cosgrove, B.; Radakovich, J.; Bosilovich, M.; et al. The global land data assimilation system. *Bull. Am. Meteorol. Soc.* **2004**, *85*, 381–394. [CrossRef]
10. Bai, P.; Liu, X.M. Intercomparison and evaluation of three global high–resolution evapotranspiration products across China. *J. Hydrol.* **2018**, *566*, 743–755. [CrossRef]
11. Liu, W.B.; Wang, L.; Zhou, J.; Li, Y.Z.; Sun, F.B.; Fu, G.B.; Li, X.P.; Sang, Y.F. A worldwide evaluation of basin–scale evapotranspiration estimates against the water balance method. *J. Hydrol.* **2016**, *538*, 82–95. [CrossRef]
12. Tapley, B.D.; Bettadpur, S.; Ries, J.C.; Thompson, P.F.; Watkins, M.M. GRACE measurements of mass variability in the Earth system. *Science* **2004**, *305*, 503–505. [CrossRef] [PubMed]
13. Pan, Y.; Zhang, C.; Gong, H.L.; Yeh, P.J.F.; Shen, Y.J.; Guo, Y.; Huang, Z.Y.; Li, X.J. Detection of human–induced evapotranspiration using GRACE satellite observations in the Haihe River basin of China. *Geophys. Res. Lett.* **2017**, *44*, 190–199. [CrossRef]
14. Eicker, A.; Forootan, E.; Springer, A.; Longuevergne, L.; Kusche, J. Does GRACE see the terrestrial water cycle "intensifying"? *J. Geophys. Res. Atmos.* **2016**, *121*, 733–745. [CrossRef]
15. Ramillien, G.; Frappart, F.; Guntner, A.; Ngo–Duc, T.; Cazenave, A.; Laval, K. Time variations of the regional evapotranspiration rate from Gravity Recovery and Climate Experiment (GRACE) satellite gravimetry. *Water Resour. Res.* **2006**, *42*, 1–8. [CrossRef]

16. Boronina, A.; Ramillien, G. Application of AVHRR imagery and GRACE measurements for calculation of actual evapotranspiration over the Quaternary aquifer (Lake Chad basin) and validation of groundwater models. *J. Hydrol.* **2008**, *348*, 98–109. [CrossRef]
17. Ferguson, C.R.; Sheffield, J.; Wood, E.F.; Gao, H.L. Quantifying uncertainty in a remote sensing–based estimate of evapotranspiration over continental USA. *Int. J. Remote Sens.* **2010**, *31*, 3821–3865. [CrossRef]
18. Swann, A.L.S.; Koven, C.D. A Direct Estimate of the Seasonal Cycle of Evapotranspiration over the Amazon Basin. *J. Hydrometeorol.* **2017**, *18*, 2173–2185. [CrossRef]
19. Castle, S.L.; Reager, J.T.; Thomas, B.F.; Purdy, A.J.; Lo, M.H.; Famiglietti, J.S.; Tang, Q.H. Remote detection of water management impacts on evapotranspiration in the Colorado River Basin. *Geophys. Res. Lett.* **2016**, *43*, 5089–5097. [CrossRef]
20. Dee, D.P.; Uppala, S.M.; Simmons, A.J.; Berrisford, P.; Poli, P.; Kobayashi, S.; Andrae, U.; Balmaseda, M.A.; Balsamo, G.; Bauer, P. The ERA–Interim reanalysis: configuration and performance of the data assimilation system. *Q. J. R. Meteorolog. Soc.* **2011**, *137*, 553–597. [CrossRef]
21. Badgley, G.; Fisher, J.B.; Jimenez, C.; Tu, K.P.; Vinukollu, R. On Uncertainty in Global Terrestrial Evapotranspiration Estimates from Choice of Input Forcing Datasets. *J. Hydrometeorol.* **2015**, *16*, 1449–1455. [CrossRef]
22. Mao, Y.N.; Wang, K.C.; Liu, X.M.; Liu, C.M. Water storage in reservoirs built from 1997 to 2014 significantly altered the calculated evapotranspiration trends over China. *J. Geophys. Res. Atmos.* **2016**, *121*, 10097–10112. [CrossRef]
23. Mao, Y.N.; Wang, K.C. Comparison of evapotranspiration estimates based on the surface water balance, modified Penman–Monteith model, and reanalysis data sets for continental China. *J. Geophys. Res. Atmos.* **2017**, *122*, 3228–3244. [CrossRef]
24. Jiang, Y.Y.; Wang, W.; Zhou, Z.H. Evaluation of MODIS MOD16 Evapotranspiration Product in Chinese River Basins. *J. Nat. Resour.* **2017**, *32*, 517–528. [CrossRef]
25. Li, X.L.; Liang, S.L.; Yuan, W.P.; Yu, G.R.; Cheng, X.; Chen, Y.; Zhao, T.B.; Feng, J.M.; Ma, Z.G.; Ma, M.G.; et al. Estimation of evapotranspiration over the terrestrial ecosystems in China. *Ecohydrology* **2014**, *7*, 139–149. [CrossRef]
26. Mu, Q.Z.; Heinsch, F.A.; Zhao, M.; Running, S.W. Development of a global evapotranspiration algorithm based on MODIS and global meteorology data. *Remote Sens. Environ.* **2007**, *111*, 519–536. [CrossRef]
27. Cleugh, H.A.; Leuning, R.; Mu, Q.Z.; Running, S.W. Regional evaporation estimates from flux tower and MODIS satellite data. *Remote Sens. Environ.* **2007**, *106*, 285–304. [CrossRef]
28. Zhang, Y.Q.; Leuning, R.; Chiew, F.H.S.; Wang, E.L.; Zhang, L.; Liu, C.M.; Sun, F.B.; Peel, M.C.; Shen, Y.J.; Jung, M. Decadal Trends in Evaporation from Global Energy and Water Balances. *J. Hydrometeorol.* **2012**, *13*, 379–391. [CrossRef]
29. Senay, G.B.; Leake, S.; Nagler, P.L.; Artan, G.; Dickinson, J.; Cordova, J.T.; Glenn, E.P. Estimating basin scale evapotranspiration (ET) by water balance and remote sensing methods. *Hydrol. Process.* **2011**, *25*, 4037–4049. [CrossRef]
30. Xue, B.L.; Wang, L.; Li, X.P.; Yang, K.; Chen, D.L.; Sun, L.T. Evaluation of evapotranspiration estimates for two river basins on the Tibetan Plateau by a water balance method. *J. Hydrol.* **2013**, *492*, 290–297. [CrossRef]
31. Feng, W.; Zhong, M.; Lemoine, J.M.; Biancale, R.; Hsu, H.T.; Xia, J. Evaluation of groundwater depletion in North China using the Gravity Recovery and Climate Experiment (GRACE) data and ground–based measurements. *Water Resour. Res.* **2013**, *49*, 2110–2118. [CrossRef]
32. Scanlon, B.R.; Longuevergne, L.; Long, D. Ground referencing GRACE satellite estimates of groundwater storage changes in the California Central Valley, USA. *Water Resour. Res.* **2012**, *48*. [CrossRef]
33. Zeng, Z.Z.; Piao, S.L.; Lin, X.; Yin, G.D.; Peng, S.S.; Ciais, P.; Myneni, R.B. Global evapotranspiration over the past three decades: estimation based on the water balance equation combined with empirical models. *Environ. Res. Lett.* **2012**, *7*, 014026. [CrossRef]
34. Wang, D.B. Evaluating interannual water storage changes at watersheds in Illinois based on long–term soil moisture and groundwater level data. *Water Resour. Res.* **2012**, *48*. [CrossRef]
35. Ministry of Water Resources of the People's Republic of China (MWR). *River Sediment Bulletin of China*; Ministry of Water Resour. of the PRC, Ed.; China Water Power Press: Beijing, China, 2013; p. 85.
36. Tan, Y.J. *Alternation of Dry and Wet Climate Zone and its Cause Analysis in China in Last 50 Years*; Nanjing University of Information Technology: Nanjing, China, 2016.

37. Forootan, E.; Safari, A.; Mostafaie, A.; Schumacher, M.; Delavar, M.; Awange, J.L. Large–Scale Total Water Storage and Water Flux Changes over the Arid and Semiarid Parts of the Middle East from GRACE and Reanalysis Products. *Surv. Geophys.* **2016**, *38*, 591–615. [CrossRef]
38. Long, D.; Shen, Y.J.; Sun, A.; Hong, Y.; Longuevergne, L.; Yang, Y.T.; Li, B.; Chen, L. Drought and flood monitoring for a large karst plateau in Southwest China using extended GRACE data. *Remote Sens. Environ.* **2014**, *155*, 145–160. [CrossRef]
39. Li, Q.; Luo, Z.C.; Zhong, B.; Zhou, H. An Improved Approach for Evapotranspiration Estimation Using Water Balance Equation: Case Study of Yangtze River Basin. *Water* **2018**, *10*, 812. [CrossRef]
40. Li, X.Y.; Long, D.; Han, Z.Y.; Scanlon, B.R.; Sun, Z.L.; Han, P.F.; Hou, A.Z. Evapotranspiration Estimation for Tibetan Plateau Headwaters Using Conjoint Terrestrial and Atmospheric Water Balances and Multisource Remote Sensing. *Water Resour. Res.* **2019**, *55*, 8608–8630. [CrossRef]
41. Save, H.; Bettadpur, S.; Tapley, B.D. High–resolution CSR GRACE RL05 mascons. *J. Geophys. Res. Solid Earth* **2016**, *121*, 7547–7569. [CrossRef]
42. Wiese, D.N.; Landerer, F.W.; Watkins, M.M. Quantifying and reducing leakage errors in the JPL RL05M GRACE mascon solution. *Water Resour. Res.* **2016**, *52*, 7490–7502. [CrossRef]
43. Landerer, F.W.; Swenson, S.C. Accuracy of scaled GRACE terrestrial water storage estimates. *Water Resour. Res.* **2012**, *48*. [CrossRef]
44. Kusche, J.; Schmidt, R.; Petrovic, S.; Rietbroek, R. Decorrelated GRACE time–variable gravity solutions by GFZ, and their validation using a hydrological model. *J. Geod.* **2009**, *83*, 903–913. [CrossRef]
45. Yi, S.; Song, C.Q.; Wang, Q.Y.; Wang, L.S.; Heki, K.; Sun, W.K. The potential of GRACE gravimetry to detect the heavy rainfall–induced impoundment of a small reservoir in the upper Yellow River. *Water Resour. Res.* **2017**, *53*, 6562–6578. [CrossRef]
46. Zhao, Y.; Zhu, J.; Xu, Y. Establishment and assessment of the grid precipitation datasets in China for recent 50 years. *J. Meteorol. Sci* **2014**, *34*, 414–420. [CrossRef]
47. Ren, Z.G.; Zhang, M.J.; Wang, S.J.; Qiang, F.; Zhu, X.F.; Dong, L. Changes in daily extreme precipitation events in South China from 1961 to 2011. *J. Geog. Sci.* **2015**, *25*, 58–68. [CrossRef]
48. Ek, M.B.; Mitchell, K.E.; Lin, Y.; Rogers, E.; Grunmann, P.; Koren, V.; Gayno, G.; Tarpley, J.D. Implementation of Noah land surface model advances in the National Centers for Environmental Prediction operational mesoscale Eta model. *J. Geophys. Res. Atmos.* **2003**, *108*. [CrossRef]
49. Beck, H.E.; van Dijk, A.I.J.M.; Levizzani, V.; Schellekens, J.; Miralles, D.G.; Martens, B.; de Roo, A. MSWEP: 3–hourly 0.25 degrees global gridded precipitation (1979–2015) by merging gauge, satellite, and reanalysis data. *Hydrol. Earth Syst. Sci.* **2017**, *21*, 589–615. [CrossRef]
50. Zhang, Z.Z.; Chao, B.F.; Chen, J.L.; Wilson, C.R. Terrestrial water storage anomalies of Yangtze River Basin droughts observed by GRACE and connections with ENSO. *Global Planet. Change* **2015**, *126*, 35–45. [CrossRef]
51. Zhong, Y.L.; Zhong, M.; Feng, W.; Zhang, Z.Z.; Shen, Y.C.; Wu, D.C. Groundwater Depletion in the West Liaohe River Basin, China and Its Implications Revealed by GRACE and In Situ Measurements. *Remote Sens.* **2018**, *10*, 493. [CrossRef]
52. Syed, T.H.; Famiglietti, J.S.; Rodell, M.; Chen, J.; Wilson, C.R. Analysis of terrestrial water storage changes from GRACE and GLDAS. *Water Resour. Res.* **2008**, *44*. [CrossRef]
53. Scanlon, B.R.; Zhang, Z.Z.; Save, H.; Wiese, D.N.; Landerer, F.W.; Long, D.; Longuevergne, L.; Chen, J.L. Global evaluation of new GRACE mascon products for hydrologic applications. *Water Resour. Res.* **2016**, *52*, 9412–9429. [CrossRef]
54. Ran, J.; Ditmar, P.; Klees, R.; Farahani, H.H. Statistically optimal estimation of Greenland Ice Sheet mass variations from GRACE monthly solutions using an improved mascon approach. *J. Geod.* **2018**, *92*, 299–319. [CrossRef] [PubMed]
55. Lv, M.X.; Ma, Z.G.; Yuan, X.; Lv, M.Z.; Li, M.X.; Zheng, Z.Y. Water budget closure based on GRACE measurements and reconstructed evapotranspiration using GLDAS and water use data for two large densely–populated mid–latitude basins. *J. Hydrol.* **2017**, *547*, 585–599. [CrossRef]
56. Rodell, M.; Beaudoing, H.K. *GLDAS Noah Land Surface Model L4 monthly 1 x 1 degree V2.1*; NASA/GSFC/HSL, Ed.; Goddard Space Flight Center Earth Sciences Data and Information Services Center (GES DISC): Greenbelt, MD, USA, 2016. [CrossRef]
57. Zhang, Q.; Singh, V.P.; Sun, P.; Chen, X.; Zhang, Z.X.; Li, J.F. Precipitation and streamflow changes in China: changing patterns, causes and implications. *J. Hydrol.* **2011**, *410*, 204–216. [CrossRef]

58. Feng, W.; Shum, C.K.; Zhong, M.; Pan, Y. Groundwater Storage Changes in China from Satellite Gravity: An Overview. *Remote Sens.* **2018**, *10*, 674. [CrossRef]
59. Rui, H.; Beaudoing, H. *Readme Document for Global Land Data Assimilation System Version 2 (GLDAS–2) Products;* DISC/HSL Group, Ed.; Goddard Space Flight Center Earth Sciences Data and Information Services Center (GES DISC): Greenbelt, MD, USA, 2011.
60. Wang, M.M.; He, G.J.; Zhang, Z.M.; Wang, G.Z.; Zhang, Z.J.; Cao, X.J.; Wu, Z.J.; Liu, X.G. Comparison of Spatial Interpolation and Regression Analysis Models for an Estimation of Monthly Near Surface Air Temperature in China. *Remote Sens.* **2017**, *9*, 1278. [CrossRef]
61. Zhang, Z.H.; Li, L.R. *Groundwater Resources of China (Liaoning Volume);* China Cartographic Publishing House: Beijing, China, 2005; p. 111.
62. Longuevergne, L.; Wilson, C.R.; Scanlon, B.R.; Cretaux, J.F. GRACE water storage estimates for the Middle East and other regions with significant reservoir and lake storage. *Hydrol. Earth Syst. Sci.* **2013**, *17*, 4817–4830. [CrossRef]
63. Wang, X.W.; de Linage, C.; Famiglietti, J.; Zender, C.S. Gravity Recovery and Climate Experiment (GRACE) detection of water storage changes in the Three Gorges Reservoir of China and comparison with in situ measurements. *Water Resour. Res.* **2011**, *47*. [CrossRef]

remote sensing

MDPI

Article

Quantitative Evaluations and Error Source Analysis of Fengyun-2-Based and GPM-Based Precipitation Products over Mainland China in Summer, 2018

Jintao Xu [1], Ziqiang Ma [2,3], Guoqiang Tang [4,5], Qingwen Ji [2], Xiaoxiao Min [1], Wei Wan [2] and Zhou Shi [1,*]

1 Institute of Agricultural Remote Sensing and Information Technology Application, College of Environmental and Resource Sciences, Zhejiang University, Hangzhou 310058, China; jintaox@zju.edu.cn (J.X.); xiaoxiaom@zju.edu.cn (X.M.)
2 Institute of Remote Sensing and Geographical Information Systems, School of Earth and Space Sciences, Peking University, Beijing 100871, China; ziqma@pku.edu.cn (Z.M.); qwenji@pku.edu.cn (Q.J.); w.wan@pku.edu.cn (W.W.)
3 State Key Laboratory of Resources and Environmental Information System, Institute of Geographical Sciences and Natural Resources Research, Chinese Academy of Sciences, Beijing 100101, China
4 University of Saskatchewan Coldwater Lab, Canmore, AB T1W 3G1, Canada; guoqiang.tang@usask.ca
5 Centre for Hydrology, University of Saskatchewan, Saskatoon, SK S7N 1K2, Canada
* Correspondence: shizhou@zju.edu.cn; Tel.: +86-188-5818-6616

Received: 13 November 2019; Accepted: 10 December 2019; Published: 12 December 2019

Abstract: Satellite-based quantitative precipitation estimates (QPE) with a fine quality are of great importance to global water cycle and matter and energy exchange research. In this study, we firstly apply various statistical indicators to evaluate and compare the main current satellite-based precipitation products from Chinese Fengyun (FY)-2 and the Global Precipitation Measurement (GPM), respectively, over mainland China in summer, 2018. We find that (1) FY-2G QPE and Integrated Multi-satellitE Retrievals for GPM (IMERG) perform significantly better than FY-2E QPE, using rain gauge data, with correlation coefficients (CC) varying from 0.65 to 0.90, 0.80 to 0.90, and 0.40 to 0.53, respectively; (2) IMERG agrees well with rain gauge data at monthly scale, while it performs worse than FY-2G QPE at hourly and daily scales, which may be caused by its algorithms; (3) FY-2G QPE underestimates the precipitation in summer, while FY-2E QPE and IMERG generally overestimate the precipitation; (4) there is an interesting error phenomenon in that both FY-based and GPM-based precipitation products perform more poorly during the period from 06:00 to 10:00 UTC than other periods at diurnal scale; and (5) FY-2G QPE agrees well with IMERG in terms of spatial patterns and consistency (CC of ~0.81). These findings can provide valuable preliminary references for improving next generation satellite-based QPE retrieval algorithms and instructions for applying these data in various practical fields.

Keywords: precipitation; evaluation; error analysis; Fengyun; quantitative precipitation estimates; GPM; IMERG

1. Introduction

As one of the most active variables in atmospheric circulation, precipitation is a critical linkage between global water and energy cycles. Obtaining spatiotemporal information on precipitation is of great importance for water resource management, climatological modeling, and many other applications [1–3]. Therefore, reliable precipitation datasets gathered from different sources, including ground stations, ground-based weather radars, and satellites, are essential [4,5].

Collecting precipitation information from ground rain gauge stations is the traditional and common method of measurement. However, the limitations are obvious due to the uneven spatial distribution of the stations. The measurements of ground stations are usually very sparse over some regions of the earth (e.g., the Tibetan plateau), which are meteorologically important [6,7]. As for ground-based weather radars, they have certain superiorities when observing precipitation in local areas. Nevertheless, due to the limitations of the scope of observation and the huge cost of equipment acquisition and maintenance, ground-based weather radars are not the first choice for large-scale precipitation observations.

However, precipitation information obtained from satellites does not meet such limitations. Satellite-based precipitation datasets can depict the spatial and temporal variability of precipitation with a considerable accuracy over regions that have few ground stations [5,8]. Over the last four decades, the progress of meteorological satellites has made it possible for scientists to acquire reliable and cost-effective precipitation datasets through a variety of sensors and inversion algorithms [9–13]. Therefore, obtaining high-resolution and accurate precipitation estimates derived from sensors on satellites at a regional or global scale has become a highly-efficient research method at present [4,14,15].

Satellite-based precipitation products provided by several institutions and organizations from all over the world are different in terms of their spatial and temporal resolution, data coverage, data continuity, and latency [16]. The products mentioned above can only be used for practical applications if there is a consistency in terms of both the spatial and temporal scales with ground-based measurements. Therefore, the validation of satellite-based precipitation products is necessary to ensure the reliability of the products. In addition, in order to provide product users with a reliable error structure and instructions for satellite precipitation products, as well as a reasonable advancement of retrieval algorithms, validation is indispensable for satellite-based data applications [17].

There have been numerous studies evaluating the performance of satellite-based precipitation products. Datasets such as Tropical Rainfall Measuring Mission (TRMM) Multisatellite Precipitation Analysis (TMPA), Integrated Multi-satellitE Retrievals for Global Precipitation Measurement (IMERG), Climate Hazards Group InfraRed Precipitation with Stations (CHIRPS), the Climate Prediction Center (CPC) MORPHing technique (CMORPH), Precipitation Estimation from Remotely Sensed Information using Artificial Neural Networks–Climate Data Record (PERSIANN-CDR), Multi-Source Weighted-Ensemble Precipitation (MSWEP), H-SAF (EUMETSAT Satellite Application Facility on Support to Operational Hydrology and Water Management) have been validated in various regions of the world [18–23]. Chen et al. [24] analyzed the similarities and differences between TMPA V6 and V7 over China, and determined that 3B42 RT V7 overestimated precipitation over the Qinghai–Tibet Plateau by approximately 139.5%. Teng et al. [25] identified overestimates outside the 95% prediction interval in TMPA data for the Xin'anjiang Reservoir, which is the largest artificial water body in southeast China. Prakash et al. [26] evaluated the accuracy of IMERG data with TMPA and Global Satellite Mapping of Precipitation (GSMaP) data in southeast India. The results showed that IMERG represented large-scale monsoon rainfall features and their variability more realistically. Tang et al. [22] evaluated IMERG from April to December 2014 at hourly scale over mainland China and found that IMERG performed with a small correlation coefficient (CC) of ~0.40 and slight overestimates by an average of ~9%. Katiraie-Boroujerdy et al. [27] found that PERSIANN-CDR agreed well with gauge-based datasets at monthly scales over Iran, with a CC of ~0.88. Rivera et al. [18] demonstrated the systematic errors that could be attributed to the varying performance of CHIRPS in different seasons over Argentina, such as the significant bias of ~65.8% over the north Patagonia region.

Although there are a large number of evaluation studies on satellite-based precipitation products, few investigations have been conducted to assess the quality of the precipitation products from Chinese Fengyun (FY) series satellites. FY series satellites are the major operational meteorological satellites of China. Currently, there are eight on-orbit FY satellites in operation, including three polar orbit satellites and five geostationary satellites, in order to provide global meteorological observation services. With the increasing influences of FY series satellites, evaluating the performance and usability of their precipitation products has become increasingly necessary.

Compared with the data obtained from polar orbit satellites, precipitation information from geostationary satellites has a fixed observation area and stable observation intervals, which can better reflect the spatial distribution of precipitation and its changes at hourly and other temporal scales in the study areas. In other words, geostationary satellites have not only the spatial continuity of most other satellites, but also the temporal continuity of ground stations. Therefore, we selected two of the main current satellite-based precipitation products from two geostationary satellites in different batches of the FY-2 series to evaluate their quality in this study. The main objects of this study are as follows: (1) To firstly evaluate and compare the precipitation products from FY-2 and GPM at meteorological scales (hourly, daily) and a climatological scale (monthly), respectively, and (2) to analyze the potential error sources of the main current satellite-based precipitation products over mainland China in summer, 2018.

2. Study Area and Datasets

2.1. Study Area

The study area is the region of mainland China with the longitude and latitude range between 73–135°E and 18–53°N, respectively (Figure 1). The spatial distribution of the Digital Elevation Model (DEM) in mainland China is also shown in Figure 1, which demonstrates that the terrain of mainland China is low in the southeast and high in the northwest, forming a three-ladder pattern. Due to the dramatic changes in terrain, the climate in mainland China is of great complexity and is mainly regulated by the monsoon system [28]. The precipitation in China shows conspicuous variability at both temporal and spatial scales.

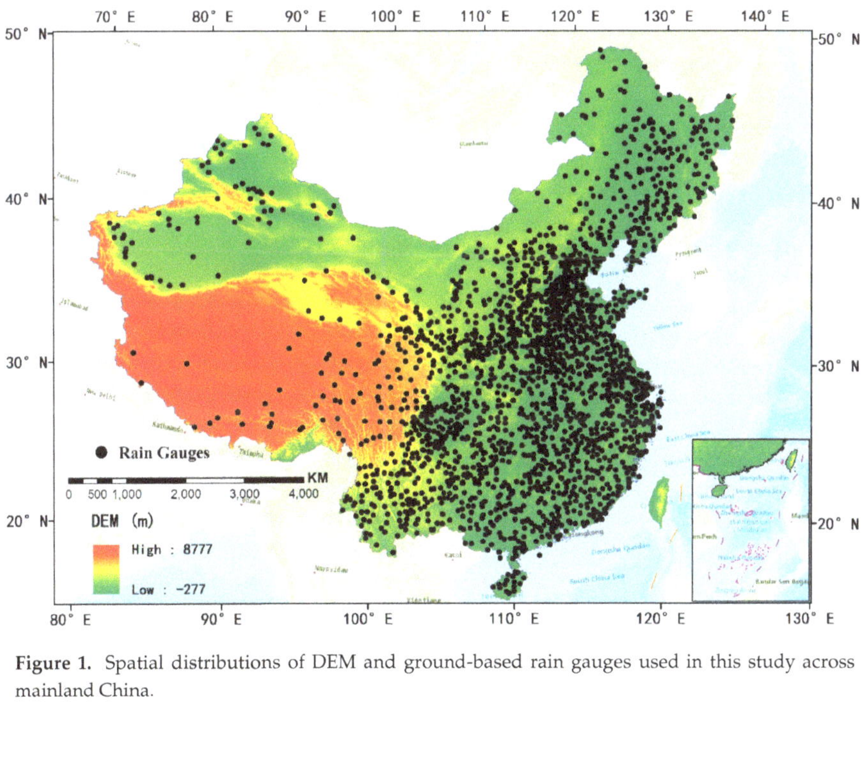

Figure 1. Spatial distributions of DEM and ground-based rain gauges used in this study across mainland China.

In this study, we chose the northern hemisphere summer (from June to August) of 2018 as the research period. The China Climate Bulletin in 2018 published by the China Meteorological Administration (CMA) shows that the annual average precipitation in China was 673.8 mm/year, which was 7% more than in other years. In particular, the average precipitation in summer was 356.4 mm/year, which was 10% above that of previous summers. Intensive typhoons and heavy rain occurred frequently in the summer of 2018. The East Asian subtropical summer monsoon was significantly stronger than usual in 2018, being the strongest since 1951.

2.2. Gauge Precipitation Measurements

The hourly rain gauge datasets from 2163 national ground stations used in this study were collected from the National Meteorological Information Center (NMIC) of CMA (http://data.cma.cn). The spatial distribution of ground stations in mainland China is shown in Figure 1. Hourly datasets from national ground stations usually include observations of air temperature, air pressure, precipitation, relative humidity, water vapor pressure, wind, and precipitation, etc. Meanwhile, the ground station datasets are quality controlled with the actual rate of each factor over 99.9%, and the accuracy of the datasets was close to 100% [29].

2.3. Satellite Precipitation Estimates

2.3.1. FY-2E Quantitative Precipitation Estimates (QPE)

The FY-2 series satellites are the principle observational platforms for covering dynamic weather events and the near-earth space environments in China. FY-2E is the third operational stationary satellite in the FY series, and was launched on 23 December 2008. Its sub-satellite point was 105°E before 1 July 2015 and has been 86.5°E over the equator to date. FY-2E is the last satellite in the first generation of Chinese operational meteorological satellites. The satellite is equipped with a five-channel (one visible channel and four infrared channels) scanning radiometer named the Visible and Infrared Spin Scan Radiometer (VISSR). The FY-2E satellite performs much better in terms of the accuracy of the inversion results of geophysical parameters, for example, precipitation, due to technical improvements, such as a reduction of the overlap of infrared spectral channels, compared with previous satellites of the FY-2 series.

FY-2E QPE data, generated by the fusion of FY-2E satellite estimate results and precipitation measurements from rain gauges, was used in this study. The QPE products have four categories at different temporal scales—hourly, three-hourly, six-hourly, and daily—with a spatial resolution of $0.1° \times 0.1°$. The latency of QPE products yielded by FY-2 series satellites is approximately one hour.

2.3.2. FY-2G QPE

FY-2G is one of the third batches of operational geostationary satellites in the FY-2 series, and was launched on 31 December 2014. The sub-satellite point of FY-2G changed from 99.5°E to 105°E, and finally became 99.2°E over the equator in April, 2018. FY-2G is the latest satellite to have Level 2 and Level 3 products since 2015. FY-2G has carried the radiometer with the best performance in operational satellites in the FY-2 series to date. Compared with satellites in the second batch, such as FY-2E, FY-2G has the ability to scan specific areas with a more flexible and higher temporal resolution. It plays a significant role in China's meteorological disaster monitoring, early warning, prevention, and reduction.

The QPE products in FY-2E and FY-2G have the same temporal scale and spatial resolution. However, the differences in the onboard sensors and fusion algorithms of FY-2E and FY-2G satellites lead to differences not only in the accuracy of precipitation estimates, but also in the numerical range and distribution. The QPE products can be downloaded from the National Satellite Meteorological Centre (NSMC, www.nsmc.org.cn).

2.3.3. IMERG

GPM is an international satellite mission. Its core observatory was launched by the National Aeronautics and Space Administration (NASA) and the Japanese Aerospace Exploration Agency (JAXA) on 27 February 2014. The first space-borne Ku/Ka-band Dual-frequency Precipitation Radar (DPR) was carried on the GPM Core Observatory, making it more sensitive to light rain rates and snowfall. IMERG is designed to intercalibrate, merge, and interpolate "all" data from satellites in the GPM constellation at fine temporal and special scales over the entire globe [11]. The version 06 IMERG Final run products were used in this study. The spatial resolution of IMERG is 0.1° × 0.1°, which is the same as FY-2 QPE products. The temporal resolution is half an hour, and the hourly data used in this study was obtained by averaging the two datafiles in the same hour.

Considering the fact that FY-2 QPE datasets merge precipitation information from ground observations, we applied the IMERG Final run dataset (V06B), which is calibrated with ground station data with a latency of about 3.5 months, as another precipitation estimate product, in this study. IMERG data could be downloaded from the Precipitation Measurement Mission's (PMM) website (https://pmm.nasa.gov/data-access/downloads/gpm).

3. Methods

3.1. Contingency Statistical Indices

Four indices are used to assess the contingency of satellite precipitation estimates. The probability of detection (POD) represents the proportion of correctly detected precipitation occurrences to the total number of events detected by satellites. The false alarm ratio (FAR) indicates the ratio of rainfall events that are falsely alarmed among the total number of satellite-detected precipitation occurrences. The frequency bias index (FBI) shows the degree of precipitation occurrence estimates from satellites. In other words, it indicates the overestimated or underestimated tendency in satellite-detected precipitation occurrences. The critical success index (CSI) denotes the fraction of rainfall events detected by satellites correctly to the total number of observed or detected rainfall events [30]. The indices mentioned above have no consideration of random assignments [31]. The equations of these indicators are given in Table 1. To discriminate between wet and dry samples, the thresholds of 1 mm day-1 for daily rain events and 0.1 mm hour-1 for hourly ones were used. [6].

Table 1. Equations and the best values of four contingency statistical indices.

Index	Equation [1]	Best Value
POD	$\frac{H}{H+M}$	1
FAR	$\frac{F}{H+F}$	0
CSI	$\frac{H}{H+M+F}$	1
FBI	$\frac{H+F}{H+M}$	1

[1] H(Hit) means that the precipitation occurrence is observed by a ground station as well as a satellite; M(Miss) denotes that the ground station observes the occurrence, while the satellite dose not detect it; F(False) indicates that the unobserved precipitation event is falsely detected by the satellite.

3.2. Statistical Indices

Four commonly used diagnostic statistics, including the correlation coefficient (CC), root mean square error (RMSE), relative bias (bias), and mean absolute error (MAE), were applied in this study to quantify the consistency between satellite precipitation products and rain gauge measurements. The four indices were also used to cross-evaluate satellite precipitation products without rain gauge measurements. The equations of these four statistical indices are shown in Table 2.

Table 2. Equations and the best values of four statistical indices.

Index	Equation [1]	Best Value
CC	$\sqrt{\frac{\sum_{i=1}^{n}\left(G_i-\overline{G}\right)^2\left(P_i-\overline{P}\right)^2}{\sum_{i=1}^{n}\left(G_i-\overline{G}\right)^2\sum_{i=1}^{n}\left(P_i-\overline{P}\right)^2}}$	1
RMSE	$\sqrt{\frac{1}{n}\sum\limits_{i=1}^{n}\left(P_i-G_i\right)^2}$	0
bias	$\frac{\sum_{i=1}^{n}(P_i-G_i)}{\sum_{i=1}^{n}G_i}\times 100\%$	0
MAE	$\frac{1}{n}\sum\limits_{i=1}^{n}\lvert P_i-G_i\rvert$	0

[1] n means the number of precipitation pairs in the analysis; G_i means ground-based precipitation measurements; $\overline{G}$ means the average ground-based precipitation measurements; P_i and $\overline{P}$ represent satellite precipitation products and their average, respectively.

4. Results

4.1. Spatial Distributions of Precipitation Estimates from FY-2E, FY-2G, and IMERG

The spatial distributions of FY-2E QPE, FY-2G QPE, and IMERG data in the summer of 2018 over mainland China are shown in Figure 2b–d, respectively, while Figure 2a displays the spatial distribution of precipitation obtained by inverse distance weighted (IDW) interpolation based on ground observations. All three satellite-based precipitation products present a distinct decreasing spatial variation of precipitation from the southeast to the northwest, which is consistent with that presented by ground observations. The spatial patterns of FY-2G QPE and IMERG are consistent with the patterns of interpolated results based on rain gauge data. However, both FY-2E and FY-2G products show an absence of data over the Tibetan Plateau and Qaidam Basin in northwest China. Moreover, both products do not provide precipitation estimates over northern parts of Heilongjiang Province, China, which exceed the extent of 50°N. Conversely, IMERG products provide full coverage precipitation estimates over mainland China.

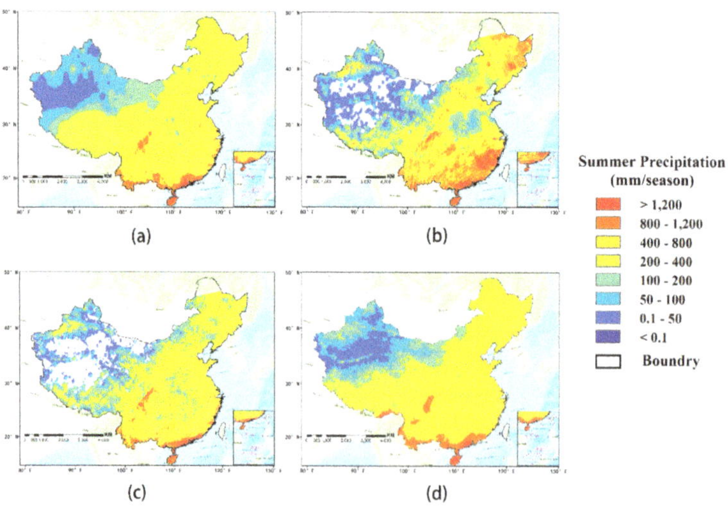

Figure 2. Spatial patterns of precipitation products estimated by (**a**) interpolated results based on rain gauge data, (**b**) Fengyun (FY)-2E quantitative precipitation estimates (QPE), (**c**) FY-2G QPE, and (**d**) Integrated Multi-satellitE Retrievals for GPM (IMERG) over mainland China in summer, 2018.

4.2. Validations of the Three Precipitation Products in the Summer, 2018

To evaluate the performances of FY-2E, FY-2G, and IMERG products, the three satellite-based precipitation products were validated separately against rain gauge data. Figure 3a–c show the validation results of FY-2E, FY-2G, and IMERG against ground observations in June (first row), July (second row), and August (third row) 2018, respectively. In general, according to the validation results, FY-2G QPE and IMERG outperform FY-2E QPE at monthly scale, with a CC of 0.65, 0.87, and 0.90 (0.90, 0.80, and 0.82) and bias of −8.13%, −3.97%, and −6.36% (8.40%, 7.84%, and 2.77%), in June, July, and August, respectively. In terms of RMSE and MAE, the results of FY-2G QPE are also lower than those of FY-2E QPE and IMERG for the entire summer of 2018, except for the worse performance compared with IMERG in June. In addition, IMERG shows small degrees of overestimation (bias of less than 10%). On the contrary, FY-2E QPE shows significant overestimation compared with ground observations, with bias of more than 30% in June and July, while FY-2G QPE also underestimated precipitation, but to a lesser degree (bias of more than −10%), for the entire summer of 2018.

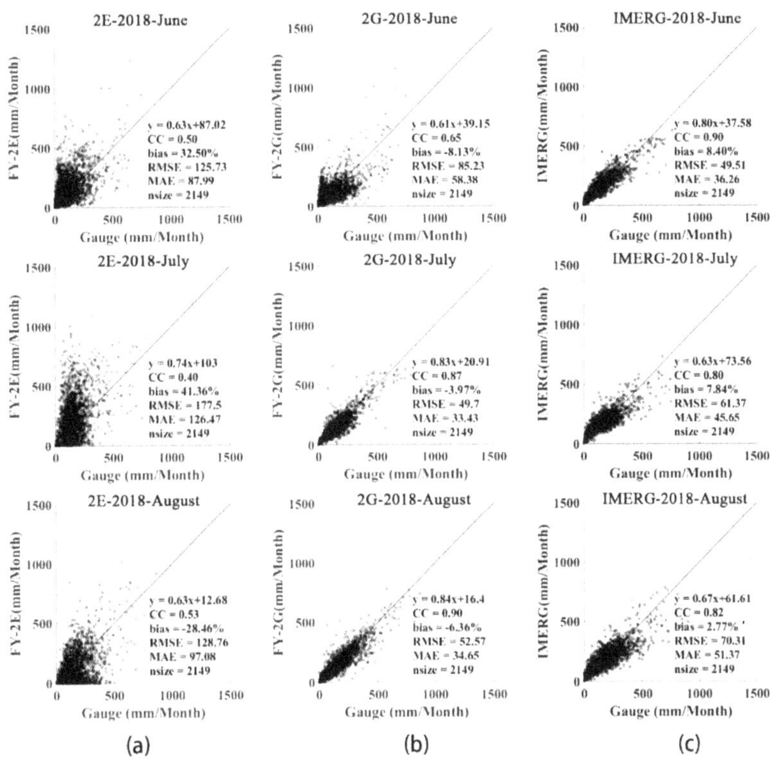

Figure 3. Validations of (**a**) FY-2E QPE, (**b**) FY-2G QPE, and (**c**) IMERG data against ground observations at monthly scale over mainland China in summer, 2018.

Figure 4a–d show Taylor diagrams of the performances of FY-2E QPE, FY-2G QPE, and IMERG against gauge precipitation measurements, in summer, June, July, and August, respectively. Taylor diagrams provide a graphical way to comprehensively evaluate the similarities between sets of patterns and observations [32]. Three classical indicators, namely, the CC, centered root-mean-square difference (CRMSD), and standard deviation (STD), are presented in a single 2D diagram, which reflect how closely the various patterns in satellite-based precipitation products match those in ground observations. If the estimated pattern is closer to the observations than other patterns in the diagram, then it means that the accuracy of the estimates is better than those of others. Taylor diagrams can convey more information more clearly than an ordinary table. They are useful because the strengths and weaknesses of the three statistical indexes are shown in the same diagram, and are thus less ambiguous [33,34].

We can conclude from the Taylor diagrams that the precipitation patterns of FY-2G QPE are the most similar to those of ground observations, since FY-2G QPE exhibits the best performances, with an RMSD value of around 48.63 mm and CC value of around 0.87 in July (Figure 4c), and an RMSD value of around 48.94 mm and CC value of around 0.90 in August (Figure 4d). In June, IMERG has the best similarity to ground observations, with RMSD and CC values of around 48.12 mm and 0.89, respectively (Figure 4b). Meanwhile, FY-2E QPE displays the largest values of RMSD, meaning that it has the lowest similarity to ground observations in all the four periods.

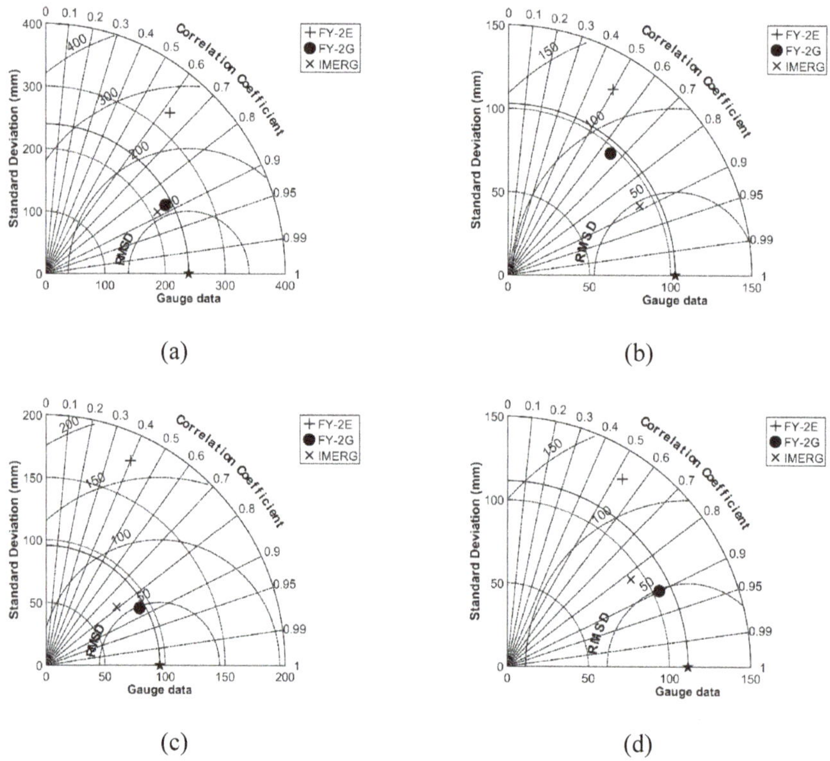

Figure 4. Taylor diagrams of performances of FY-2E QPE, FY-2G QPE, and IMERG against ground observations in terms of the centered root-mean-square difference, correlation coefficient, and standard deviation in (**a**) summer, (**b**) June, (**c**) July, and (**d**) August, 2018.

4.3. Validations of the Three Precipitation Products Based on Statistical Indices at Hourly Scale

Figure 5a–d illustrate the spatial patterns of CC, RMSE, bias, and MAE of FY-2E QPE (first column), FY-2G QPE (second column), and IMERG (third column), respectively, against ground observations at hourly scale, over mainland China in summer, 2018. It is obvious that FY-2G QPE outperforms FY-2E QPE and IMERG, with the best spatial patterns and numerical ranges of all the four indices, while IMERG performs better than FY-2E QPE. The CC of FY-2E QPE in mainland China is generally lower than 0.3, while the CC values of IMERG vary from 0 to 0.5, and are rarely larger than 0.5. Among the IMERG data, the best performing CC values are mainly distributed in the middle part of mainland China. As for FY-2G QPE, the CC values are larger than 0.6 over more than half of the area of China, especially in the eastern and central parts of mainland China. All three satellite-based precipitation products perform poorly in the southern and northwestern provinces of China. In terms of bias, FY-2G QPE also has the best performance, with the lowest bias over the majority of mainland China. The bias values of IMERG are greater than 10% over half of mainland China, especially in northwestern China, where the bias values are generally more than 50%.

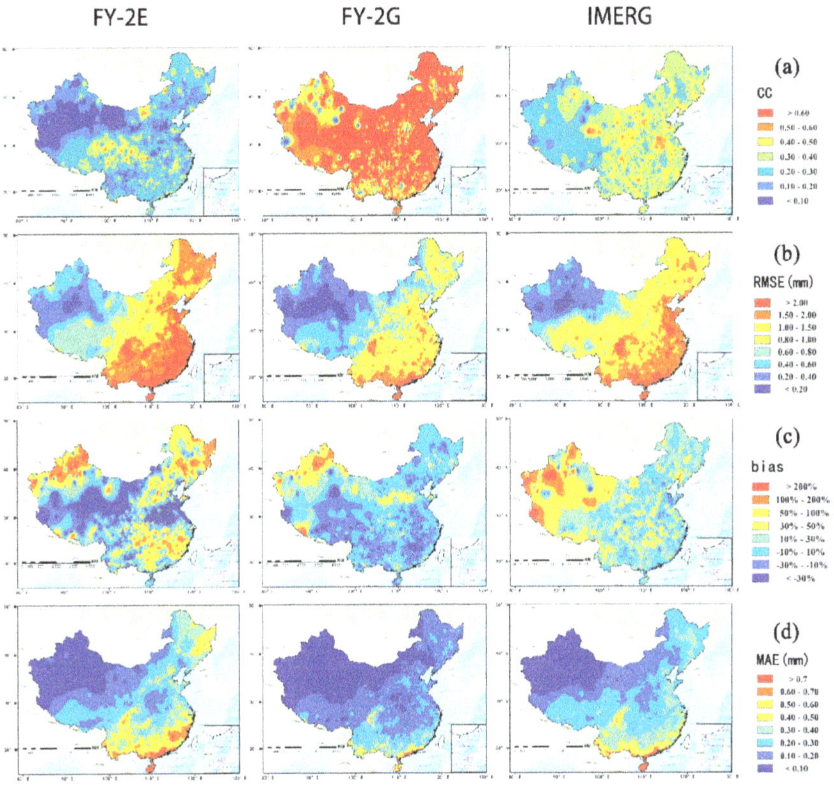

Figure 5. Spatial patterns of performances for FY-2E QPE, FY-2G QPE, and IMERG in terms of (**a**) the correlation coefficient (CC), (**b**) root mean square error (RMSE), (**c**) bias, and (**d**) mean absolute error (MAE) against ground observations at hourly scale, respectively.

Averaged values of the four statistical indices of the three products at hourly scale in June, July, August, and summer are displayed in Table 3. FY-2G QPE has the largest values of CC of 0.45, 0.66, 0.66, and 0.59 in June, July, August, and summer, respectively. The averaged RMSE and MAE values of all the three products are nearly smaller than 1.80 and 0.40 mm, respectively. IMERG shows overestimation in June (14.59%), July (11.34%), and August (10.07%), while FY-2G QPE underestimates the precipitation in all three months (−7.45% in June, −2.28% in July, and −4.34% in August, respectively). The averaged bias values of FY-2E QPE show significant variation. FY-2E QPE greatly overestimates precipitation in June (35.35%) and July (36.07%), while underestimates precipitation in August (−25.42%).

Table 3. Averaged statistical indices for FY-2E QPE, FY-2G QPE, and IMERG at hourly scale over mainland China in summer, 2018.

Data Type	Index	June	July	August	Summer
FY-2E QPE	CC	0.23	0.25	0.23	0.24
	RMSE (mm)	1.51	1.84	1.70	1.69
	bias (%)	35.35	36.07	−25.42	15.76
	MAE (mm)	0.33	0.40	0.31	0.35
FY-2G QPE	CC	0.45	0.66	0.66	0.59
	RMSE (mm)	1.14	1.13	1.21	1.16
	bias (%)	−7.45	−2.28	−4.34	−4.66
	MAE (mm)	0.20	0.18	0.19	0.19
IMERG	CC	0.36	0.36	0.37	0.36
	RMSE (mm)	1.26	1.54	1.62	1.48
	bias (%)	14.59	11.34	10.07	12.00
	MAE (mm)	0.25	0.31	0.32	0.29

Figure 6 displays the temporal patterns of performances at hourly scale for the three types of products compared to ground measurements. The statistical indices were calculated by the following steps: firstly, the gauge-based data and satellite-based data were extracted for 24 h; secondly, the statistical indices were calculated for each hour; and finally, the results from all stations across the country were averaged. Generally, both the performances of FY-based and GPM-based precipitation products are poorer during the period from 06:00 to 10:00 Coordinated Universal Time (UTC) than other periods in one day. Specifically, in Figure 6a, CC reaches its highest value during the periods of 00:00–3:00 and 18:00–24:00, and obtains its lowest value at about 09:00 (meant 9:00–10:00 UTC, which is the same as below), during the entire day. At about 15:00–17:00, the CC values of IMERG exhibit a decreasing trend, which does not appear in either FY-2E or FY-2G products. The variation of RMSE is contrary to that of CC, which means that a higher CC value always indicates a lower RMSE value (Figure 6b). It is clear that the RMSE of IMERG at 03:00 is the lowest in the 24-h period, at which time the curves of RMSE for FY-2E QPE and FY-2G QPE are smoother. As for the variations of bias (Figure 6c), FY-2E QPE and IMERG show overestimates (i.e., bias greater than 0%) in most of the periods, while FY-2G QPE generally underestimates precipitation for the entire day. Regarding the variations in MAE (Figure 6d), all three precipitation products show similar trends.

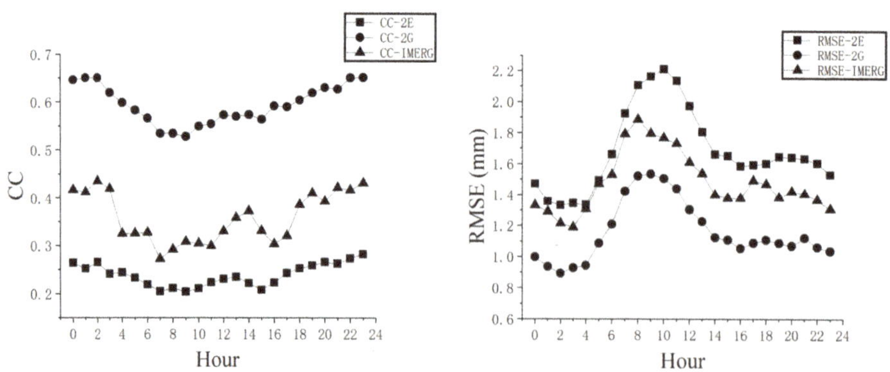

(a) (b)

Figure 6. *Cont.*

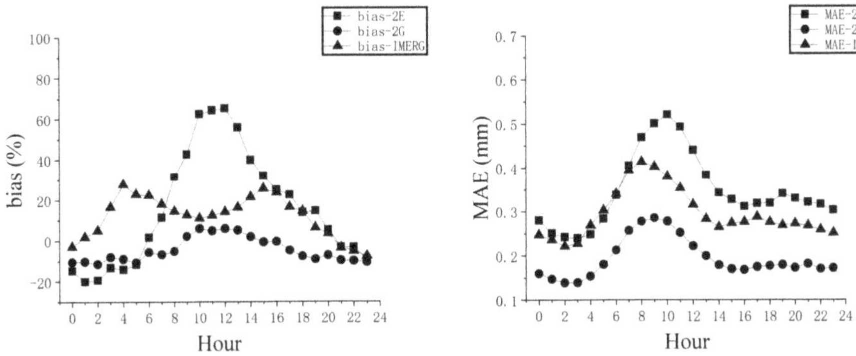

(c) (d)

Figure 6. Temporal patterns of performances of FY-2E QPE, FY-2G QPE, and IMERG in terms of (**a**) the CC, (**b**) RMSE, (**c**) bias, and (**d**) MAE against ground observations, respectively.

4.4. Contingency Indices of the Three Precipitation Products at Hourly and Daily Scales

Figure 7a–d display the spatial distributions of the contingency indices (POD, FAR, CSI, and FBI, respectively), generated by IDW interpolation based on validation results of corresponding rain gauge data, over mainland China during summer, 2018. Generally, the POD values of FY-2G QPE (>0.70) are much better than those of FY-2E QPE and IMERG, across mainland China. The POD values of IMERG are around 0.4 to 0.7 over most areas, while the POD values of FY-2E QPE are the smallest in most parts of mainland China (<0.5), especially in the northwest (<0.3) (Figure 7a). The FAR values of FY-2E QPE are above 0.5 over most regions and are larger than 0.8 in northwestern China, which is similar to the case of the IMERG products. As for FY-2G QPE, the FAR values (<0.6) are smaller than both values of FY-2E QPE and IMERG. Regarding the distributions of CSI (Figure 7c), FY-2G QPE shows a better performance than FY-2E QPE and IMERG, with values of around 0.4 to 0.7 over mainland China. The CSI values of FY-2E QPE and IMERG show similar spatial distributions. Both CSI values of FY-2E QPE and IMERG are lower than 0.4 overall, and lower than 0.2 in northwestern China. The FBI values of IMERG are higher than 1.8 in more than half of the areas, which indicates high overestimates in precipitation over such regions. The FBI values of FY-2G QPE are also greater than 1.2 over most parts of mainland China, reflecting overestimates in these areas, but to a lower degree compared with estimates of IMERG data. FY-2E QPE tends to overestimate precipitation in south, northeast, and northwest China, with FBI values larger than 1.6, and underestimate precipitation in the west and east coast regions of China, with values smaller than 1.

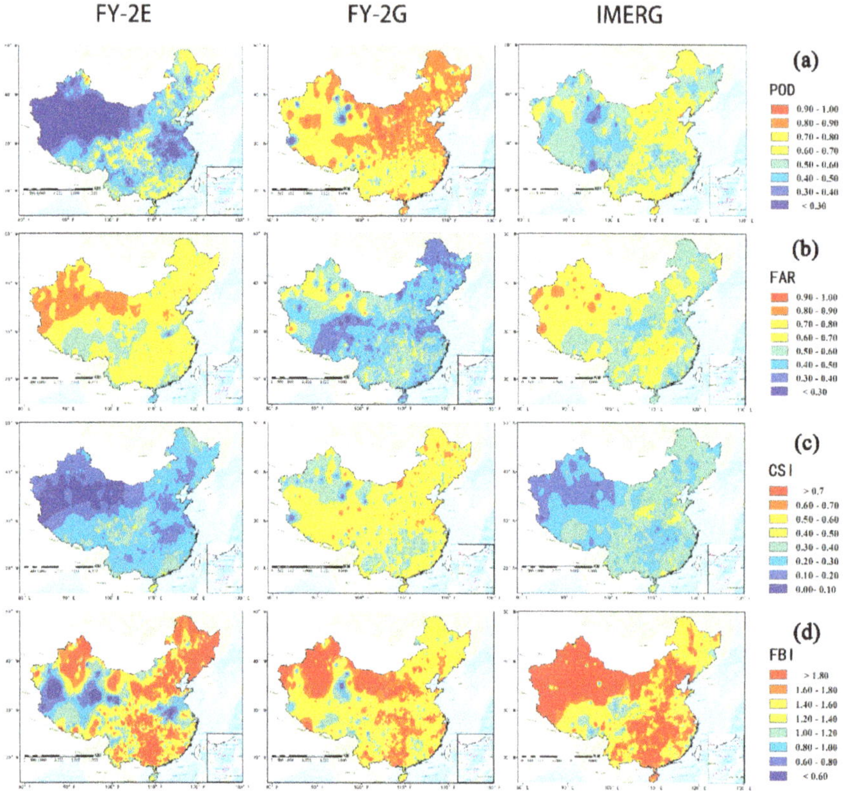

Figure 7. Spatial patterns of the performance of FY-2E QPE, FY-2G QPE, and IMERG in terms of the (**a**) probability of detection (POD), (**b**) (false alarm ratio) FAR, (**c**) critical success index (CSI), and (**d**) frequency bias index (FBI) compared to ground observations at hourly scale, respectively.

Averaged values of contingency indicators of the three products at hourly scale in June, July, August, and summer are exhibited in Table 4. FY-2G QPE shows the best values of POD in all the four periods compared with the other two products (around 0.61 in June, 0.84 in July, 0.84 in August, and 0.77 in summer). FY-2E QPE and IMERG have higher averaged values of FAR than those of FY-2G QPE, which are relevant to the lower values of CSI of both FY-2E QPE and IMERG. The values of CSI of FY-2E QPE are the lowest in each month, as well as for the entire summer. The averaged FBI values of all three precipitation products are much greater than one, which indicates that each of the three products show a larger proportion of false alarms than false negatives.

Table 4. Averaged contingency indices for the FY-2E QPE, FY-2G QPE, and IMERG at hourly scale over mainland China in summer, 2018.

Data Type	Index	June	July	August	Summer
FY-2E QPE	POD	0.49	0.53	0.47	0.50
	FAR	0.70	0.65	0.62	0.66
	CSI	0.23	0.26	0.26	0.25
	FBI	1.79	1.61	1.28	1.56
FY-2G QPE	POD	0.61	0.84	0.84	0.77
	FAR	0.56	0.46	0.44	0.48
	CSI	0.36	0.49	0.51	0.45
	FBI	1.39	1.59	1.54	1.51
IMERG	POD	0.59	0.63	0.61	0.61
	FAR	0.60	0.61	0.59	0.60
	CSI	0.31	0.31	0.32	0.31
	FBI	1.60	1.70	1.64	1.64

Figure 8a displays the temporal variations of POD of FY-2E QPE, FY-2G QPE, and IMERG. The values of POD of FY-2G QPE are the largest during the entire day, with values ranging from 0.75 to 0.80. The values of POD of IMERG are smaller than those of FY-2G QPE at each hour, with values varying from 0.55 to 0.70. The temporal variations of POD of IMERG are not smooth. IMERG shows a peak around 09:00 and valleys at 02:00, 13:00, and 17:00. FY-2E QPE shows the smallest values of POD (<0.57) compared with those of FY-2G QPE and IMERG, which suggests that FY-2E QPE could not detect rainfall events reasonably and effectively during the summer. Figure 8b shows the temporal variations of FAR. In general, FY-2G QPE shows the lowest FAR values at each time during the entire day, while the FAR values of IMERG are smaller than those of FY-2E QPE overall. As for the performances of FY-2G QPE, the FAR values exceed 0.50 from 05:00 to 14:00. Generally, the variations of CSI (Figure 8c) still demonstrate that FY-2G QPE outperforms IMERG and FY-2E QPE, with the largest CSI values during the entire day, while the CSI values of FY-2E QPE are the smallest. All of the three satellite-based precipitation products have values of FBI larger than one (Figure 8d), which indicates that all products tend to overestimate precipitation occurrences over the study area.

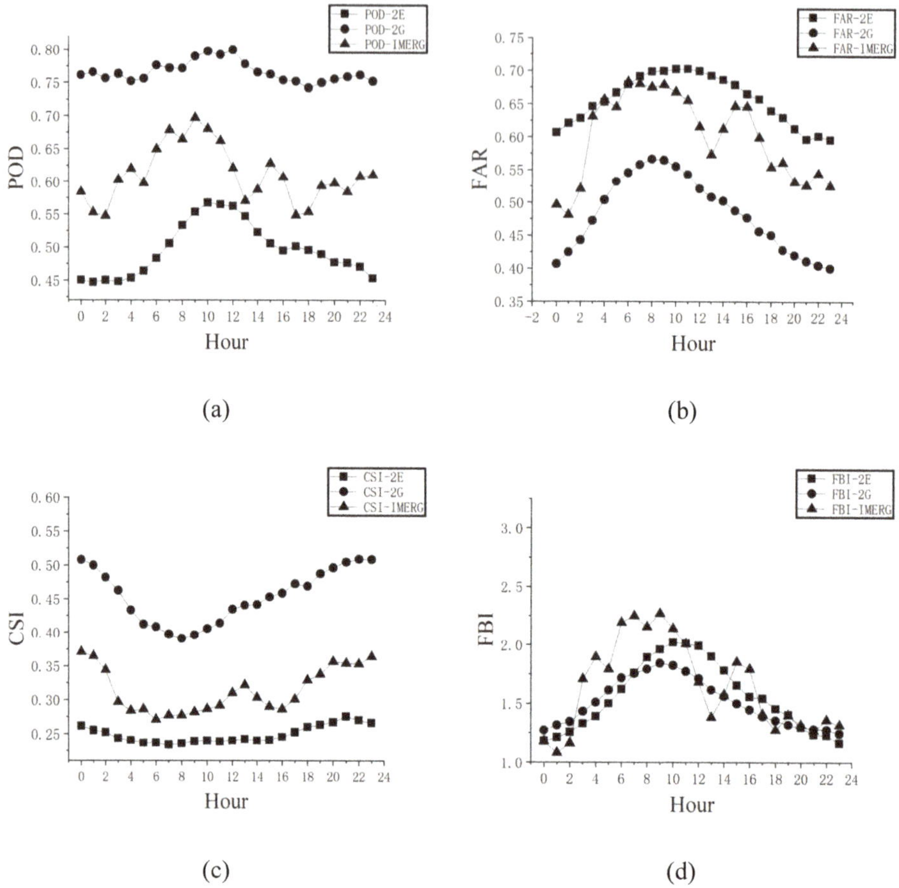

Figure 8. Temporal patterns of performances of FY-2E QPE, FY-2G QPE, and IMERG in terms of (**a**) POD, (**b**) FAR, (**c**) CSI, and (**d**) FBI against ground observations, respectively.

Figure 9 illustrates the numerical distributions of contingency statistical indices for FY-2E QPE, FY-2G QPE, and IMERG, at daily scale. In terms of POD (Figure 9a), the performance of FY-2G QPE is close to that of IMERG, with mean values of around 0.87 and 0.82, respectively, while the mean value of POD for FY-2E QPE is around 0.62. For the distributions of FAR (Figure 9b), the mean value of FY-2G QPE is the smallest (around 0.25), while the mean values of IMERG and FY-2E QPE are both around 0.4. In spite of the well-performing median, FY-2E QPE shows the worst POD and FAR distributions, since the range of whiskers is too large compared with that of the other two products. Regarding CSI (Figure 9c), it shows similar distributions and numerical characteristics to those of POD, which indicates that FY-2G QPE outperforms IMERG and FY-2E QPE, with the largest mean value of around 0.6. For FBI (Figure 9d), the mean values of all three precipitation products are larger than 1, which indicates that each of the three products shows a tendency to overestimate precipitation occurrences at diurnal scale. The mean values of FBI of the FY-2 series satellite precipitation products are closer to one than IMERG, indicating a smaller degree of overestimation. Note that some FBI values of FY-2G QPE are smaller than one, which indicates that FY-2G QPE underestimates precipitation occurrences in some areas.

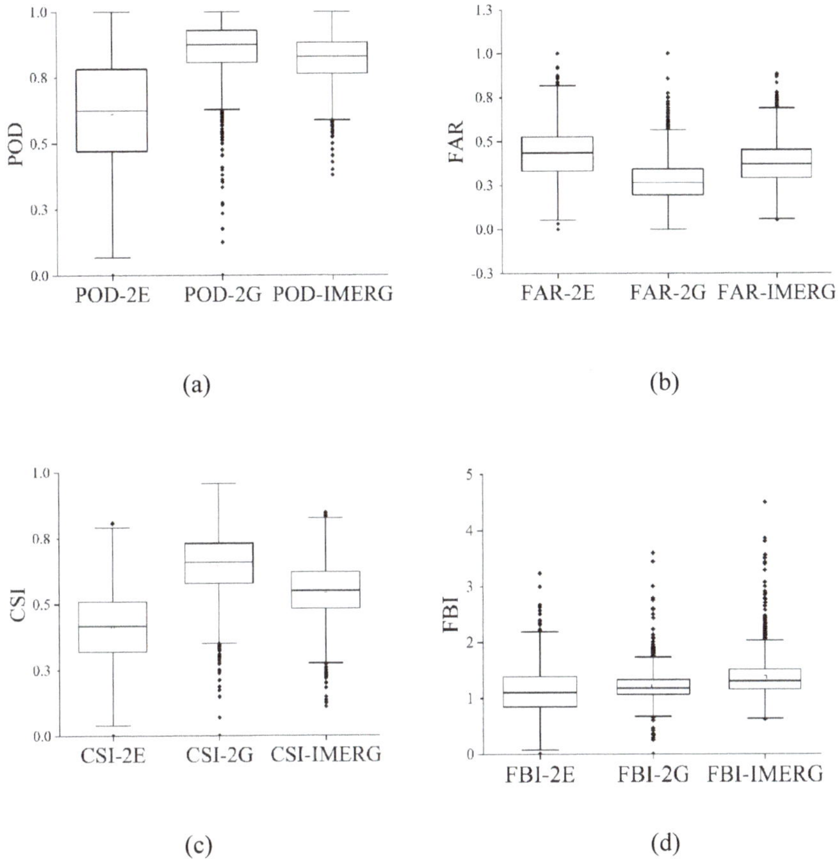

Figure 9. The numerical distributions of contingency statistics for FY-2E QPE, FY-2G QPE, and IMERG in terms of (**a**) POD, (**b**) FAR, (**c**) CSI, and (**d**) FBI, respectively.

4.5. Cross-Evaluation of FY-2 Precipitation Products Based on IMERG

Figure 10a–c demonstrate the inter-comparison results for FY-2E, FY-2G precipitation products, and IMERG, in terms of the total precipitation in summer, 2018. The number of pixels involved in cross evaluation between FY-2E QPE and IMERG is different from the number between FY-2G QPE and IMERG, which is mainly caused by the different ratios of data absence of FY-2E QPE and FY-2G QPE in northwestern mainland China. It is obvious that the correlations between FY-2G QPE and IMERG (CC of ~0.81) are much larger than those between FY-2E QPE and IMERG (CC of ~0.29), which is mainly caused by some significant overestimates of FY-2E QPE for the total precipitation in summer, when the precipitation is relatively small, compared with IMERG data. Additionally, the values of CC and other indicators between FY-2E QPE and FY-2G QPE are relatively poor, which indicates that the estimates of FY-2E are somewhat unreliable. Overall, according to the inter-comparisons displayed in Figure 10, FY-2G QPE agreed better with IMERG than FY-2E QPE in terms of spatial patterns and consistency.

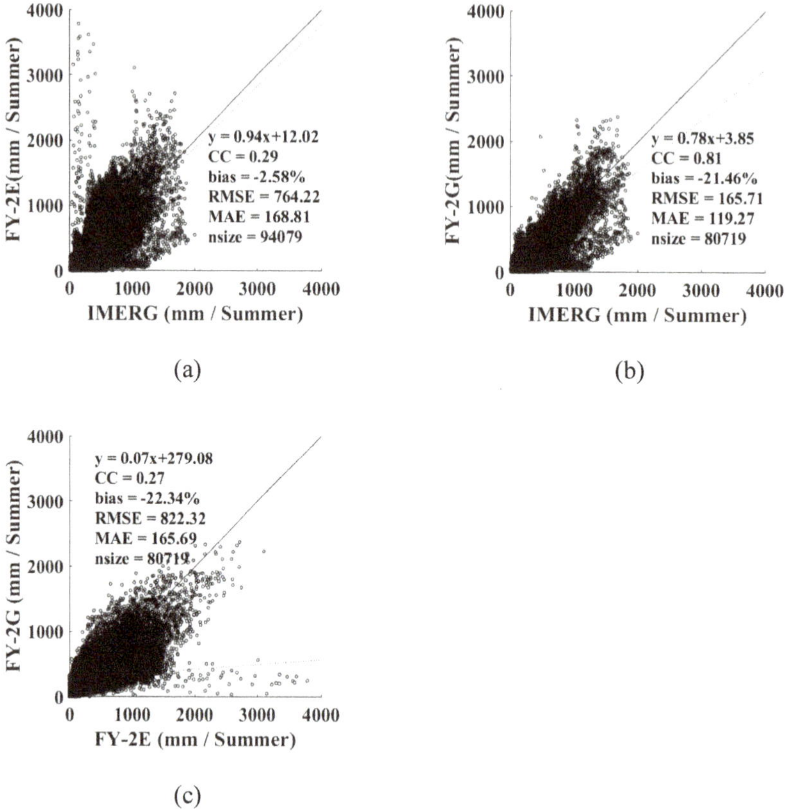

Figure 10. The inter-comparisons of (**a**) FY-2E QPE and IMERG, (**b**) FY-2G QPE and IMERG, and (**c**) FY-2G QPE and FY-2E QPE in terms of total precipitation in summer, 2018.

5. Discussion

5.1. The Advantages and Disadvantages of FY-2E QPE, FY-2G QPE, and IMERG

As mentioned above, we found that FY-2G QPE generally outperformed IMERG in terms of the statistical metrics over mainland China in summer, 2018. One of the possible reasons for this could be the different correction strategies; for example, the fusion method of FY-2 QPE considers not only the intensity, but also the directionality, of precipitation in the estimate fields. The FY-2 QPE fusion methods assume that the error field of the satellite-based precipitation estimate is related to not only the distance to the ground stations, but also the directionality of precipitation.

The unsatisfying performances of FY-2E QPE are significant, and may be related to the service life designed for FY-2E. FY-2E was launched in 2008 and was discontinued in early 2019, with a running time of about 11 years [35]. Therefore, the summer of 2018 coincides with the late stage of its operation. This could also explain the fact that there are some striped textures of precipitation spatial distributions derived from FY-2E in southern China in Figure 2b, which shows the spatial discontinuity of the satellite-based precipitation products. With the inevitable degradation of the sensors aboard the satellite, performance degradation is understandable.

According to the results demonstrated above, we found that FY-2 series satellites QPE and IMERG have advantages and disadvantages across the study area (Table 5). In mainland China, FY-2G QPE is more suitable in operational applications than IMERG, not only in terms of data accuracy, but also for the latency of the products (1 h for FY, 4 h for IMERG Early-run, 14 h for IMERG Late-run, and 3.5 months for IMERG Final-run), although there is no data coverage in the northern part of Heilongjiang Province (>50°N), while the time span of FY-2 series satellites is not long enough. In contrast, the IMERG Final-run precipitation product has been calculated back to 2000. Therefore, it is appropriate for IMERG to be used for long-term studies related to precipitation with fine spatiotemporal resolutions. Regarding the spatial coverage of these precipitation products, IMERG is more applicable for global-scale research due to the wide coverage of its products. Nonetheless, users should still pay great attention to the not so satisfying performance of IMERG at hourly and diurnal scales. Furthermore, some algorithms and methods, such as downscaling and retrospective studies, could be applied to yield long-term precipitation estimates with finer spatiotemporal resolutions in the future [36–39].

Table 5. Summary of advances and weaknesses of the three products over mainland China.

Data Type	Advances	Weaknesses
FY-2E QPE	Low latency	Poor data quality Short time span Limited coverage
FY-2G QPE	Best data quality Low latency	Short time span Limited coverage
IMERG Final-run	Fine data quality High temporal resolution Long time span Wide coverage	High latency Not so satisfying performance at hourly and diurnal scales

5.2. Possible Error Source Analysis of the GPM IMERG Product

For decades, numerous researchers have focused on the errors of satellite-based precipitation products at multiple scales all over the world, leading to the continuous improvement of these products to [40–45]. In this study, we have proposed some possible error sources of the GPM IMERG product, hoping to provide preliminary references for improving satellite-based QPE for the next generation. As can be seen from Figures 5 and 7, the large FAR (>0.7) of IMERG is mainly distributed in northwestern China, where the values of CC are relatively small compared with the other regions over mainland China. Additionally, the bias is generally greater than 50%. The dominant arid and semi-arid climate means that the area exhibits little precipitation over the entire year. The small amount of rainfall in summer makes it difficult to obtain correct detections [46,47]. Moreover, the ground observations obtained from meteorological stations for calibrating the satellite-based precipitation estimates are limited. These two issues may lead to a high false alarm ratio and significant overestimates over northwestern China.

In spite of the good performance compared with ground observations at monthly scale, the hourly and daily performance of IMERG shown by various indicators is not so satisfying. The characteristics of IMERG algorithms, including calibration algorithms and retrieval algorithms, might be related to this phenomenon. We know from the Algorithm Theoretical Basis Document (V06) of IMERG that the calibration strategy of IMERG Final-run products still has much room to improve [12]. The half-hourly precipitation estimates are simply multiplied by the monthly calibration ratios against monthly ground observations to yield half-hourly Final-run products. Although this will result in monthly estimates matching the gauge values more closely, IMERG Final-run datasets show an unsatisfying performance at meteorological scales (e.g., hourly or daily scale). We could assume that by using gauge calibrations at finer temporal resolutions, such as a daily scale, IMERG would likely yield satisfying performances at meteorological scales in terms of diagnostic indicators, with decreasing proportions of false negatives and false alarms. As for the retrieval algorithms of IMERG, the databases, including the a-priori database of cloud and precipitation profiles for inverting the passive microwave-based satellite precipitation estimates and the cloud feature database for inverting the infrared-based satellite precipitation estimates, might not be robust enough in China, considering the complex terrains and climatic factors [12,48–51]. In Figure 9, the POD patterns of IMERG are similar to those of FY-2G QPE, while the performances of its FAR and CSI patterns are not good. The phenomenon is caused by the larger proportion of false alarms of IMERG than those of FY-2G QPE. The high probability of false alarm occurrence indicated that the ability of IMERG in detecting the precipitation clouds at meteorological scales is comparatively weak, which may be related to the not well-matched feature database for precipitation retrieval algorithms over mainland China. In addition, significant overestimates and false alarms of IMERG in some areas may also result in large surrounding values for IMERG products. Meanwhile, the inconsistency between IMERG and FY-2G QPE would be significantly aggravated, as shown in Figure 10b.

6. Conclusions

Evaluations of satellite-based quantitative precipitation estimates are of great importance when applying these datasets in related fields, such as hydrology, meteorology, and agriculture. In this study, we firstly evaluated and compared the main current satellite-based precipitation products from Chinese Fengyun (FY)-2 and the Global Precipitation Mission (GPM), respectively, over mainland China in summer, 2018. The main conclusions are as follows:

(1) The three products (FY-2E QPE, FY-2G QPE, and IMERG) demonstrate similar spatial precipitation patterns; for example, a general decreasing trend from the southeast to northwest over mainland China;

(2) Compared with rain gauge measurements, FY-2G QPE and IMERG perform better among the three products, with the CC varying from 0.65 to 0.90 and 0.80 to 0.90 in summer, 2018, followed by FY-2E QPE (CC of ~0.40 to 0.53);

(3) IMERG agrees well with rain gauge data at monthly scale, while it performs worse than FY-2G QPE at hourly and daily scales, which might be caused by the algorithm characteristics of IMERG Final-run products;

(4) Compared with ground observations, FY-2G QPE exhibits underestimates in capturing the precipitation at both a monthly and hourly scale, while FY-2E QPE and IMERG generally tend to overestimate the precipitation in summer, 2018;

(5) The performances of both FY-based and GPM-based precipitation products are poorer during the period from 06:00 to 10:00 UTC than other periods at diurnal scale, which might have resulted from the satellite-based precipitation retrieval algorithms and the impact of regional meteorological and climatological influences. Further study is required to investigate the underlying reasons for this phenomenon;

(6) FY-2G QPE agrees well with IMERG in terms of spatial patterns and consistency (CC of ~0.81), which means that these two products have similar capacities to capture the spatial patterns of precipitation events.

The findings presented in this study could provide valuable preliminary references for improving the current satellite-based QPE retrieval algorithms for the next generation.

Author Contributions: Conceptualization, J.X., Z.M., and Z.S.; methodology, J.X., Z.M., and G.T.; software, X.M., W.W., and Q.J.; validation, J.X., G.T., and Q.J.; formal analysis, J.X., Q.J., and X.M.; investigation, J.X., Z.M., G.T., and Q.J.; resources, J.X., X.M., and W.W.; data curation, J.X. and W.W.; writing—original draft preparation, J.X. and Z.M.; writing—review and editing, J.X. and Z.M.; visualization, J.X., Q.J., X.M., and W.W.; supervision, Z.M., G.T., and Z.S.; project administration, Z.M. and Z.S.; funding acquisition, Z.M. and Z.S.

Funding: This research is financially supported by the National Key Research and Development Program (2017YFD0700501); the State Key Laboratory of Resources and Environmental Information System, National Natural Science Foundation of China (41571339, 41901343); the Open Fund of the State Key Laboratory of Remote Sensing Science (OFSLRSS201909), the Key R&D Program of Ministry of Science and Technology (2018YFC1506500); and the China Postdoctoral Science Foundation (2018M630037 and 2019T120021).

Acknowledgments: The authors' great gratitude is extended to the NSMC for providing FY-2 QPE and to the CMA for providing ground observations. We also appreciate the NASA/Goddard Space Flight Center's Mesoscale Atmospheric Processes Laboratory for providing the IMERG data.

Conflicts of Interest: The authors declare no conflicts of interest.

References

1. Kidd, C.; Huffman, G. Global Precipitation Measurement. *Meteorol. Appl.* **2011**, *18*, 334–353. [CrossRef]
2. Hegerl, G.; Black, E.; Allan, R.; Ingram, W.J.; Polson, D.; Trenberth, K.E.; Chadwick, R.; Arkin, P.A.; Sarojini, B.; Becker, A.; et al. Challenges in Quantifying Changes in the Global Water Cycle. *Bull. Am. Meteorol. Soc.* **2014**, *96*, 1097–1115. [CrossRef]
3. Ruhi, A.; Messager, M.L.; Olden, J.D. Tracking the Pulse of the Earth's Fresh Waters. *Nat. Sustain.* **2018**, *1*, 198–203. [CrossRef]
4. Hirpa, F.A.; Gebremichael, M.; Hopson, T. Evaluation of High-Resolution Satellite Precipitation Products over Very Complex Terrain in Ethiopia. *J. Appl. Meteorol. Clim.* **2010**, *49*, 1044–1051. [CrossRef]
5. Jongjin, B.; Jongmin, P.; Dongryeol, R.; Minha, C. Geospatial Blending to Improve Spatial Mapping of Precipitation with High Spatial Resolution by Merging Satellite-based and Ground-based data. *Hydrol. Process.* **2016**, *30*, 2789–2803. [CrossRef]
6. Ebert, E.E.; Janowiak, J.E.; Kidd, C. Comparison of Near-Real-Time Precipitation Estimates from Satellite Observations and Numerical Models. *Bull. Am. Meteorol. Soc.* **2007**, *88*, 47–64. [CrossRef]
7. Ma, Z.; Xu, Y.; Peng, J.; Chen, Q.; Wan, D.; He, K.; Shi, Z.; Li, H. Spatial and Temporal Precipitation Patterns Characterized by TRMM TMPA over the Qinghai-Tibetan Plateau and Surroundings. *Int. J. Remote Sens.* **2018**, *39*, 3891–3907. [CrossRef]
8. Li, X.; Zhang, Q.; Xu, C.-Y. Assessing the Performance of Satellite-based Precipitation Products and its Dependence on Topography over Poyang Lake Basin. *Theor. Appl. Climatol.* **2014**, *115*, 713–729. [CrossRef]
9. Levizzani, V.; Bauer, P.; Joseph Turk, F. *Measuring Precipitation from Space—EURAINSAT and the Future*; Springer: Dordrecht, The Netherlands, 2007; pp. 49–58.
10. Kucera, P.A.; Ebert, E.E.; Turk, F.J.; Levizzani, V.; Kirschbaum, D.; Tapiador, F.J.; Loew, A.; Borsche, M. Precipitation from Space: Advancing Earth System Science. *Bull. Am. Meteorol. Soc.* **2013**, *94*, 365–375. [CrossRef]
11. Hou, A.Y.; Kakar, R.K.; Neeck, S.; Azarbarzin, A.A.; Kummerow, C.D.; Kojima, M.; Oki, R.; Nakamura, K.; Iguchi, T. The Global Precipitation Measurement Mission. *Bull. Am. Meteorol. Soc.* **2014**, *95*, 701–722. [CrossRef]
12. Huffman, G.J.; Bolvin, D.T.; Braithwaite, D.; Hsu, K.; Joyce, R.; Kidd, C.; Nelkin, E.J.; Sorooshian, S.; Tan, J.; Xie, P. *Algorithm Theoretical Basis Document (ATBD) Version 06: NASA Global Precipitation Measurement (GPM) Integrated Multi-SatellitE Retrievals for GPM (IMERG)*; NASA: Greenbelt, MD, USA, 2018. Available online: https://pmm.nasa.gov/sites/default/files/document_files/IMERG_ATBD_V06.pdf (accessed on 11 December 2019).
13. Shukla, S.; McNally, A.; Husak, G.; Funk, C. A Seasonal Agricultural Drought Forecast System for Food-insecure Regions of East Africa. *Hydrol. Earth Syst. Sci.* **2014**, *18*, 3907–3921. [CrossRef]
14. Anjum, M.N.; Ding, Y.; Shangguan, D.; Ijaz, M.W.; Zhang, S. Evaluation of High-Resolution Satellite-Based Real-Time and Post-Real-Time Precipitation Estimates during 2010 Extreme Flood Event in Swat River Basin, Hindukush Region. *Adv. Meteorol.* **2016**, *2016*, 2604980. [CrossRef]
15. Derin, Y.; Yilmaz, K.K. Evaluation of Multiple Satellite-Based Precipitation Products over Complex Topography. *J. Hydrometeorol.* **2014**, *15*, 1498–1516. [CrossRef]
16. Beck, H.E.; Vergopolan, N.; Pan, M.; Levizzani, V.; Van Dijk, A.I.J.M.; Weedon, G.P.; Brocca, L.; Pappenberger, F.; Huffman, G.J.; Wood, E.F. Global-scale Evaluation of 22 Precipitation Datasets Using Gauge Observations and Hydrological Modeling. *Hydrol. Earth Syst. Sci.* **2017**, *21*, 6201–6217. [CrossRef]
17. Porcù, F.; Milani, L.; Petracca, M. On the Uncertainties in Validating Satellite Instantaneous Rainfall Estimates with Raingauge Operational Network. *Atmos. Res.* **2014**, *144*, 73–81. [CrossRef]
18. Rivera, J.A.; Marianetti, G.; Hinrichs, S. Validation of CHIRPS precipitation dataset along the Central Andes of Argentina. *Atmos. Res.* **2018**, *213*, 437–449. [CrossRef]
19. Darand, M.; Amanollahi, J.; Zandkarimi, S. Evaluation of the Performance of TRMM Multi-satellite Precipitation Analysis (TMPA) Estimation over Iran. *Atmos. Res.* **2017**, *190*, 121–127. [CrossRef]

20. Gadelha, A.N.; Coelho, V.H.R.; Xavier, A.C.; Barbosa, L.R.; Melo, D.C.D.; Xuan, Y.; Huffman, G.J.; Petersen, W.A.; Almeida, C.D.N. Grid Box-level Evaluation of IMERG over Brazil at Various Space and Time Scales. *Atmos. Res.* **2019**, *218*, 231–244. [CrossRef]
21. Sun, R.; Yuan, H.; Liu, X.; Jiang, X. Evaluation of the Latest Satellite–gauge Precipitation Products and their Hydrologic Applications over the Huaihe River basin. *J. Hydrol.* **2016**, *536*, 302–319. [CrossRef]
22. Tang, G.; Ma, Y.; Long, D.; Zhong, L.; Hong, Y. Evaluation of GPM Day-1 IMERG and TMPA Version-7 Legacy Products over Mainland China at Multiple Spatiotemporal Scales. *J. Hydrol.* **2015**, *533*, 152–167. [CrossRef]
23. Feidas, H.; Porcu, F.; Puca, S.; Rinollo, A.; Lagouvardos, C.; Kotroni, V. Validation of the H-SAF Precipitation Product H03 over Greece Using Rain Gauge Data. *Theor. Appl. Climatol.* **2018**, *131*, 377–398. [CrossRef]
24. Chen, S.; Hong, Y.; Cao, Q.; Gourley, J.J.; Kirstetter, P.E.; Yong, B.; Tian, Y.D.; Zhang, Z.X.; Shen, Y.; Hu, J.J.; et al. Similarity and Difference of the Two Successive V6 and V7 TRMM Multisatellite Precipitation Analysis Performance over China. *J. Geophys. Res. Atmos.* **2013**, *118*, 13060–13074. [CrossRef]
25. Teng, H.; Shi, Z.; Ma, Z.; Li, Y. Estimating Spatially Downscaled Rainfall by Regression Kriging using TRMM Precipitation and Elevation in Zhejiang Province, Southeast China. *Int. J. Remote Sens.* **2014**, *35*, 7775–7794. [CrossRef]
26. Prakash, S.; Mitra, A.K.; AghaKouchak, A.; Liu, Z.; Norouzi, H.; Pai, D.S. A Preliminary Assessment of GPM-based Multi-satellite Precipitation Estimates over a Monsoon Dominated Region. *J. Hydrol.* **2018**, *556*, 865–876. [CrossRef]
27. Katiraie-Boroujerdy, P.-S.; Akbari Asanjan, A.; Hsu, K.-L.; Sorooshian, S. Intercomparison of PERSIANN-CDR and TRMM-3B42V7 Precipitation Estimates at Monthly and Aaily Time Scales. *Atmos. Res.* **2017**, *193*, 36–49. [CrossRef]
28. Ma, S.; Zhou, T.; Dai, A.; Han, Z. Observed Changes in the Distributions of Daily Precipitation Frequency and Amount over China from 1960 to 2013. *J. Clim.* **2015**, *28*, 6960–6978. [CrossRef]
29. Shen, Y.; Xiong, A.; Wang, Y.; Xie, P. Performance of High Resolution Satellite Precipitation Products over China. *J. Geophys. Res. Atmos.* **2010**, *115*. [CrossRef]
30. Wilks, D.S. *Statistical Methods in the Atmospheric Sciences*; Academic Press: Oxford, UK, 2011.
31. Nurmi, P. *Recommendations on the Verification of Local Weather Forecasts*; ECMWF Technical Memorandum. No. 430; ECMWF: Reading, UK, 2003.
32. Taylor, K.E. Summarizing Multiple Aspects of Model Performance in a Single Diagram. *J. Geophys. Res. Atmos.* **2001**, *106*, 7183–7192. [CrossRef]
33. Guo, H.; Chen, S.; Bao, A.; Behrangi, A.; Hong, Y.; Ndayisaba, F.; Hu, J.; Stepanian, P.M. Early assessment of Integrated Multi-satellite Retrievals for Global Precipitation Measurement over China. *Atmos. Res.* **2016**, *176–177*, 121–133. [CrossRef]
34. Ning, S.; Wang, J.; Jin, J.; Ishidaira, H. Assessment of the Latest GPM-Era High-Resolution Satellite Precipitation Products by Comparison with Observation Gauge Data over the Chinese Mainland. *Water* **2016**, *8*, 481. [CrossRef]
35. National Satellite Meteorological Centre (China). Announcement on the Discontinued Operations of FY-2E. 2019. Available online: http://www.nsmc.org.cn/NSMC/Contents/103440.html (accessed on 5 December 2019). (In Chinese)
36. Ma, Z.; Shi, Z.; Zhou, Y.; Xu, J.; Yu, W.; Yang, Y. A Spatial Data Mining Algorithm for Downscaling TMPA 3B43 V7 Data over the Qinghai–Tibet Plateau With the Effects of Systematic Anomalies Removed. *Remote Sens. Environ.* **2017**, *200*, 378–395. [CrossRef]
37. He, K.; Ma, Z.; Zhao, R.; Biswas, A.; Teng, H.; Xu, J.; Yu, W.; Shi, Z. A Methodological Framework to Retrospectively Obtain Downscaled Precipitation Estimates over the Tibetan Plateau. *Remote Sens.* **2018**, *10*, 1974. [CrossRef]
38. Ma, Z.Q.; Zhou, L.Q.; Yu, W.; Yang, Y.Y.; Teng, H.F.; Shi, Z. Improving TMPA 3B43 V7 Data Sets Using Land-Surface Characteristics and Ground Observations on the Qinghai–Tibet Plateau. *IEEE Geosci. Remote Sens. Lett.* **2018**, *15*, 178–182. [CrossRef]
39. Ma, Z.; He, K.; Tan, X.; Xu, J.; Fang, W.; He, Y.; Hong, Y. Comparisons of Spatially Downscaling TMPA and IMERG over the Tibetan Plateau. *Remote Sens.* **2018**, *10*, 1883. [CrossRef]
40. Tian, Y.; Peters-Lidard, C.D.; Choudhury, B.J.; Garcia, M. Multitemporal Analysis of TRMM-based Satellite Precipitation Products for Land Data Assimilation Applications. *J. Hydrometeorol.* **2007**, *8*, 1165–1183. [CrossRef]

41. Pfeifroth, U.; Mueller, R.; Ahrens, B. Evaluation of Satellite-based and Reanalysis Precipitation Data in the Tropical Pacific. *J. Appl. Meteorol. Clim.* **2013**, *52*, 634–644. [CrossRef]
42. Gehne, M.; Hamill, T.M.; Kiladis, G.N.; Trenberth, K.E. Comparison of Global Precipitation Estimates across a Range of Temporal and Spatial Scales. *J. Clim.* **2016**, *29*, 7773–7795. [CrossRef]
43. Herold, N.; Behrangi, A.; Alexander, L.V. Large Uncertainties in Observed Daily Precipitation Extremes over Land. *J. Geophys. Res. Atmos.* **2017**, *22*, 668–681. [CrossRef]
44. Sun, Q.H.; Miao, C.Y.; Duan, Q.Y.; Ashouri, H.; Sorooshian, S.; Hsu, K.L. A Review of Global Precipitation Data Sets: Data Sources, Estimation, and Intercomparisons. *Rev. Geophys.* **2018**, *56*, 79–107. [CrossRef]
45. Chen, H.; Yong, B.; Gourley, J.J.; Liu, J.; Ren, L.; Wang, W.; Hong, Y.; Zhang, J. Impact of the Crucial Geographic and Climatic Factors on the Input Source Errors of GPM-based Global Satellite Precipitation Estimates. *J. Hydrol.* **2019**, *575*, 1–16. [CrossRef]
46. Gao, F.; Zhang, Y.; Chen, Q.; Wang, P.; Yang, H.; Yao, Y.; Cai, W. Comparison of Two Long-term and High-resolution Satellite Precipitation Datasets in Xinjiang, China. *Atmos. Res.* **2018**, *212*, 150–157. [CrossRef]
47. Marra, F.; Morin, E. Autocorrelation Structure of Convective Rainfall in Semiarid-arid Climate Derived from High-resolution X-Band Radar Estimates. *Atmos. Res.* **2018**, *200*, 126–138. [CrossRef]
48. Hong, Y.; Hsu, K.-L.; Sorooshian, S.; Gao, X. Precipitation Estimation from Remotely Sensed Imagery Using an Artificial Neural Network Cloud Classification System. *J. Appl. Meteorol. Clim.* **2004**, *43*, 1834–1853. [CrossRef]
49. Bitew, M.M.; Gebremichael, M. Evaluation through independent measurements: Complex terrain and humid tropical region in Ethiopia. In *Satellite Rainfall Applications for Surface Hydrology*; Springer: Dordrecht, The Netherlands, 2010; pp. 205–214.
50. Maggioni, V.; Meyers, P.; Robinson, M. A Review of Merged High Resolution Satellite Precipitation Product Accuracy during the Tropical Rainfall Measuring Mission (TRMM)-Era. *J. Hydrometeorol.* **2016**, *17*, 1101–1117. [CrossRef]
51. Passive Microwave Algorithm Team Facility. Global Precipitation Measurement (GPM) Mission Algorithm Theoretical Basis Document GPROF2017 Version 1. 2017. Available online: https://pmm.nasa.gov/sites/default/files/document_files/ATBD_GPM_GPROF_June1_2017.pdf (accessed on 15 October 2019).

remote sensing

MDPI

Article

Evaluation of the Radar QPE and Rain Gauge Data Merging Methods in Northern China

Qingtai Qiu [1], Jia Liu [1,*], Jiyang Tian [1], Yufei Jiao [1], Chuanzhe Li [1], Wei Wang [1,2] and Fuliang Yu [1]

1 State Key Laboratory of Simulation and Regulation of Water Cycle in River Basin, China Institute of Water Resources and Hydropower Research, Beijing 100038, China; qqt31415926@163.com (Q.Q.); tjyshd@126.com (J.T.); yufeij2018@163.com (Y.J.); azhe051@163.com (C.L.); WangWei_hydro@163.com (W.W.); yufl@iwhr.com (F.Y.)

2 College of Hydrology and Water Resources, Hohai University, Nanjing 210098, China

* Correspondence: hettyliu@126.com; Tel.: +86-15010443860

Received: 30 December 2019; Accepted: 21 January 2020; Published: 22 January 2020

Abstract: Radar-rain gauge merging methods have been widely used to produce high-quality precipitation with fine spatial resolution by combing the advantages of the rain gauge observation and the radar quantitative precipitation estimation (QPE). Different merging methods imply a specific choice on the treatment of radar and rain gauge data. In order to improve their applicability, significant studies have focused on evaluating the performances of the merging methods. In this study, a categorization of the radar-rain gauge merging methods was proposed as: (1) Radar bias adjustment category, (2) radar-rain gauge integration category, and (3) rain gauge interpolation category for a total of six commonly used merging methods, i.e., mean field bias (MFB), regression inverse distance weighting (RIDW), collocated co-kriging (CCok), fast Bayesian regression kriging (FBRK), regression kriging (RK), and kriging with external drift (KED). Eight different storm events were chosen from semi-humid and semi-arid areas of Northern China to test the performance of the six methods. Based on the leave-one-out cross validation (LOOCV), conclusions were obtained that the integration category always performs the best, the bias adjustment category performs the worst, and the interpolation category ranks between them. The quality of the merging products can be a function of the merging method that is affected by both the quality of radar QPE and the ability of the rain gauge to capture small-scale rainfall features. In order to further evaluate the applicability of the merging products, they were then used as the input to a rainfall-runoff model, the Hybrid-Hebei model, for flood forecasting. It is revealed that a higher quality of the merging products indicates a better agreement between the observed and the simulated runoff.

Keywords: weather radar quantitative precipitation estimation; rain gauge; radar-rain gauge merging; leave-one-out cross validation; verification

1. Introduction

Precipitation is a key driving component for hydrological water cycle processes at regional and global scales. A catchment reacts very specifically to intense rainfall due to its steep slopes and shallow soils, and precipitation data with a high spatial and temporal distribution is critical for forecasting flash flooding events [1]. These events invariably have the characteristics of high intensity and sudden occurrence, and under climate change, the needs for high resolution and accurate rainfall data have increased, particularly because effective hydrological forecasting depends greatly on precipitation accuracy [2–4]. Rain gauges can measure precipitation very accurately at a point scale, but we would need a dense network of instruments to ascertain the rainfall intensity at local/regional scales because of its high variability. In a short-time flash flood simulation, the spatial representatives of the rain gauge and the accurate representation of spatial rainfall variability in the surrounding area need to be considered [5,6]. High-resolution precipitation based on rain gauge data are usually geometrically interpolated from a limited number of observation points using geographic information systems (GIS).

Compared with a rain gauge, a quantitative precipitation estimation (QPE), based on the weather radar, has a primary advantage because it provides very high spatial and temporal resolution rainfall information, making it very suitable for hydrological modeling [7,8]. With the development of weather radar over the past 60 years, QPE, with its very high spatial and temporal resolutions, can accurately detect the location of precipitation, and can be applied to practical hydrological operations such as flood forecasting [9,10]. However, an error-free radar QPE is not possible due to various sources of error, such as indirect precipitation measurement, the Z-R relationship, being above the ground, beam shielding, and ground clutter, which result in range degradation [8,11,12]. Preserving the high spatial accuracy of rainfall in radar QPEs remains a challenge for meteorologists. This has been the case since the 1940s, when the potential for measuring precipitation with high spatial and temporal distributions based on weather radar was realized [13]. With the advantage of radar to estimate the spatial pattern and rain gauge data to obtain the correct point value, a combined product based on radar QPE and rain gauge data has significant potential for achieving superior rainfall estimations [14,15].

The concept of achieving high-resolution precipitation estimations by merging QPE and rain gauge data has resulted in proposals of numerous merging methods, and different ways of categorizing these methods have been applied [16]. An additional correction factor is the most commonly investigated and is currently being used by many national meteorological services due to its simplicity [17]. With the development of these interpolation methods, some studies have attempted to interpolate point rain gauge values with a variogram, which represents the spatial association of radar fields [18]. Sharon et al. (2015) found a clear difference between geostatistical and non-geostatistical methods, where the geostatistical methods attempt to use the variogram to represent the spatial bias and error variance of the rainfall field [19]. In a review, McKee (2015) adopted a viewpoint proposed by Wang (2013) that such merging methods generally achieve merging precipitation through either bias minimization methods or error variance reduction methods [20,21]. An integration method was recently proposed with the aim of minimizing data uncertainty [22]. When considering these merging methods, a better, application-oriented categorization is necessary.

Despite the research on this study, most of the studies have focused on evaluating the feasibility of the applied merging techniques and measuring the performance of the merged rainfall estimates against the rain gauge observation and radar estimates. Few studies have attempted to compare the results from various merging categories and have instead focused on large scale applications [23,24]. Although the impact of limited rain gauge data cannot be neglected in the merging performance when using rain gauge data for ground truthing, many studies have shown that more rain gauges across the catchment can increase the chances of capturing rainfall features, while fewer rain gauges may miss small convective cells [25]. To identify the commonly used merging techniques with better performances, many inter-comparison studies have focused on the performances of these methods, including the applied merging details of the type of method, spatiotemporal resolutions, and the better performance methods identified in previous work. Generally, the performances of different merging methods in most studies are assessed based on accuracy measures by comparing merged estimates against rain gauge observation through cross validation [26], but recently, some studies have attempted to evaluate the radar-rain gauge merging methods by comparing the hydrological performances resulting from these methods [27].

High-resolution precipitation data have been used in various types of hydrological studies, and the improvement of simulated hydrological dynamics using radar-based QPE has been highlighted [28–30]. It should be noted that in spite of the residual errors often remaining, these merging products have significant uses in hydrological applications, particularly when forecasting flash floods or extreme events [31]. When merging for flood forecasting, the application of high resolution and accurate precipitation at fine spatiotemporal scales presents some challenges, such as (1) preserving small-scale features (e.g., convective), (2) density of rain gauges across mountain basins, and (3) fitness of the hydrology model for the local catchment. With regard to applying flood forecasting at piedmont plain scales, it is therefore critical to consider these factors when examining the performances of different interpolated precipitation models and their ability to deal with challenges in flood forecasting [32].

In this study, the potential of flood forecasting with high-resolution precipitation was described, including its variability and uncertainty regarding less clarity. For hourly precipitation, few studies have focused on different interpolations regarding possible covariates over the catchments in semi-humid and semi-arid climates. Evaluating the performance of both radar-based and rain-gauge-based precipitation produced in the hydrological model can thus not only help to understand its physical processes, but also its function as an indirect measure to assess the accuracy of the rainfall input.

Although many merging methods of different categories are available, little research has been conducted to compare their performances and the applications driving hydrological models. In addition to choosing a reliable radar-rain gauge merging method to obtain high resolution and accurate precipitation data for the study area, the objective of this research was also to assess the detailed performances of different quality merging data in flash flood forecasting. In this study, we aimed to assess how different rain gauge observations, merged with radar data, leads to both better high-resolution precipitation resolutions and improved hydrological applications, thereby further enhancing the potential benefit of flash flood forecasting.

2. Methods

2.1. Radar-Rain Gauge Merging Methods

The potential of high spatial and temporal resolution precipitation based on weather radar is known. Hence, different merging methods have been proposed, and are generally classified as bias reduction and error variance minimization [33]. Identifying the spatial correlation in the error structure model is the most important step in the merging process. A categorization similar to a starting point and refined based upon a theoretical categorization was adopted by Wang et al. (2013), who also proposed the following (Table 1): First, a radar bias adjustment methods focusing on bias adjustment of radar estimates; second, radar-rain gauge integration methods, undertaking a true integration of both radar precipitation and rain gauge data; and, third, rain gauge interpolation methods using different interpolation methods with the radar spatial association as additional information [21].

Table 1. Categories of radar-rain gauge merging, the merging methods, and their abbreviations used to derive the data.

Category	Merging Method	Abbreviation
Radar bias adjustment category	Mean field bias	MFB
	Regression inverse distance weighting	RIDW
Radar-rain gauge integration category	Collocated co-kriging	CCoK
	Fast Bayesian regression kriging	FBRK
Rain gauge interpolation category	Regression kriging	RK
	Kriging with external drift	KED

2.1.1. Radar Bias Adjustment Category

The methods in this category attempt to predict the ungauged location value by the bias that computes radar accumulation and the rain gauge accumulation. In this category, the rain gauge observation is assumed as the true rainfall value, the radar precipitation is used as the entire background, and the radar values at the gauged locations are used to compare with the bias adjustment [34,35]. The ungauged locations value is adjusted by multiplying or adding the gauge-radar comparison correction factor, which is given over a long or short time period [36].

(1) Mean field bias QPE (MFB)

Mean field bias (MFB) adjustment is the simplest method in this category and assumes that the radar estimates are affected by a spatially uniform multiplicative error, which can be a bad electronic calibration or an offset in the Z-R relation used to convert radar reflectivity to rainfall value [13,34]. It is acknowledged that MFB adjustment is the most common and simplest technique in radar meteorology and the correction factor is simply obtained by comparing a spatially averaged ratio of rain gauge with the radar accumulations at gauged locations over the given time period. In this method, a simple multiplicative factor is used to correct the radar domain uniformly. In the current study, the adjustment factor is estimated as

$$C_{MFB} = \frac{\sum_{i=1}^{N} G_i}{\sum_{i=1}^{N} R_i} \tag{1}$$

where N is the number of valid rain gauge, G_i is the rain gauge observations, and R_i is the radar estimated values at the rain gauge located pixel.

(2) Regression inverse distance weighting (RIDW)

In this method, the rain gauge observations are used as the true rainfall value to correct the entire radar background field by multiplying a dynamic adjustment correction factor. In geostatistics, a random process $R(s, t)$ consists of two parts, where the first deterministic part $D(s, t)$ corresponds to the trend component, and the other stochastic residual part $\varepsilon(s, t)$ corresponds to local fluctuations of the trend. In this study, in addition to the observation vector $g(s, t)$ measured by the rain gauge, the radar-based QPE at rain gauge locations over the whole period were also considered and used at each interpolation point. In this context, the radar external variable was used as a linear function to model the trend $D(s, t)$ [23], so that

$$\mathrm{R}(s,t) = D(s,t) + \varepsilon(s,t) \tag{2}$$

$$\mathrm{D}(s,t) = a(t)r(s,t) \tag{3}$$

where s is the location of a given point at time t. $r(s,t)$ is the radar value at location s and time t. The coefficient is computed as the slope of a linear regression of all pairs of points composed of the gauge values on the y-axis and the values of the radar pixel on the x-axis. $a(t)$ is assumed to be constant spatially.

$$\varepsilon(s,t) = \mathrm{g}(s,t) - D(s,t) = g(s,t) - a(t)r(s,t) \tag{4}$$

$$\hat{R}_{RIDW}(s,t) = D_p(s,t) + \hat{\varepsilon}_{RIDW} \tag{5}$$

2.1.2. Radar-Rain Gauge Integration Category

The methods in this category aim to minimize the estimation uncertainty by conducting an actual integration of both rain gauge and radar data. As well as differing with some local bias adjustment aiming at reducing local biases between radar and rain gauge observation, the integration category also attempts to minimize overall estimation uncertainty [22]. The merging methods in this category do not simply retain the radar as background or impact the local magnitude at the rain gauge location, but depend on their relative uncertainties and estimate the rainfall value at given location grid in the weighted average of both resources [37]. Two main methods in this category are applied to obtain a minimum uncertainty estimation in different ways, which are:

(1) Collocated co-kriging (CCoK)

To achieve the final aim of reducing uncertainty as much as possible, this method integrates both data instead of using only radar or rain gauge precipitation as the background. The co-kriging (CoK) belongs to this category because it minimizes the variance of estimation by solving a single kriging system, including both radar and rain gauge data. Although CoK can be seen as a radar-based interpolation, it is a liner combination of a multivariate variant, merging radar and rain gauge data [38]. These results, however, from several full forms of CoK with different secondary variables, show a significantly larger kriging system and always lead to a numerically unstable co-kriging matrix, with a significant difference between the primary and secondary data. To avoid these uncertainties, collocated co-kriging (CCoK) has been proposed as a reduced form of CoK and applied in merging rainfall with this variation [39,40]. Compared with CoK, CCoK employs the radar value at the rain gauge location and only incorporates the cross-covariance between the radar value and observation at the rain gauge location. Consequently, the kriging system has been efficiently reduced.

The CCoK estimates rainfall and can be defined as

$$R_{CCoK}(s_0, t) = \sum_{i=1}^{n} \lambda_i^{CCoK} R_g(s_i, t) + \lambda_2' R_r(s_0, t) \tag{6}$$

where $R_g(s_i, t)$ are the rainfall values of the n known rain gauges at time t, λ_i^{CCoK} are the weights of the rain gauges, $R_r(s_0, t)$ is the radar QPE value at the target point, and λ_2' is the weight of related to the radar field.

In this method, the constraint of both data weight can be defined as

$$\sum_{i=1}^{n} \lambda_i^{CCoK} + \lambda_2' = 1 \tag{7}$$

The full radar covariance is, hence, not required. Instead, the covariance of rain gauges and the cross-covariance between radar and gauges are necessary. In this paper, we used rain gauge data and radar data as the primary variable and the secondary variable, respectively, to integrate the precipitation estimation, and approximate the cross-correlation from the radar spatial correlation using the alternative Markov approach [41]. Instead of the multiple (cross-) spatial correlation and large equations using the full COK, for this study, the CCoK used a simplified approximation with its advantage easily applied [40].

$$\sum_{i=1}^{n} \lambda_i^{CCoK} C_{GG}(x_i, x_0) + \lambda_2' C_{GR}(x_i, x_o) = C_{GG}(x_i, x_o) \tag{8}$$

$$\sum_{i=1}^{n} \lambda_{ki}^{CCoK} C_{RG}(x_0, x_i) + \lambda_2' C_{RR}(0) = C_{RG}(0) \tag{9}$$

where $C_{GG}(h)$ is the covariance of the rain gauges, $C_{RG}(h)$ is the cross-covariance between radar and rain gauges, and $C_{RR}(0)$ is the radar covariance at h = 0.

(2) Fast Bayesian regression kriging (FBRK)

In this category, we integrated both rainfall data with the purpose of obtaining the estimation at the minimum uncertainty. For this purpose, methods in a Bayesian framework are widely used, and we applied the fast Bayesian regression kriging (FBRK), method, as proposed by Yang and Ng [42], to merge different data types [42]. We explicitly considered the difference in the errors from the raw input data and aimed to estimate an accurate rainfall field and obtain better precipitation data. Unlike most other existing kriging-based merging methods, the likelihood function in the FBRK is modified. Further accounting for the differences, Yang and Ng [42] applied the FBRK in three different data types, i.e., the residuals of the regression model were used to regress radar estimations, and rain gauge observations were interpolated by the ordinary kriging.

$$\hat{I}_0 = a_r R_0 + \beta_r + e_0 \tag{10}$$

where $\hat{I}_0$ is the FBRK interpolated rainfall intensity at x_0, R_0 is the radar measured intensity at the same location, a_r and β_r are the regression coefficients, and e_0 is a random error term whose mean and standard deviation are computed following the kriging equations below:

$$u_{e0} = \sum_{i=1}^{M} \lambda_i e_i \tag{11}$$

$$\sigma_0^2 = \left(1 - \sum_{i=1}^{M} \lambda_i\right) \gamma_\infty + \sum_{i=1}^{M} \lambda_i r_{0,i} \tag{12}$$

$$r_{i,j} = \tau^2 + s^2\left[1 - exp\left(-\frac{|x_i, x_j|}{d_\infty}\right)\right] \quad (13)$$

$$\begin{pmatrix} r_{1,1} & \cdots & r_{M,1} \\ \vdots & \ddots & \vdots \\ r_{1,M} & \cdots & r_{M,M} \end{pmatrix}\begin{pmatrix} \lambda_1 \\ \vdots \\ \lambda_M \end{pmatrix} = \begin{pmatrix} r_{0,1} \\ \vdots \\ r_{0,M} \end{pmatrix} \quad (14)$$

where u_{e0} and σ_o are the expected value and standard deviation of e_0, respectively, e_i is the residual of the ith of the M rain gauge and crowdsourced observations, and λ_i is its associated weight.

2.1.3. Rain Gauge Interpolation Category Using the Spatial Association of Radar as an Addition

Unlike the bias adjustment methods using the entire radar field as background with these integration methods integrating both data sources, the methods in this category simply used the spatial association as an external drift to interpolate the rain gauge values. Ochoa-Rodriguez et al. proposed that the merging methods in this category are all geostatistical and kriging-based [33]. The kriging-based interpolation approaches predict the ungauged located values with the linear weights of observations at gauged locations by minimizing the variance of the error. As such, the main component of the methods is that the rainfall filed can be characterized as a Gaussian random variable, and because of this, the methods predict ungauged values with the liner combination of gauged value by deriving the weights through minimizing the variance. This differs to the classical geostatistics by assuming a Gaussian distribution and stationarity of the spatial covariance, with the distribution of precipitation skewed over the domain [43]. The transformation of applying both rain gauge and radar data into a more Gaussian distribution is termed trans-gaussian kriging [44]. It is based on the quantitative spatial variability of both data, and a more Gaussian distribution always has a better Gaussianity in the residuals. Two widely used methods of this category were applied in this study, which are:

(1) Regression kriging (RK)

As one of the kriging family hybrid interpolation methods, regression kriging (RK) is a spatial interpolation technique that integrates a linear regression and the regression residuals with simple kriging. The advantage of this method is that all points are used to interpolate the residual with a global neighborhood, which can extend to a broader range. RK uses arbitrarily complex regression on auxiliary information (radar data) and chooses simple kriging to interpolate the residuals acquired from the regression model. In this study, we defined the following successive steps to implement this method: (a) The linear regression step, (b) the residuals variogram computation step, and (c) the kriging-based interpolation of the residual steps.

For steps a and b, the trend and residuals computation based on Equation (3) are also valid for RK. At the beginning of step c, the covariance matrix C_{aa} of computing the covariance of the residuals at the target location is:

$$C_{aa} = \begin{pmatrix} C_Z^2 & C_{12} & \cdots & C_{1N} \\ C_{21} & C_Z^2 & \cdots & C_{2N} \\ \vdots & \vdots & \ddots & \vdots \\ C_{N2} & C_{N2} & \cdots & C_Z^2 \end{pmatrix} \quad (15)$$

where element C_{aa} of the matrix is computed by the covariance between the observation locations I and j, where C_Z^2 is the variance of the observations. In this method, the square-root transformation of the data is used in the process of applying kriging. This transformation shows a trending increase on the Gaussianity of the overall residuals, although some analysis of the effect sometimes show a few that are limited. Based on the linear kriging and linear combination, the weights used to compute residuals at the target location could thus be given as:

$$\hat{e}_{RK}(s_0, t) = \sum_{i=1}^{N} \lambda_i \in (s, t) \quad (16)$$

which is then added to the trend $m_p(S_0, t)$ to obtain the expected value of the precipitation depth.

Finally, the expected precipitation value at the interpolation location can be computed as:

$$\hat{p}_{RK}(s_o, t) = m_p(s_0, t) + \hat{e}_{RK}(s_0, t) \tag{17}$$

(2) Kriging with external drift (KED)

Kriging with external drift (KED) is an extension of universal kriging interpolation, in which the interpolated variable, in this case, is computed as the sum of stochastic term and a deterministic term. Kriging with external drifts allows the incorporation of several additional variables that are used as background information to interpolate the primary variable [45]. In this study, we focused on merging rain gauge and radar data, and therefore, radar data were considered as the only additional information in this method. The basic assumption of KED is that the expected value of the estimated variable $G(x)$ has a linear relationship with an additional variable $R(x)$:

$$G(x) = a + b \cdot R(x) \tag{18}$$

where $G(x)$ is the rain gauge value at location x, $R(x)$ is the radar rainfall estimate at the gauged location x, and a and b are linear coefficients that are determined.

The external drift can clearly indicate the full spatial variability of the radar QPE data, especially in the events that the rain gauge-radar consistency is high. Thus, the estimation at given location x_0 is derived from a linear estimator, and the weights are computed as follows:

$$\sum\nolimits_{i=1}^{n} \lambda_i^{KED} = 1 \tag{19}$$

$$\sum\nolimits_{i=1}^{n} \lambda_i^{KED} R(x_i) = R(x_0) \tag{20}$$

As mentioned above, data transformation is used to deal with the rainfall showing non-Gaussian features and the problematic cases that lack enough rain gauges to obtain a reliable variogram [46]. In this method, normal score transformation, which can associate every given probability quantile to the corresponding quantiles of a standard normal probability distribution, is used to transform the data to obtain a continuous, strictly cumulative distribution [44].

2.2. Meteorological Evaluation

2.2.1. Leave-One-Out Cross Validation (LOOCV)

No independent precipitation observations exist at a high resolution. Hence, to validate the merging methods, a leave-one-out cross validation (LOOCV) was used to assess the performance of the rainfall-merged techniques. In this method, the rain gauge point observations are assumed to be directly measured as true values, which are used to access the performance by a comparison with other given grid point rainfall products that are computed by different rainfall-merged methods. For the scale mismatch between rain gauge and merging grid cell, the gauged location value is substituted by merged rainfall data from the nearest grid center. It is notable that LOOCV only assesses the accuracy of estimations at the rain gauge locations, but the LOOCV statistics allow us to compare the performance of different merging methods systematically.

In this study, we undertook the evaluation on an hourly basis for each rain gauge location and for each of the merging methods.

The following indicators were used to quantitatively compare the different radar-rain gauge merged products and the rain gauge observations for each time step:

Bias: The systematic errors assessment is calculated from the mean of difference between the observed and predicted rainfall values.

$$\text{Bias} = \frac{1}{n}\sum_{i=1}^{n}\left(R_i - \hat{R}_i\right) \tag{21}$$

RMSE: The root mean square (RMSE) represents the standard deviation of the differences between the observed and predicted rainfall values and is widely used in verification.

$$\text{RMSE} = \sqrt{\frac{1}{n}\sum_{i=1}^{n}\left(R_i - \hat{R}_i\right)^2} \tag{22}$$

MRTE: The mean-root-transformed error (MRTE) can mitigate the dominant of errors from large precipitation amounts for the given lower weight:

$$\text{MRTE} = \frac{1}{n}\sum_{i=1}^{n}\left(\sqrt{R_i} - \sqrt{\hat{R}_i}\right)^2 \tag{23}$$

where R_i is the rain gauge observed value, n is the number of the rain gauges, and $\hat{R}_i$ is the estimated value at the rain gauge location. For each whole event, the bias was computed as an average of the entire period and entire spatial range. The bias value can range from $-\infty$ to $+\infty$, with optimal value equal to 0. The RMSE and MRTE can range from 0 to $+\infty$, with optimal value equal to 0.

Finally, the assessment of individual rain gauges only provides a performance at the point scale. For the catchment that has fewer rain gauges to validate the performance, the evaluation over the whole catchment is necessary. Thus, in order to assess the performance of the merging methods at larger areas, a hydrological application approach was implemented.

2.2.2. Hybrid Hydrological Model (Hybrid-Hebei Model)

In this study, all the rainfall runoff forecasts were produced with the semi-distributed rainfall-runoff Hybrid-Hebei model. This model is a semi-distributed model with a lumped conceptual from the Hebei model and a spatially distributed feature based on GIS [47]. The model is in operational use in semi-arid and semi-humid regions. The runoff of the semi-distributed Hebei model is divided into surface runoff and underground runoff in each 1-km^2 grid cell. When the precipitation intensity is greater than the infiltration intensity, the landmark runoff confluence is generated and, conversely, the infiltration component generates underground runoff after considering the soil water demand. Finally the confluence generates the outlet flow of the basin. The structure of the Hebei model in each 1-km^2 grid cell is shown in Figure 1.

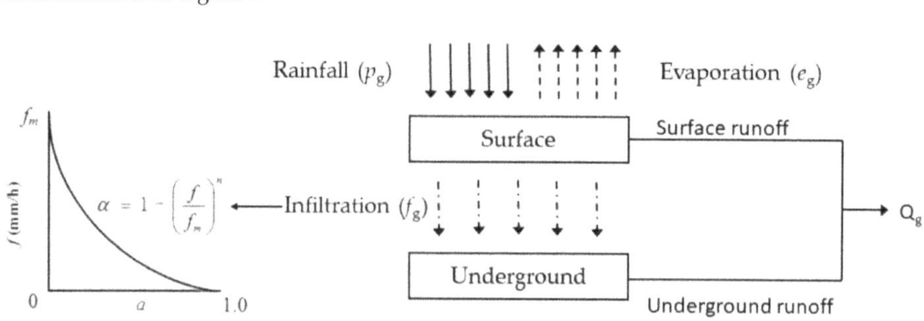

Figure 1. Structure of the Hybrid-Hebei model in each grid cell.

In semi-humid and semi-arid areas, the middle zone of the soil vadose zone is relatively thick, the infiltration rate generated by precipitation often fails to reach the diving surface, and the infiltration rate gradually decreases during the infiltration process. Therefore, the main factor affecting the infiltration rate is the water content of the surface soil. In the semi-distributed Hebei model, considering the complex changes of the underlying surface and the significant difference in infiltration capacity in the semi-arid and semi-humid areas, the infiltration curve in the model is a parabolic infiltration curve controlled by surface soil moisture. This is based on measured data of the Tuanshan gully in northern Shaanxi, China.

The model's infiltration curve within the grid is:

$$f_g = \left(p_g - \frac{i^{(1+n)}}{(1+n)f_m^n}\right)e^{-um} + f_c \tag{24}$$

where f_g is infiltration rate within the grid, unit: mm/h; p_g is rainfall intensity within the grid, mm/h; n is the index; f_c is stable infiltration rate, mm/h; u is the index; m is the surface soil moisture, mm; and f_m^n is the infiltration capacity within the grid, mm/h.

The structure of the lumped Hebei model was described in a previous study [47]. Compared to the original lumped Hebei model, the Hybrid-Hebei model provides improved hydrograph simulations. It can be coupled with high-resolution precipitation to achieve a superior runoff simulation. In this study, the Nash efficiency coefficient (NSE) was used to evaluate the ensemble runoff simulation.

$$\text{NSE} = 1 - \frac{\sum_{t=1}^{T}\left(Q_0^t - Q_m^t\right)^2}{\sum_{t=1}^{T}\left(Q_0^t - \overline{Q}_o\right)^2} \tag{25}$$

where Q_0^t is the observation at time t, Q_m^t is the estimated value at time t, and $\hat{Q}_o$ is the mean value of the whole time T.

For floods in mountainous catchments, the peak flow is an important index, and the relative error (RE) of the peak flow is adopted:

$$\text{RE} = \frac{Q_m - Q_p}{Q_p} \tag{26}$$

where Q_p is the observed peak flow and Q_m is the estimated peak flow.

3. Study Area and Data

3.1. Study Area and Events

The two river catchments of Fuping and Zijinguan were selected as the study areas. These catchments belong to the south and north reaches of the Daqinghe catchment located in Northern China and have semi-humid and semi-arid climatic conditions. The drainage area of Fuping (from latitude 39°22′ to latitude 38°47′N and from longitude 113°40′ to longitude 114°18′E) is 2210 km^2, and the area of Zijingguan (from latitude 39°13′ to latitude 39°40′N and from longitude 114°28′ to longitude 115°11′E) is 1760 km^2. The elevation above sea level in the Fuping and Zijingguan catchments varies from 254 m to approximately 2456 m and 519 m to 2105 m, respectively (Figure 2). Both catchments react very specifically to intense rainfall for the steep terrain and the low vegetation coverage, and the rivers of the two catchments flow from west to east. The river flow is measured at the catchment outlets. There are eight rain gauges in the Fuping catchment and eleven rain gauges in the Zijingguan catchment. An S-band Doppler weather radar with a scan radius of 250 km is located in Shijiazhuang city, which is approximately 100 km to the southeast of the two catchments and both catchments can be completely covered by the radar. In this study, four storm events from the Zijingguan catchment and four storm events from the Fuping catchment were selected to assess the performance of different merging methods. When we chose the storm events, a 24-h time window was used. Within the 24-h window, the storms which showed representative rainfall evenness in space and time were chosen, which then formed the eight storm events. The start times, durations, and peak flows of these events are shown in Table 2.

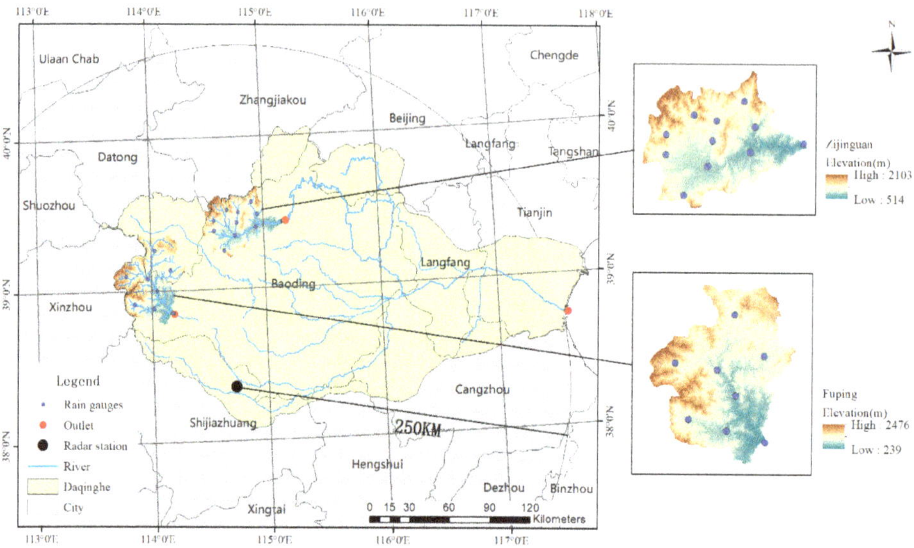

Figure 2. The information of the Zijingguan and Fuping catchments.

Table 2. Durations and rainfall totals for the eight selected storm events.

Catchment	Event ID	Date	Start Time	Duration	Rain Gauges	Accumulated Rainfall (mm)	Peak Flow (m^3s^{-1})
Zijingguan	Z1	22/05/2007	00:00	17 h	11	39.52	6.8
	Z2	10/08/2008	00:00	10 h		45.53	2.5
	Z3	21/07/2012	04:00	14 h		155.43	2580.0
	Z4	19/07/2016	05:00	19 h		74.29	53.4
Fuping	F1	29/07/2007	20:00	24 h	8	63.38	29.7
	F2	30/07/2012	08:00	24 h		50.48	70.7
	F3	01/09/2012	08:00	18 h		40.30	13.7
	F4	25/07/2016	00:00	11 h		10.8	2020

3.2. Weather Radar and Data

The radar data used in this study were retrieved from the S-band single-polarization Doppler weather radar located at Shijiazhuang city in Northern China (Figure 1). The detailed information of this radar can be seen in Table 3. The radar is operated by the China Meteorological Administration (CMA) and has an optimal detection range of 230 km. The radar obtains per base reflectivity data on every six-minute volume scan and completely covers the two study areas [48]. A radar QPE Group System (QPEGS) was developed by the CMA, providing hourly QPE data at a high spatial resolution. For different sources of errors, such as radar calibration, variation of the vertical reflectivity profile, attenuation, and anomalies, quality control of radar data, such as the removal of ground clutter, is necessary and carried out by the OPEGS. A maximum-pixel-value method was also used to generate the "mixed height radar reflectivity" in each volume gridded value, as in the hybrid scan reflectivity proposed by the National Severe Storms Laboratory (NSSL). For the location of the selected radar in this study, a threshold value (38 dBZ) of radar reflectivity was set to differentiate between the convective and stratiform rainfall types considering the rainfall features in eastern China [49]. The accumulated clutter was always magnified during the hourly accumulated rainfall collection period, so a simple clutter filter was used to remove the static clutter. The power law Marshall–Palmer relationship converted the radar reflectivity (Z) to rain rata (R), and a Z-R relationship calculated the accumulated rainfall for convective and stratiform rainfall types [50]. In the QPEGS, the common definition of "a" and "b" is shown in Equations (25) and (26) [48]. To ensure that the spatial resolution of radar data can reflect the precipitation, a higher radar precipitation, rather than a low threshold of 0.1 mm, was considered for the estimation of a 1 km × 1 km radar grid.

For convective:

$$Z = 300R^{1.4} \tag{27}$$

For stratiform:

$$Z = 200R^{1.6} \tag{28}$$

Table 3. The parameters and hardware functions of the Shijiazhuang SA Doppler Radar.

Radar Name	Name of Radar Site	Frequency (GHz)	Beam Width (°)	Antenna Diameter (m)	Pulse Width (μs)	Antenna Gain (db)	Peak Power (kw)
Shijiazhuang	SA	2.7~3.0	1	11.8 *	1.57	≥44	650

* The antenna diameter includes the radome.

4. Results and Discussion

We produced a series of sets of radar-rain gauge merged data by combining the radar and rain gauge precipitation using different merging methods. During the computing period with these merging methods, some applications were defined. For example, a value of 2 (common default value) was given in the RIDW method, and a minimum of three stations were used in the kriging-based methods. Spherical-model semivariograms defined the variograms that were ill defined, with an insufficient number of points. The last previously computed valid variogram was instead used when the present condition was not valid, and the regression and integration were computed on stations only located in the basins [51,52]. The evaluation of the quality and the reliability of the merged data in this study was assessed in two stages: The quantitative evaluation of radar-rain gauge merging methods based on LOOCV and the hydrological model performance driven from the merged data as Hybrid-Hebei model input.

4.1. Evaluation of Radar-Rain Gauge Merging Methods

To assess the performance of the different merging methods, we computed the present performance indicators for the eight events. For each indicator, hourly rainfall values were averaged for each event (Table 4, Figure 3), and the calculated weight of each rain gauge was determined by the area weight divided by the Thiessen polygon. As indicated, the QPE-only-based radar data (OR) clearly showed the weakest performance in all performance indicators, which confirms the necessity of bias correction using rain gauge observations. For the BIAS indicator, the median of BIAS should be close to zero with minimum dispersion because the cross-validation errors of a good estimator should be unbiased. The results shown in Figure 3 indicate that all merging methods were relatively unbiased (median errors were close to zero in all events), but the range of BIAS varied between the different methods. The main difference between these methods can be seen in the range of the indicators, whereby the integration category has the smallest range of BIAS rather than the interpolation category and bias adjustment category [53]. The results show that the kriging-based methods all performed well, with the bias reduced. A comparison of the results of the three categories on the RMSE and MRTE (Figure 3) is noticeable. Irrespective of the two merging methods chosen in the integration or interpolation category, one of these had a clear improvement over the bias adjustment and another has a slight improvement. The performance and added value of the associated merging methods generally increase with the improvement of correlation between rain gauge and radar estimation for medium or large geographic domains, but is invisible for the smaller domains for the limited rain gauges observation and the uncertainty of precipitation under rapid spatial transformation [33].

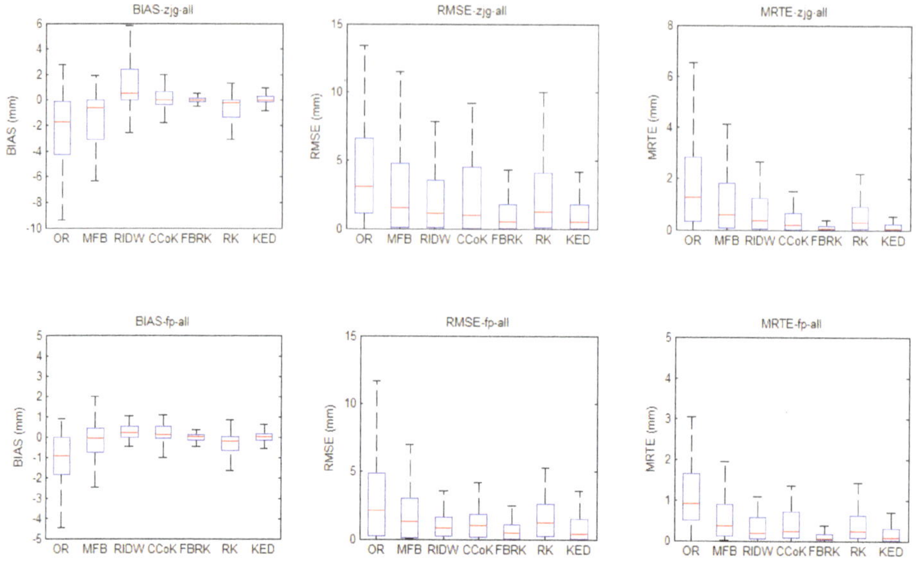

Figure 3. Boxplots of the indicators BIAS, root mean square error (RMSE), and mean-root-transformed error (MRTE) of the chosen two basins. For each method, the central bar is the median, the bounds of the box are the first and third quartiles, and the whiskers include 1.5-times the interquartile range from the box. Note that only the hourly rainfall values in the domain >0.1 mm are provided in this figure.

Table 4. The indicators performance based on leave-one-out cross validation (LOOCV) in the two catchments.

Basin	Indicator	OR	MFB	RIDW	CCoK	FBRK	RK	KED
Zijingguan	BIAS	−2.84	−1.69	1.61	0.58	0.24	−0.71	0.34
	RMSE	4.84	4.49	3.28	3.3	1.31	2.92	1.41
	MRTE	1.86	1.22	1.03	0.55	0.21	0.78	0.22
Fuping	BIAS	−1.08	−0.36	0.33	0.20	−0.08	−0.28	−0.11
	RMSE	3.78	2.14	1.64	2.22	1.18	2.72	1.21
	MRTE	1.63	0.98	0.49	0.57	0.19	0.73	0.22

Figure 3 provides a graphical representation of the results of the different merging methods on the two basins and indicates that the FBRK results performed best, followed by the KED, CCoK, RK, RIDW, and MFB. The improvement of the FBRK data over the KED, however, was relatively minor compared to their improvements over other merging methods. The two methods that performed best were the FBRK and KED, which belong to the integration category and interpolation category, respectively. It has been pointed out in previous studies that the methods in the bias adjustment category were not found to have a better performance in any inter-comparison studies [32,54]. The results of these methods make us consider that correction can be used to estimate all points inside the study area and can deal with anisotropy and the spatial evolution of precipitation with assuming translation invariance and small basins. The results are not clear, however, for partial methods and indicators. For example, the RIDW performed better overall than the CCoK and RK when applied in Fuping. In terms of the BIAS, the RIDW showed a similar performance with the FBRK or KED and had a better performance than CCoK or RK in terms of RMSE and MTRE indicators. This is due to the fact that the quality of the kriging-based methods and bias adjustment methods depend significantly on the geometry of the rain gauge network distribution and, in particular, on the low-density rain gauge network [32].

Through the overall comparison of the performance of the two basins using different merging methods, we found that different merging methods not only produced similar results, but also showed different performances. This is because there is an information gap, such as the area of the basin, the distribution information of the rain gauges, and the distance from the radar. Therefore, we need to analyze and discuss each basin in order to determine suitable merging methods that suit different basins and the reasons for their suitability, as well as the reasons why other methods do not perform well.

In Table 5, the indicator performances of the merging methods for the four storm events of Zijingguan are shown. Figures 4 and 5 show the boxplot of the three indicators for the four events in different merging methods and the scatterplot of the rain gauge observations with the predicted rainfall values from the six merging methods. The merging methods generally outperformed the radar-only estimations and the quality of radar data determined the quality of the merging products. The BIAS shows a clear underestimation when comparing the radar QPE with the rain gauge observation and confirms the need for correction of the radar QPE using rain gauges. The results for the BIAS show that for the performance of the rainfall data compared with the gauge observation, OR and MFB showed a strong negative value and the RIDW had a strong positive value. The other methods had a bias value of approximately 0. Furthermore, the values of the RMSE and MRTE strengthen the observation. It can also be seen that the interpolation methods and integration methods performed better than the bias adjustment methods. This could be due to the simplicity of bias adjustment methods and the complex formulation of the other two methods. It seems that a more complex implementation of merging methods always achieves a better result.

The FBRK method and the KED method provided the best performance and second-best performance, respectively, for the three indicators, whereby they provided the best and second-best values over all four events in the Zijingguan catchment. The scatter diagram (Figure 5) demonstrates that the FBRK and KED had a significant relationship between the merging data and rain gauge data, thereby indicating the high potential of merging skills in the applications. To the authors' knowledge, in all merging methods studied, the BAY-based method and the KED are the most popular merging methods that generally perform best [32], and this is in concordance with our results. The overall indicators and events show that the FBRK has a slightly better performance than KED. The third and fourth best performing methods were the CCoK and the RK methods. It is noted that the BIAS value-based RK had a slightly negative bias. In terms of all indicators, CCoK outperformed for three events, but for the Z3 event, the RK performed better than CCoK. The RK preserved the relative spatial rainfall resolution of the radar data, but its value estimates tended to be under the range of the gauge observations. We highlight that the RIDW method was applied with the default value of 2 [52]. The parameter, however, is often applied across large areas (particularly including a degree of spatial varying of rain gauge-radar biases), along with small-scale features that are spatially variable. This means that a default value may fail to quantitatively correct the rainfall data. An adjustment of the parameter may achieve a better performance of the RIDW method. The MFB performed worst in all methods and events. The MFB is, however, the most commonly used and investigated method among all merging methods, and it scales the original radar data to match the rainfall accumulations recorded by rain gauges. Considering this, the MFB can potentially provide a better representation of small-rainfall features compared with the other five merging methods [33]. The simple use and way in which the radar QPE is employed throughout the merging process, however, suggests that the MFB may fail to satisfactorily correct the rainfall features.

Table 5. The indicator performance based on LOOCV in Zijingguan.

Event	Indicator	OR	MFB	RIDW	CCoK	FBRK	RK	KED
Z1	BIAS	−0.84	−0.49	0.25	0.12	0.07	−0.16	−0.10
	RMSE	2.33	1.05	1.08	0.87	0.41	1.05	0.53
	MRTE	0.73	0.34	0.23	0.16	0.04	0.18	0.06
Z2	BIAS	−1.195	−0.78	0.61	−0.14	−0.06	−0.31	0.09
	RMSE	5.69	2.26	2.15	1.86	0.85	1.97	1.07
	MRTE	1.87	0.67	0.52	0.34	0.09	0.40	0.14
Z3	BIAS	−4.40	−2.36	2.34	1.17	0.54	−1.01	0.76
	RMSE	9.11	7.97	5.35	6.17	2.96	5.78	3.23
	MRTE	2.69	1.90	1.27	1.53	0.50	1.34	0.69
Z4	BIAS	−2.68	−1.01	0.71	−0.33	−0.16	−0.66	0.21
	RMSE	6.61	4.12	3.12	1.31	0.97	2.10	0.99
	MRTE	2.12	1.68	1.11	0.66	0.20	0.87	0.31

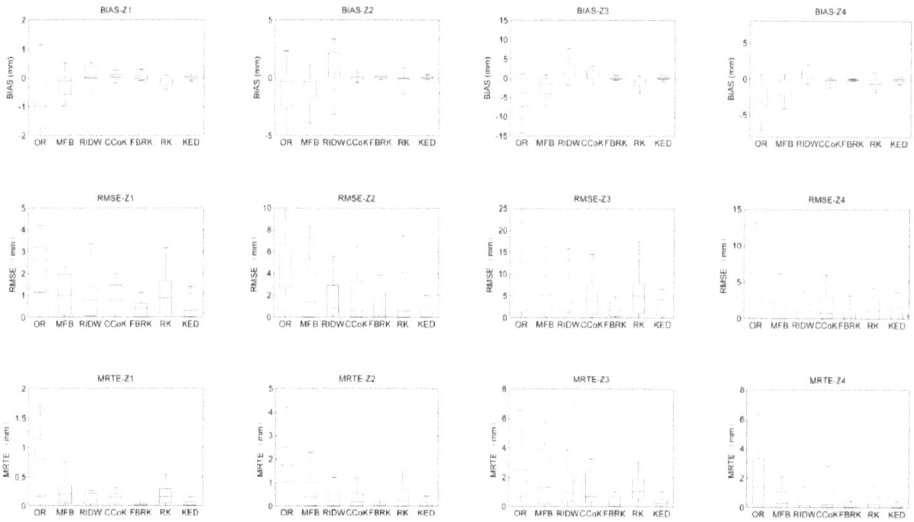

Figure 4. Boxplot for the radar only and six merging methods values of three indicators in four events. For each method, the central bar is the median, the bounds of the box are the first and third quartiles, and the whiskers include 1.5-times the interquartile range from the box. Note that only the hourly rainfall values in the domain >0.1 mm are provided in this figure.

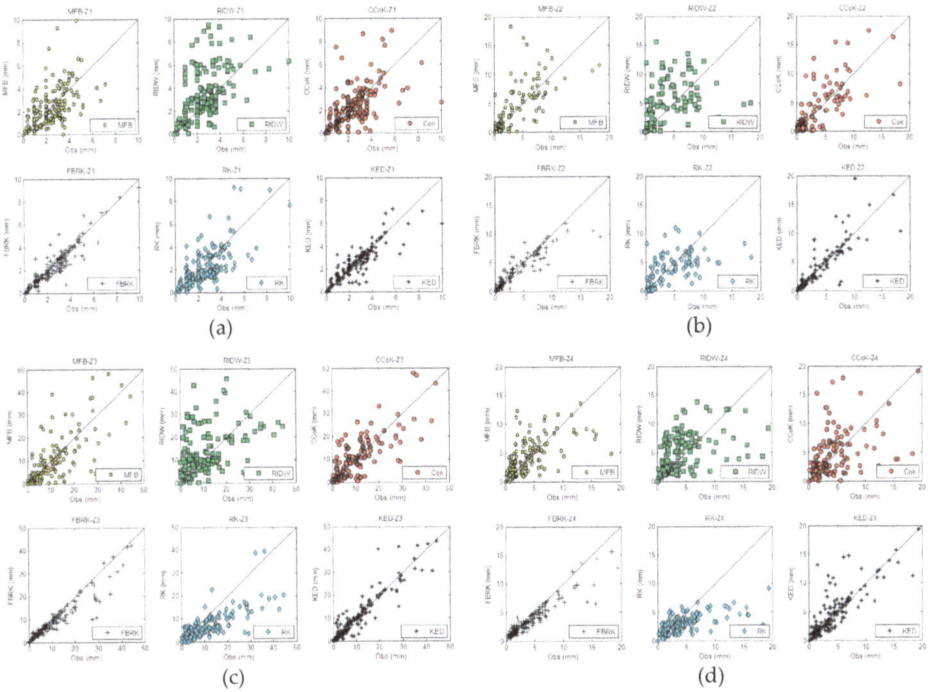

Figure 5. Scatterplots of data between the rain gauge observations and the radar-rain gauge merged products (the mean field bias (MFB), regression inverse distance weighting (RIDW), collocated co-kriging (CCok), fast Bayesian regression kriging (FBRK), regression kriging (RK), and kriging with external drift (KED)) of the Zijingguan basin. Continuous 1/1 slope lines are shown for the purpose of visualization when comparing different merging methods. In this figure, (**a**) the scatter distribution of the chosen merging methods in Z1 event; (**b**) the scatter distribution of the chosen merging methods in Z2 event; (**c**) the scatter distribution of the chosen merging methods in Z3 event; (**d**) the scatter distribution of the chosen merging methods in Z4 event.

Table 6 shows the indicator performances of the merging methods for the four storm events of Fuping. Figures 6 and 7 show detailed information regarding the six merging methods applied in Fuping. The results of the performances for the three categories are the same as those of the Zijingguan events. When comparing the indicator performances of the two basins, the OR performed better in Fuping, inferring that there was a better QPE for close to the radar station. As expected, the methods belonging to the bias adjustment category had a better performance than their applications in Zijingguan. In contrast to the results of the chosen methods applied in Zijingguan, it is clear that the RIDW belonging to the bias adjustment category outperformed the CCoK belonging to the integration category and the RK belonging to the interpolation category. It is well known that the smaller the value in the indicators and the smaller the range in the boxplot, the better the scatter correction. The better QPE is the main reason leading to the superior performance of the RIDW, and with a better QPE, the RIDW can preserve the original structure of the radar rainfall, especially the small-scale features. The RK method is highly reliant upon rain gauge numbers because it simply utilizes the spatial information of the radar field at the rain gauge locations to interpolate. As well as the ability of RK to reproduce rainfall features, it is highly dependent upon the density of rain gauges at the small scale. The RK method is likely to be used more in the limited rain gauges, with the variogram generation based on the point rain gauge value. The relatively poor performance of the CCoK and RK results (Figure 6) were likely due to the assumption that the Gaussian distribution in dynamic merging methods would compute results that may be limited in simulating rainfall values [43]. For CCoK, the method takes spatial information from the radar in the basin. According to the performance results (Table 6), the better QPE and limited rain gauges combined to produce a satisfactory performance.

Table 6. The indicator performance based on LOOCV in Fuping.

Event	Indicator	OR	MFB	RIDW	CCoK	FBRK	RK	KED
F1	BIAS	−1.87	−0.93	0.41	0.76	−0.23	−0.68	−0.31
	RMSE	5.43	3.33	2.35	2.12	1.42	2.67	1.66
	MRTE	3.27	1.40	1.06	0.73	0.19	1.28	0.43
F2	BIAS	−0.93	−0.53	0.17	0.21	0.08	−0.24	0.09
	RMSE	2.79	1.89	1.36	1.84	0.61	1.45	0.96
	MRTE	1.55	0.62	0.33	0.45	0.11	0.34	0.20
F3	BIAS	−0.95	−0.53	0.21	0.26	0.03	−0.26	0.07
	RMSE	2.71	1.32	0.89	1.01	0.51	0.92	0.65
	MRTE	0.71	0.47	0.18	0.28	0.06	0.19	0.12
F4	BIAS	−0.59	−0.25	0.17	0.24	−0.04	0.28	0.05
	RMSE	4.18	2.02	1.68	1.87	0.43	1.78	0.64
	MRTE	0.88	0.36	0.21	0.26	0.06	0.26	0.11

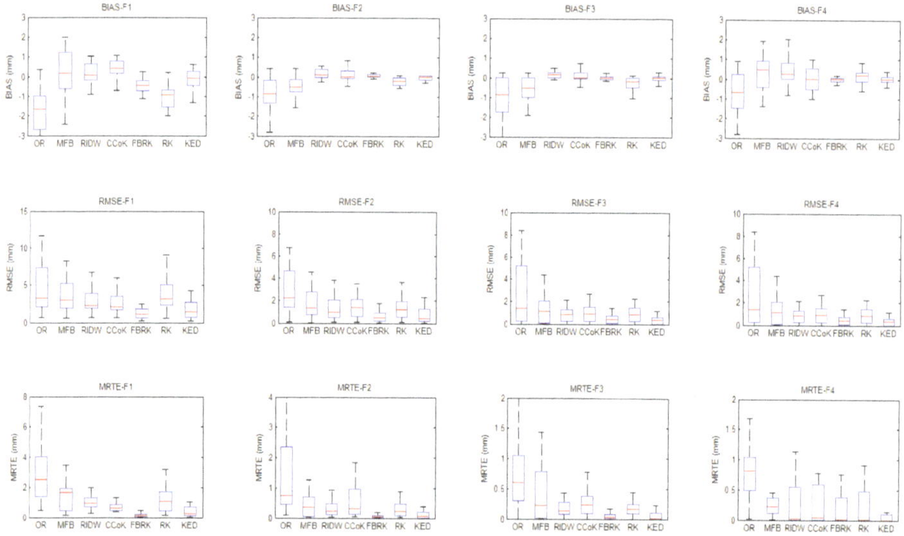

Figure 6. Boxplot for the radar only and six merging methods values of three indicators in four events. For each method, the central bar is the median, the bounds of the box are the first and third quartiles, and the whiskers include 1.5-times the interquartile range from the box. Note that only the hourly rainfall values in the domain >0.1 mm are shown in this figure.

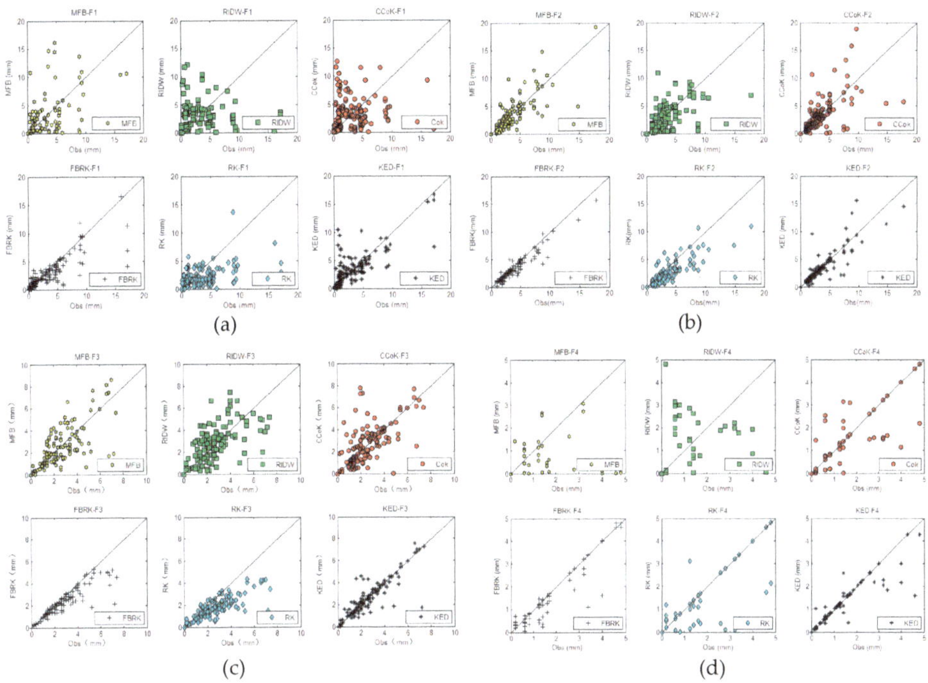

Figure 7. Scatterplots of data between the rain gauge observations and the radar-rain gauge merged products (the MFB, RIDW, CCoK, FBRK, RK, and KED) of the Fuping basin. Continuous 1/1 slope lines are shown for the purpose of visualization when comparing different merging methods. In this figure, (**a**) the scatter distribution of the chosen merging methods in F1 event; (**b**) the scatter distribution of the chosen merging methods in F2 event; (**c**) the scatter distribution of the chosen merging methods in F3 event; (**d**) the scatter distribution of the chosen merging methods in F4 event.

Through the comparison of the three evaluation indicators with the two basins, the two classical approaches (FBRK and KED), based on inotropic and variograms, are the best suited of the merging methods, which work well with the high spatial and temporal variability of precipitation. The performance of the different merging methods is clearly shown in this section. Multiple factors, which affect the application of radar-rain gauge merging methods used in the basins, were identified in this study. Obviously, the most important single factor affecting the performance is the quality of QPE that indicates the ability of radar to sample the rainfall conditions [55]. As the application in the Fuping shows, the RIDW, based a better QPE, outperformed the CCoK and RK, although the limited rain gauges significantly affected the performances of these methods. The lack of rain gauges created a lack of consistency provided by radar and rain gauges. As some studies have found, merging performance generally improves with increasing consistency between radar and rain gauge measurements, particularly in the integration and interpolation categories [19].

For small-scale basins, the preservation of small-scale rainfall features is critical to apply the methods. Different merging methods mean different choices regarding how the radar and rain gauge data are treated and applied during the merging process [56]. In this study, methodological choices focused on improving the quality of radar and rain gauge estimates through different methods. For small-scale basins, the radar QPE was the main data source, providing spatial details of the rainfall. Thus, the ability of these merging methods to preserve local rainfall features was highly dependent on the proportion of, and way of, employing radar data through the merging process. As described in Section 2.1.1, the bias adjustment category scaled the original radar estimation by multiplying the

QPE accumulations with factors to match the rain gauge record. As such, the original structure of the radar rainfall field was essentially preserved. However, the MFB did not achieve a satisfactory performance in the rainfall rates correction associated with small-scale features. Because MFB is usually applied uniformly to large areas, it often ignores the spatial variability in radar QPEs when applied at the small scale [57]. As Table 6 demonstrates, RIDW achieved a satisfactory performance with a high-quality radar QPE. As discussed in the previous section, a local regression of the radar data on the rain gauge data could contribute positively to the RIDW performance. The methods in the integration category combine the two datasets based on their relative uncertainty (see Section 2.1.2). As such, with limited rain gauges or high spatial rainfall variability, integration methods take more information from the radar data. This means that it is possible for integration methods to obtain a satisfactory reproduction of small-scale features, as in the assessment result of FBRK (Table 5). To a certain extent, however, CCoK prefers the interpolation category, in both the performance of comparison from the results and the Gaussian assumption. Unlike the other two categories, the methods in the interpolation category are highly reliant on rain gauge observations, simply considering the gauged radar data in the interpolation process. As per the result comparison discussed in the previous sections, the performance of interpolation-based methods is clearly associated with the density of rain gauges, which decide the ability to capture the small-scale rainfall features. Conversely, interpolation methods with Gaussian assumptions always smooth the high nonlinearities in small-scale features [58]. This is why the error of the RK prediction value is more obvious when the precipitation is large. For KED, the spatial details are reconstructed after the merging process. The density of rain gauges employed in radar-rain gauge merging has an impact on the performance of the merging methods. The impact of limited rain gauge availability on merging performance is closely linked to the reliance of a given merging method upon rain gauge data, as well as to the way in which radar data is employed in the merging process. Having a sufficient number of rain gauges in the study area may increase the ability of the rain gauges to capture the relevant precipitation features. However, the two catchments in this manuscript had a limited number of rain gauges due to the lack of monitoring network. With the increasing of monitoring stations, a further work should be implemented to study the influence of the rain gauge density on the merging performance.

In the six chosen merging method performances, we conclude the strengths and weaknesses of the three categories. The bias adjustment category has the advantage of ease of use, which leads to its wide application. For the methods in this category, small-scale rainfall features are generally preserved, although the correction may fail to correct the rain rates. The integration category allows for the consideration of rain gauge and radar uncertainties. The complex computation of the integration category, such as requiring solving matrix systems, leads to it being applied and tested the least. The interpolation category ranks between the other two categories in both complexity and performance, which is why the methods in the interpolation category are becoming increasingly popular [19].

4.2. Hydrological Model Performance Evaluation

The LOOCV analysis result does not allow a direct comparison because the left rain gauge is just used to compare and not used to compute the product [52]. Some authors have proposed that a higher quality of the merging products can be indicated from agreements between the simulated and observed runoff using the merging products as the input [53]. All authors, however, pointed out that the calibration of a hydrological model is a demanding task and subject to various uncertainties, particularly for mountain flood simulations, whereby it is not easy to find the adequate parameters for considering uncertainties in the model structure and parameterization. In this study, a set of calibrated and validated model parameters were, therefore, used with the Hybrid-Hebei rainfall-runoff model in the same two basins [47].

Table 7 shows the performance of merging methods in the runoff simulation. As expected, the merging methods performed a different goodness-of-fit with the runoff simulation. For the NSE, all merging methods performed normally, and the extent of failure was revealed more clearly in RE. The values of NSE were increased from 0.21 (the MFB in F4) to 0.62 (the KED in F2). Concerning peak flows, the values of MFB and RK were negative overall, indicating a general tendency of underestimation together with the performance of rainfall estimates in the previous section. All the other methods showed a higher estimate. It can be noticed that the RK method seemed to have a better performance for the peak flows than FBRK and KED methods for the Z3, F2, F3, and F4 storm events. In this study, the purpose of inputting the QPEs to the hydrological model was not to rank the merging methods, but to test the applicability of the merging rainfall products for flood forecasting. The performance of the rainfall-runoff model is subject to its parameter calibration. Since only one certain set of parameters was used for the Hybrid-Hebei model, the differences of QPEs from different merging methods may be obscured in the runoff simulations. As indicated in Table 7, the MFB performed worse than any of the methods chosen. The best results were the FBRK and KED methods. The performance of the FBRK and KED methods were almost identical when considering the simulation of NSE and RE in the entire events. This may be because the model calibration seems to have compensated for the differences in the rainfall garnering in the merging process. Compared with the CCoK and RK, the CCoK showed a better performance in most events, and therefore, the result indicates that the performance of the hydrological output is highly dependent on the accuracy of the rainfall product. The RIDW method ranked between the RK and MFB in both Fuping and Zijingguan. With one exception (Z3), however, the performances of these merging methods were different from other events.

Table 7. The indicator performance based on the Hybrid-Hebei model in two catchments.

Event	Indicator	MFB	RIDW	CCoK	FBRK	RK	KED
Z1	NSE	0.37	0.44	0.51	0.57	0.46	0.60
	RE	−0.47	0.57	0.24	0.21	−0.38	0.26
Z2	NSE	0.47	0.41	0.55	0.59	0.46	0.61
	RE	−0.38	0.33	0.36	0.28	−0.29	0.24
Z3	NSE	0.36	0.52	0.42	0.53	0.41	0.49
	RE	−0.52	0.54	0.35	0.28	−0.16	0.41
Z4	NSE	0.26	0.37	0.41	0.52	0.38	0.50
	RE	−0.68	0.61	0.55	0.38	−0.69	0.28
F1	NSE	0.32	0.38	0.48	0.58	0.42	0.55
	RE	−0.47	0.68	0.56	0.42	−0.48	0.38
F2	NSE	0.41	0.49	0.51	0.61	0.49	0.62
	RE	−0.55	0.95	0.77	0.55	−0.48	0.48
F3	NSE	0.36	0.42	0.46	0.54	0.40	0.51
	RE	−0.49	0.78	0.63	0.37	−0.33	0.65
F4	NSE	0.21	0.38	0.41	0.53	0.35	0.55
	RE	−0.68	0.44	0.36	0.68	−0.57	0.48

Figure 8 shows the 24-h accumulated rainfall in the Zijingguan catchment based on different merging methods in the Z3 event. Figure 9 shows the hourly precipitation distribution and simulated stream flow hydrograph of all merging methods in this event, and the hydrograph flood process was extended to 36 h. As indicated in Figure 9, the entire runoff simulation process lines of merging methods had the same trend as the measured flow process line. The runoff hydrograph indicates that the catchment in the study area was relatively fast. The occurrence time of flood peak was the same in all simulation processes, and the peak staggering time was no more than a maximum of one hour. This indicates that the Hybrid-Hebei model can successfully simulate runoff under different precipitation products. For all merging methods, it is can be seen from Table 7 and Figure 9 that FBRK continued to rank highest in the hydrological verification. The RIDW outperformed at NSE because the runoff simulation hydrograph of RIDW was the only one with a double-peak flow as the measured runoff streamline. RIDW, however, was evaluated as only better than the MFB in LOOCV. KED performed the third best in the NSE, however, the extent of failure was found in the peak flow. The CCoK performed worse, and a one-hour peak flow lag was found (Figure 9). RK and MFB both showed an average performance in NSE and RE, and both showed a clear underestimation in peak flow.

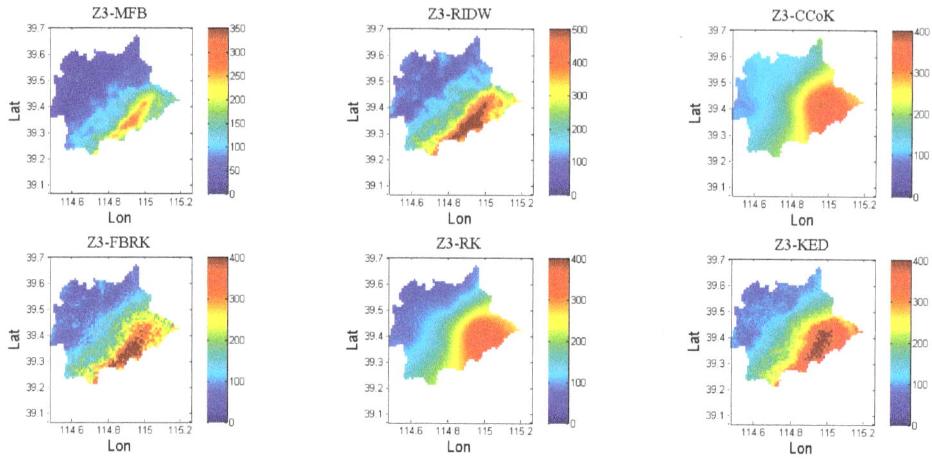

Figure 8. Twenty-four-hour accumulated rainfall based on different merging methods (mm).

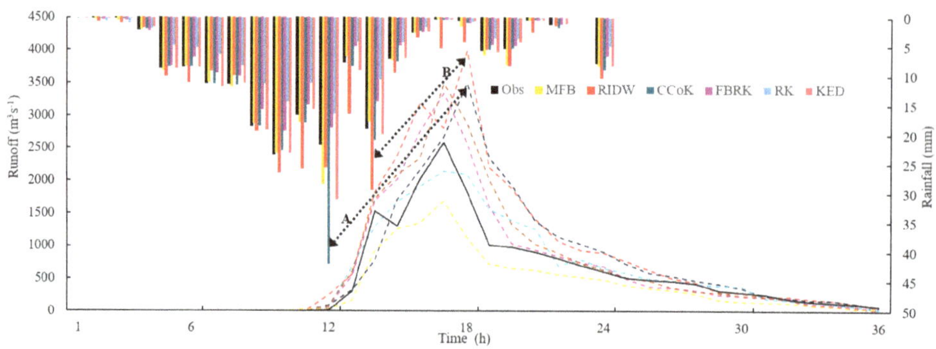

Figure 9. Hydrographs showing simulated stream discharge at the discharge station in Zijingguan.

Based on the result of NSE (Table 7), the results of runoff simulations were normal. This is particularly because these merging methods do not have a large difference in accumulated precipitation and temporal distribution in Z3 (Figure 9). When considering the peak flow, however, the discrepancy was pronounced in both value and occurrence time. Furthermore, for RIDW and CCoK, the peak flow occurrence time of the other four methods was consistent with the measurement, and the difference between them was the value of peak flow. The lagged phase of the double peak flows in RIDW and peak flow in CCoK was approximately one hour. Compared to line A and line B, clear rainfall random errors in RIDW and CCoK were found, and the response of the runoff model performed differently. It is known that the transformation of precipitation to runoff is a smoothing operation in both space and time. For this catchment, a quick response time leads to a failure in the "smoothing effect." If the rainfall accumulation is larger than the response time of catchment, the performance of these methods will be improved [53].

In this section, the fitness of the simulated flow driven by different merging products was used to assess the performance of these products. The hydrological performance of the merging methods was not as good as expected. It is notable that, in runoff simulations, these methods all significantly ranked the lowest in the bias adjustment category. This may be because a spatially differentiated correction was not adopted in this category, and the transformation of rain to runoff is a temporal and spatial process. If the correlation length of random errors is close to the catchment's response time, random errors will be averaged out to a lower extent [53]. For the chosen methods, both FBRK and KED exhibited a low variation of approximately 0.6. The median ranking of RIDW in LOOCV was different to its performance in runoff simulations than CCoK and RK. It should be noted that the potential of giving the merging products as hydrological input is also a function of further multiple factors, including the methodological choices in the merging process, the climatological conditions in the basins, structural model errors, and the cross uncertainties in the entire merging and hydrological application [59]. Moreover, in order to assess the performance of hydrological variables while considering spatial observations, temporal observations must be taken into consideration because products generated by a distributed hydrological model or semi-distributed model, such as stream flow, are found to be sensitive to different high-resolution precipitation [60].

5. Conclusions

In this paper, radar-rain gauge merging categories were conducted. Eight different storm events were chosen from two catchments in semi-humid and semi-arid areas of Northern China to test six different radar-rain gauge-merging methods that belong to three categories using a LOOCV and a rainfall-runoff model (Hybrid-Hebei model). We generated six merged radar-rain gauge rainfall products and compared their performances at gauged location estimations to further their effectiveness as inputs to a semi-distributed rainfall-runoff model of the two study catchments, the Zijingguan and the Fuping catchments in the Northern China. Their relative performances were assessed based on the LOOCV and compared. Two main conclusions can be drawn:

(1) The merging methods have significant potential to improve the quality of rainfall estimates. The integration category performed best in most cases. The bias adjustment category always performed significantly worse. The interpolation category ranked between the aforementioned. The degree of improvement can be a function of merging method that is affected by the quality of both the data and the ability to capture small-scale rainfall features and methodological factors. The total bias of the merging products is because of components of merging methods or other uncertainties. This means that the use of merging methods, without considering the small-scale rainfall features, can be misleading. The quality and representativeness of the radar and rain gauge data should be carefully considered with refinements to mathematical techniques.

(2) In this study, we assumed that a higher quality of the merging products would be indicated from agreements between the simulated and observed runoff using the merging products as the input of the rainfall-runoff model. As expected, the results revealed that a higher quality of merging products indicated a better agreement between the observed and simulated runoff. However, the precipitation estimation random errors will be averaged out to a lower extent when the correlation length of random errors is close to the catchment's response time. Thus, it is hard to know if the streamflow simulation errors were due to precipitation estimation random errors or the rainfall-runoff model's structural errors.

It should be noted that the computational requirements and runtimes are a significant challenge in the merging process. In general, the bias adjustment methods are the least complex and are easy to compute. The interpolation methods are computed relying on the solution of the kriging system, which increases the computational complexity by adding the variables. The integration methods are the most complex and will continuously increase with radar QPE of higher spatial resolution.

In conclusion, this synthetic study demonstrated the potential benefit of the radar-rain gauge-merged rainfall precipitation at a high spatial resolution. The performance in gauged locations evaluation and hydrological application based on the different merging methods was also demonstrated. It is should be noted that the quality of radar QPE will be improved in the future with the increasing available of dual-polarization radars [61]. As discussed in Section 4.1, the quality of QPE plays a critical role on the performance of different merging methods, in which the spatial information of QPE is employed in different merging techniques. It is recommended that the three merging categories are tested in combination with the higher quality QPEs, and it is critical to study how the quality of QPE affects the performance of these merging methods [44]. Furthermore, with the increasing of monitoring stations, a further work should be implemented to study the affection of different density of rain gauges on the merging performance in the future. Notably, the conditions and assumptions of this study, including the hydrology parameters chosen and the Gaussian assumptions in the kriging, are merely simplifications of reality. The difference between the theoretical study and simulated data in this study is that the rainfall observations from radar or rain gauge in reality are even more complicated due to dynamic spatial changes.

Author Contributions: Conceptualization, Q.Q. and J.L.; methodology, Q.Q.; software, Q.Q.; validation, Q.Q., C.L. and Y.J.; formal analysis, J.L. and J.T.; investigation, W.W.; resources, C.L.; data curation, F.Y.; writing—original draft preparation, Q.Q. and J.L.; writing—review and editing, Q.Q. and J.T.; visualization, Q.Q.; supervision, Q.Q. and Y.J.; project administration, F.Y.; funding acquisition, C.L. All authors have read and agreed to the published version of the manuscript.

Funding: This research was funded by the National Natural Science Foundation of China (51822906), the National Key Research and Development Project (2017YFC1502405), the Major Science and Technology Program for Water Pollution Control and Treatment (2018ZX07110001), and the IWHR Research & Development Support Program (WR0145B732017).

Acknowledgments: Thanks to Alexandre Wadoux for help on KED computation; thanks to Xinyi Li for help on scatter figure.

Conflicts of Interest: The authors declare no conflict of interest.

References

1. Varlas, G.; Anagnostou, M.; Spyrou, C.; Papadopoulos, A.; Kalogiros, J.; Mentzafou, A.; Michaelides, S.; Baltas, E.; Karymbalis, E.; Katsafados, P. A multi-platform hydrometeorological analysis of the flash flood event of 15 November 2017 in Attica, Greece. *Remote Sens.* **2019**, *11*, 45. [CrossRef]
2. Salvadore, E.; Bronders, J.; Batelaan, O. Hydrological modelling of urbanized catchments: A review and future directions. *J. Hydrol.* **2015**, *529*, 62–81. [CrossRef]
3. Westra, S.; Fowler, H.; Evans, J.; Alexander, L.; Berg, P.; Johnson, F.; Kendon, E.; Lenderink, G.; Roberts, N. Future changes to the intensity and frequency of short-duration extreme rainfall. *Rev. Geophys.* **2014**, *52*, 522–555. [CrossRef]
4. Molnar, P.; Fatichi, S.; Gaál, L.; Szolgay, J.; Burlando, P. Storm type effects on super Clausius–Clapeyron scaling of intense rainstorm properties with air temperature. *Hydrol. Earth Syst. Sci.* **2015**, *19*, 1753–1766. [CrossRef]
5. Lobligeois, F.; Andréassian, V.; Perrin, C.; Tabary, P.; Loumagne, C. When does higher spatial resolution rainfall information improve streamflow simulation? An evaluation using 3620 flood events. *Hydrol. Earth Syst. Sci.* **2014**, *18*, 575–594. [CrossRef]
6. Nikolopoulos, E.I.; Anagnostou, E.N.; Borga, M.; Vivoni, E.R.; Papadopoulos, A. Sensitivity of a mountain basin flash flood to initial wetness condition and rainfall variability. *J. Hydrol.* **2011**, *402*, 165–178. [CrossRef]
7. Van de Beek, C.; Leijnse, H.; Stricker, J.; Uijlenhoet, R.; Russchenberg, H. Performance of high-resolution X-band radar for rainfall measurement in The Netherlands. *Hydrol. Earth Syst. Sci.* **2010**, *14*, 205–221. [CrossRef]
8. Berne, A.; Krajewski, W.F. Radar for hydrology: Unfulfilled promise or unrecognized potential? *Adv. Water Resour.* **2013**, *51*, 357–366. [CrossRef]
9. He, X.; Refsgaard, J.C.; Sonnenborg, T.O.; Vejen, F.; Jensen, K.H. Statistical analysis of the impact of radar rainfall uncertainties on water resources modeling. *Water Resour. Res.* **2011**, *47*, W09526. [CrossRef]
10. Chang, W.-Y.; Vivekanandan, J.; Ikeda, K.; Lin, P.-L. Quantitative precipitation estimation of the epic 2013 Colorado flood event: Polarization radar-based variational scheme. *J. Appl. Meteorol. Clim.* **2016**, *55*, 1477–1495. [CrossRef]
11. McKee, J.L.; Binns, A.D. A review of gauge-radar merging methods for quantitative precipitation estimation in hydrology. *Can. Water Resour. J.* **2016**, *41*, 186–203. [CrossRef]

12. Gou, Y.; Ma, Y.; Chen, H.; Yin, J. Utilization of a C-band polarimetric radar for severe rainfall event analysis in complex terrain over eastern China. *Remote Sens.* **2019**, *11*, 22. [CrossRef]
13. Wilson, J.W. Integration of radar and raingage data for improved rainfall measurement. *J. Appl. Meteorol. Clim.* **1970**, *9*, 489–497. [CrossRef]
14. Arsenault, R.; Brissette, F. Determining the optimal spatial distribution of weather station networks for hydrological modeling purposes using RCM datasets: An experimental approach. *J. Hydrometeorol.* **2014**, *15*, 517–526. [CrossRef]
15. Bárdossy, A.; Das, T. Influence of rainfall observation network on model calibration and application. *Hydrol. Earth Syst. Sci.* **2008**, *12*, 77–89. [CrossRef]
16. Habib, E.; Haile, A.T.; Tian, Y.; Joyce, R.J. Evaluation of the high-resolution CMORPH satellite rainfall product using dense rain gauge observations and radar-based estimates. *J. Hydrometeorol.* **2012**, *13*, 1784–1798. [CrossRef]
17. Goudenhoofdt, E.; Delobbe, L. Generation and verification of rainfall estimates from 10-yr volumetric weather radar measurements. *J. Hydrometeorol.* **2016**, *17*, 1223–1242. [CrossRef]
18. Wang, L.-P.; Ochoa-Rodríguez, S.; Onof, C.; Willems, P. Singularity-sensitive gauge-based radar rainfall adjustment methods for urban hydrological applications. *Hydrol. Earth Syst. Sci.* **2015**, *19*, 4001–4021. [CrossRef]
19. Jewell, S.A.; Gaussiat, N. An assessment of kriging-based rain-gauge–radar merging techniques. *Q. J. R. Meteor. Soc.* **2015**, *141*, 2300–2313. [CrossRef]
20. McKee, J.L. Evaluation of Gauge-Radar Merging Methods for Quantitative Precipitation Estimation in Hydrology: A Case Study in the Upper Thames River Basin. Master's Thesis, The University of Western Ontario, London, ON, Canada, 2015.
21. Wang, L.P.; Ochoa-Rodríguez, S.; Simões, N.E.; Onof, C.; Maksimović, C. Radar-raingauge data combination techniques: A revision and analysis of their suitability for urban hydrology. *Water Sci. Technol.* **2013**, *68*, 737–747. [CrossRef]
22. Todini, E. A Bayesian technique for conditioning radar precipitation estimates to rain-gauge measurements. *Hydrol. Earth Syst. Sci.* **2001**, *5*, 187–199. [CrossRef]
23. Goovaerts, P. *Geostatistics for Natural Resources Evaluation*; Oxford University Press on Demand: Oxford, UK, 1997.
24. Sinclair, S.; Pegram, G. Combining radar and rain gauge rainfall estimates using conditional merging. *Atmos. Sci. Lett.* **2005**, *6*, 19–22. [CrossRef]
25. Villarini, G.; Mandapaka, P.V.; Krajewski, W.F.; Moore, R.J. Rainfall and sampling uncertainties: A rain gauge perspective. *J. Geophys. Res. Atmos.* **2008**, *113*. [CrossRef]
26. Schuurmans, J.; Bierkens, M.; Pebesma, E.; Uijlenhoet, R. Automatic prediction of high-resolution daily rainfall fields for multiple extents: The potential of operational radar. *J. Hydrometeorol.* **2007**, *8*, 1204–1224. [CrossRef]
27. Cho, Y.; Engel, B.A. NEXRAD quantitative precipitation estimations for hydrologic simulation using a hybrid hydrologic model. *J. Hydrometeorol.* **2017**, *18*, 25–47. [CrossRef]
28. Cole, S.J.; Moore, R.J. Hydrological modelling using raingauge-and radar-based estimators of areal rainfall. *J. Hydrol.* **2008**, *358*, 159–181. [CrossRef]
29. Reichert, P.; Mieleitner, J. Analyzing input and structural uncertainty of nonlinear dynamic models with stochastic, time-dependent parameters. *Water Resour. Res.* **2009**, *45*, W10402. [CrossRef]
30. Anagnostou, M.; Nikolopoulos, E.; Kalogiros, J.; Anagnostou, E.; Marra, F.; Mair, E.; Bertoldi, G.; Tappeiner, U.; Borga, M. Advancing precipitation estimation and streamflow simulations in complex terrain with X-band dual-polarization radar observations. *Remote Sens.* **2018**, *10*, 1258. [CrossRef]
31. Rico-Ramirez, M.; Liguori, S.; Schellart, A. Quantifying radar-rainfall uncertainties in urban drainage flow modelling. *J. Hydrol.* **2015**, *528*, 17–28. [CrossRef]
32. Nanding, N.; Rico-Ramirez, M.A.; Han, D. Comparison of different radar-raingauge rainfall merging techniques. *J. Hydroinform.* **2015**, *17*, 422–445. [CrossRef]
33. Ochoa-Rodriguez, S.; Wang, L.P.; Willems, P.; Onof, C. A review of radar-rain gauge data merging methods and their potential for urban hydrological applications. *Water Resour. Res.* **2019**, *55*, 6356–6391. [CrossRef]

34. Borup, M.; Grum, M.; Linde, J.J.; Mikkelsen, P.S. Dynamic gauge adjustment of high-resolution X-band radar data for convective rain storms: Model-based evaluation against measured combined sewer overflow. *J. Hydrol.* **2016**, *539*, 687–699. [CrossRef]
35. Wood, S.; Jones, D.; Moore, R. Static and dynamic calibration of radar data for hydrological use. *Hydrol. Earth Syst. Sci.* **2000**, *4*, 545–554. [CrossRef]
36. Smith, J.A.; Baeck, M.L.; Meierdiercks, K.L.; Miller, A.J.; Krajewski, W.F. Radar rainfall estimation for flash flood forecasting in small urban watersheds. *Adv. Water Resour.* **2007**, *30*, 2087–2097. [CrossRef]
37. Kitzmiller, D.; Miller, D.; Fulton, R.; Ding, F. Radar and multisensor precipitation estimation techniques in National Weather Service hydrologic operations. *J. Hydrol. Eng.* **2013**, *18*, 133–142. [CrossRef]
38. Seo, D.J.; Krajewski, W.F.; Bowles, D.S. Stochastic interpolation of rainfall data from rain gages and radar using cokriging: 1. Design of experiments. *Water Resour. Res.* **1990**, *26*, 469–477. [CrossRef]
39. Chiles, J.-P.; Delfiner, P. *Geostatistics: Modeling Spatial Uncertainty*; John Wiley & Sons: Hoboken, NJ, USA, 2009; Volume 497.
40. Velasco-Forero, C.A.; Sempere-Torres, D.; Cassiraga, E.F.; Gómez-Hernández, J.J. A non-parametric automatic blending methodology to estimate rainfall fields from rain gauge and radar data. *Adv. Water Resour.* **2009**, *32*, 986–1002. [CrossRef]
41. Shmaryan, L.; Journel, A. Two Markov models and their application. *Math. Geol.* **1999**, *31*, 965–988. [CrossRef]
42. Yang, P.; Ng, T.L. Fast Bayesian Regression Kriging Method for Real-Time Merging of Radar, Rain Gauge, and Crowdsourced Rainfall Data. *Water Resour. Res.* **2019**, *55*, 3194–3214. [CrossRef]
43. Cecinati, F.; Wani, O.; Rico-Ramirez, M.A. Comparing Approaches to Deal with Non-Gaussianity of Rainfall Data in Kriging-Based Radar-Gauge Rainfall Merging. *Water Resour. Res.* **2017**, *53*, 8999–9018. [CrossRef]
44. Erdin, R.; Frei, C.; Künsch, H.R. Data transformation and uncertainty in geostatistical combination of radar and rain gauges. *J. Hydrometeorol.* **2012**, *13*, 1332–1346. [CrossRef]
45. Haberlandt, U. Geostatistical interpolation of hourly precipitation from rain gauges and radar for a large-scale extreme rainfall event. *J. Hydrol.* **2007**, *332*, 144–157. [CrossRef]
46. Wadoux, A.M.-C.; Brus, D.J.; Rico-Ramirez, M.A.; Heuvelink, G.B. Sampling design optimisation for rainfall prediction using a non-stationary geostatistical model. *Adv. Water Resour.* **2017**, *107*, 126–138. [CrossRef]
47. Tian, J.; Liu, J.; Yan, D.; Ding, L.; Li, C. Ensemble flood forecasting based on a coupled atmospheric-hydrological modeling system with data assimilation. *Atmos. Res.* **2019**, *224*, 127–137. [CrossRef]
48. Shen, Y.; Hong, Z.; Pan, Y.; Yu, J.; Maguire, L. China's 1 km Merged Gauge, Radar and Satellite Experimental Precipitation Dataset. *Remote Sens.* **2018**, *10*, 264. [CrossRef]
49. Zhong, L.; Liu, L.; Gu, S. An algorithm of identifying convective and echo in mixed precipitation and its application in estimating rainfall intensity. *Plateau Meteorol.* **2007**, *26*, 593–602. [CrossRef]
50. Zhang, J.; Langston, C.; Howard, K. Brightband identification based on vertical profiles of reflectivity from the WSR-88D. *J. Atmos. Ocean. Technol.* **2008**, *25*, 1859–1872. [CrossRef]
51. Manz, B.; Buytaert, W.; Zulkafli, Z.; Lavado, W.; Willems, B.; Robles, L.A.; Rodríguez-Sánchez, J.P. High-resolution satellite-gauge merged precipitation climatologies of the Tropical Andes. *J. Geophys. Res. Atmos.* **2016**, *121*, 1190–1207. [CrossRef]
52. Foehn, A.; Hernández, J.G.; Schaefli, B.; De Cesare, G. Spatial interpolation of precipitation from multiple rain gauge networks and weather radar data for operational applications in Alpine catchments. *J. Hydrol.* **2018**, *563*, 1092–1110. [CrossRef]
53. Heistermann, M.; Kneis, D. Benchmarking quantitative precipitation estimation by conceptual rainfall-runoff modeling. *Water Resour. Res.* **2011**, *47*, W06514. [CrossRef]
54. Goudenhoofdt, E.; Delobbe, L. Evaluation of radar-gauge merging methods for quantitative precipitation estimates. *Hydrol. Earth Syst. Sci.* **2009**, *13*, 195–203. [CrossRef]
55. Erdin, R. *Geostatistical Methods for Hourly Radar-Gauge Combination: An Explorative, Systematic Application at Meteoswiss*; MeteoSchweiz: Zürich, Switzerland, 2013.
56. García-Pintado, J.; Barberá, G.G.; Erena, M.; Castillo, V.M. Rainfall estimation by rain gauge-radar combination: A concurrent multiplicative-additive approach. *Water Resour. Res.* **2009**, *45*, W01415. [CrossRef]
57. Ochoa-Rodriguez, S.; Wang, L.; Bailey, A.; Schellart, A.; Willems, P.; Onof, C. Evaluation of radar-rain gauge merging methods for urban hydrological applications: Relative performance and impact of gauge density.

In Proceedings of the UrbanRain15 Proceedings "Rainfall in Urban and Natural Systems", Pontresina, Switzerland, 1–5 December 2015.

58. Berndt, C.; Rabiei, E.; Haberlandt, U. Geostatistical merging of rain gauge and radar data for high temporal resolutions and various station density scenarios. *J. Hydrol.* **2014**, *508*, 88–101. [CrossRef]
59. Kavetski, D.; Kuczera, G.; Franks, S.W. Bayesian analysis of input uncertainty in hydrological modeling: 1. Theory. *Water Resour. Res.* **2006**, *42*. [CrossRef]
60. Koch, J.; Cornelissen, T.; Fang, Z.; Bogena, H.; Diekkrüger, B.; Kollet, S.; Stisen, S. Inter-comparison of three distributed hydrological models with respect to seasonal variability of soil moisture patterns at a small forested catchment. *J. Hydrol.* **2016**, *533*, 234–249. [CrossRef]
61. Huuskonen, A.; Saltikoff, E.; Holleman, I. The operational weather radar network in Europe. *Bull. Am. Meteorol. Soc.* **2014**, *95*, 897–907. [CrossRef]

remote sensing

MDPI

Article

Convective/Stratiform Precipitation Classification Using Ground-Based Doppler Radar Data Based on the K-Nearest Neighbor Algorithm

Zhida Yang, Peng Liu and Yi Yang *

Research and Development Center of Earth System Model (RDCM), College of Atmospheric Sciences, Lanzhou University, Lanzhou 730000, China; yangzhd17@lzu.edu.cn (Z.Y.); liup16@lzu.edu.cn (P.L.)
* Correspondence: yangyi@lzu.edu.cn

Received: 2 August 2019; Accepted: 25 September 2019; Published: 29 September 2019

Abstract: Stratiform and convective rain types are associated with different cloud physical processes, vertical structures, thermodynamic influences and precipitation types. Distinguishing convective and stratiform systems is beneficial to meteorology research and weather forecasting. However, there is no clear boundary between stratiform and convective precipitation. In this study, a machine learning algorithm, K-nearest neighbor (KNN), is used to classify precipitation types. Six Doppler radar (WSR-98D/SA) data sets from Jiangsu, Guangzhou and Anhui Provinces in China were used as training and classification samples, and the 2A23 product of the Tropical Precipitation Measurement Mission (TRMM) was used to obtain the training labels and evaluate the classification performance. Classifying precipitation types using KNN requires three steps. First, features are selected from the radar data by comparing the range of each variable for different precipitation types. Second, the same unclassified samples are classified with different k values to choose the best-performing k. Finally, the unclassified samples are put into the KNN algorithm with the best k to classify precipitation types, and the classification performance is evaluated. Three types of cases, squall line, embedded convective and stratiform cases, are classified by KNN. The KNN method can accurately classify the location and area of stratiform and convective systems. For stratiform classifications, KNN has a 95% probability of detection, 8% false alarm rate, and 87% cumulative success index; for convective classifications, KNN yields a 78% probability of detection, a 13% false alarm rate, and a 69% cumulative success index. These results imply that KNN can correctly classify almost all stratiform precipitation and most convective precipitation types. This result suggests that KNN has great potential in classifying precipitation types.

Keywords: precipitation classification; K-nearest neighbor; Doppler radar; Tropical Precipitation Measurement Mission (TRMM)

---

1. Introduction

Precipitation can be divided into stratiform precipitation and convective precipitation [1]. A convective precipitation system generally has the characteristics of strong upward motion, small areal coverage and high precipitation intensity, while a stratiform precipitation system has the characteristics of weak upward motion, large areal coverage and weak precipitation intensity. The classification of precipitation can be used in meteorological research, weather forecasting and meteorological disaster prevention. First, there are different precipitation growth mechanisms and different physical principles between convective and stratiform precipitation. Research on convective and stratiform classification can provide a better understanding of the physical mechanisms of clouds. Additionally, convective systems have an important influence on the thermal balance of the atmosphere [2,3], and thermodynamic differences can lead to different latent heat distributions, moisture cycling, and cold rain and warm rain processes, which can have different effects on cloud lifetime and Earth's climate [1,4]. The energy of a convective system is often expressed in terms of the apparent heat sources and apparent moisture sinks. An apparent heat source is difficult to measure but can be estimated by precipitation [5,6], and different types of precipitation can reflect different thermodynamic structures. Finally, precipitation estimation plays an important role in understanding the hydrological cycle reducing uncertainties in global climate change model predictions for future environmental scenarios, weather forecasting and disaster prevention [7–9]. In the traditional method of precipitation estimation, a Z-R relationship is adopted, and the estimated precipitation is calculated through the relation between the radar echo intensity and precipitation. However, different types of clouds have different structures and precipitation growth mechanisms, and the use of a single Z-R relation cannot provide a good estimate of precipitation. It is helpful to improve the accuracy of precipitation estimation by classifying precipitation clouds and adopting different Z-R relations for different types of clouds [10,11]. However, there is no clear boundary between stratiform precipitation and convective precipitation, and it is difficult to distinguish one from the other directly from radar data. As a result, a number of methods have been developed to classify precipitation types.

In an early study, ground rain gauges were used for classification. This method classifies rainfall as convective when the gauge data exceed some background level by a certain amount [12,13]. The background-exceedance technique (BET) uses radar reflectivity to identify the convective core in a certain plane and set the radius of influence; the area inside the influence radius is considered the convective rain zone, and the area outside the influence radius is the stratiform rain zone [14]. Steiner, et al. [15] modified the BET method using a variable radius of influence and a variable threshold instead of a fixed radius of influence and threshold in the BET method. The authors proposed the method in 1995, and the method was named SHY95 using the initials of the three authors and the year that the method was proposed. An extended SHY95 method was applied by DeMott, et al. [16], who used a two-dimensional BET at each height level within a volume of radar reflectivity to extend this approach to three dimensions. They suggested that using low-level data may lead to the misclassification of convective cells that tilted strongly with height and showed that using three-dimensional data can improve the accuracy of precipitation classification. Biggerstaff and Listemaa [5] modified the classification results of SHY95 by considering the vertical structure of the radar reflectivity factor based on the SHY95 method and found that the method yielded higher accuracy than SHY95. Bringi, et al. [17] classified precipitation types by calculating the standard deviation of the drop size distribution (DSD). When the standard deviation of the DSD is smaller than a certain standard, it is classified as stratiform precipitation, and when the standard deviation of the DSD is larger than this standard, it is classified as convective precipitation. Instead of using the traditional method based on the BET, Anagnostou [18] proposed an algorithm for classifying stratiform and convective clouds using an artificial neural network (ANN). The cloud-top height, reflectivity at a height of 2 km and 4 other features were used in the ANN training. Compared with other traditional algorithms based on the BET, the ANN exhibited better performance. The DSD has also been used to classify precipitation [19]. Based on a large number of rain events and by computing the Z-R relationship, the average DSD and the corresponding parameters, microphysical analysis can be performed; the rain distribution and precipitation type can be adequately characterized by a gamma DSD. Zhang and Qi [20] developed a method that automatically corrects for large errors due to the bright band in a real-time national radar quantitative precipitation estimation product, and the performances were good [21–23]. Yang, et al. [24] applied the fuzzy logic (FL) method for precipitation classification research using the 2-km height echo reflectivity, vertical integral liquid water content and other characteristics for classification, and the FL classification results were more natural and realistic than those of other methods. Yang, et al. [25] used FL to classify precipitation types and estimate precipitation. The results showed that compared with the Z-R relationship, FL can reduce the underestimation of precipitation and improve the accuracy of estimating precipitation using radar data.

Some studies have used satellite data to classify precipitation types. Adler and Negri [26] used infrared satellite data and applied a variant of the BET to classify convective and stratiform precipitation. Unlike convective cores denoted by radar reflectivity in the BET, they used the minimum cloud-top temperatures to identify the convective core area. The radius of influence of each core was dependent of the magnitude of the infrared brightness temperature of the core [26,27]. Goldenberg, et al. [28] used an infrared cloud-top temperature method similar to the BET to classify convective and stratiform precipitation for a tropical cloud cluster. Awaka, et al. [29] used TRMM precipitation radar data. Two algorithms, the vertical contour mode (V-method) and horizontal contour mode (H-method), were used in the study to classify precipitation types. If the classification results of the two algorithms are the same, the classification result is determined, and if the classification results differ, fusion-based classification results are used. The V-method can be used to detect the bright band. Once the bright band is detected, the precipitation type is classified as stratiform precipitation. Then, the V-method continues to detect convective precipitation according to the radar reflectivity. If the precipitation type is neither stratiform nor convective, it is classified as another type of precipitation. The H-method is based on Houze's classification model [15] using the horizontal echo intensity at a height of 2 km to assess the type of precipitation.

The precipitation process involves complex thermodynamic mechanisms and cloud microphysical mechanisms during sedimentation. These principles are difficult to fully explore. Thus, it is difficult to classify precipitation types based on these mechanisms. Machine learning can be used to build models and capture the characteristics of data such that changes in the data can be predicted and the data can be classified into different categories based on the relevant characteristics. When using machine learning to classify precipitation types, it is not necessary to understand the precipitation mechanisms of convective precipitation or stratiform precipitation, and the representative and appropriate variables for classification and labeling can be selected to achieve optimal classification. Machine learning is a discipline that uses experience to improve the performance of a system by means of calculations [30]. In computer systems, experience often exists in the form of data; thus, the main area of machine learning research involves computer algorithms that generate models from data. The main types of machine learning include supervised learning, unsupervised learning and semisupervised learning. Supervised learning is a method of adjusting the parameters of a model with a set of known classes of samples to achieve the required performance. Supervised learning includes decision trees, boosting and bagging algorithms, support vector machines, etc. Semisupervised learning refers to the fact that data sets contain both identified and unidentified data, and unidentified data are obtained using the identified data. Semisupervised learning usually includes semisupervised Support Vector Machine (SVM), semisupervised clustering, etc. In unsupervised learning, training samples do not have known characteristic information. Unsupervised learning reflects the inherent nature and laws of data by learning unlabeled training samples, providing a basis for further data analysis. This approach is commonly used in clustering.

The K-nearest neighbor (KNN) method is a type of supervised learning algorithm that has been widely used in pattern recognition and classification. KNN relies on the nearest k samples instead of all the samples for classification and is most suitable for classifying samples with overlap and unclear boundaries. KNN was proposed by Fix and Hodges [31], and Cover and Hart [32] further developed and improved the algorithm. KNN has fewer tunable parameters and provides faster calculations for small data sets than other methods. Thus, this approach has advantages in solving classification problems involving precipitation types. Machine learning is seldom used to classify precipitation types. KNN is a mature classification algorithm with many advantages and has been used in many fields, but there is no relevant study to prove the applicability of KNN in the classification of precipitation types, and the present study attempts to use and explore the applicability of KNN to classify precipitation types.

This paper consists of the following: Section 2 introduces radar data and satellite data used in this paper, Section 3 describes the implementation process of the KNN algorithm and the performance of the selected variables under different conditions, Section 4 presents the results of the KNN classification of precipitation type, and the final section provides a summary and conclusion.

2. Data Description

The Doppler radar data used in this study are from the six S-band China Next-Generation Weather Radars (CINRAD/SA), and the site information and usage period of the radars are shown in Table 1. The radars are 10-cm wavelength Doppler radars with a 1° half-power beam width. The radar data consist of volume scans of the radar reflectivity, average radial velocity and spectral width. The radars are operated in 360° azimuthal volume scan mode with steps in elevation angles from 0.5° to 19.5° during periods of precipitation. The number of elevation steps and temporal resolution of the data depend on the operational mode of each radar. The radial bin spacing is 250 m. The radar data used in this paper are interpolated by the Barnes interpolation algorithm to a horizontal grid with a resolution of 1 km × 1 km [33] and a vertical resolution of 500 m over a depth of 18 km in the Cartesian coordinate system. The origin of the coordinates is the position of the radar. The data are quality controlled.

The precipitation radar (PR) is mounted on the TRMM satellite. The system takes 92.5 min to scan Earth, and it can scan Earth 16 times a day. The scanning range is from 38°N to 38°S and 180°W to 180°E, and the scanning swath width is 247 km. The spatial resolution is 5 km. As TRMM uses a low-altitude orbit, the PR can provide measurements of 3D rainfall distributions with unprecedented accuracy in the tropics and subtropics. The products of TRMM have been widely used in a variety of studies, such as the study of precipitation distribution patterns in tropical and subtropical regions [34], to improve the accuracy of precipitation prediction [35]. Research on precipitation structure and properties [36] has demonstrated the reliability of TRMM and its products. In addition to the basic information provided by the PR, the 2A23 product includes rain characteristics observed by the PR. Based on the high vertical resolution of the PR data, the 2A23 product can accurately detect the bright band (BB) occurrence and its height. The following variables are used in this paper: rain flag, which indicates the possibility of precipitation in a grid, the rain type, which is the classification of the precipitation type, including stratiform, convective and others, and the height of the bright band, which indicates whether a BB is detected in a grid and the height of the BB if there is one. As warm and cold rain precipitation are not directly classified and interpreted in the 2A23 product, the classification results do not include the classification of cold or warm rain. The precipitation radar has a wavelength of 2.2 cm, and the ground-based radar used in this paper has a wavelength of 10 cm. Therefore, the precipitation radar will be subject to more two-way path attenuation. In addition, the scanning angle, signal frequency and sensitivity of ground-based radar differ from the PR. The main purpose of this paper is to classify the types of precipitation, taking the 2A23 precipitation classification product of the PR instead of the echo reflectivity data of the PR as the training sample label for KNN and evaluating the training results; these differences are not taken into consideration and have no effect on the results.

The 2A23 product has a horizontal resolution of 5 km, and the horizontal resolution of the radar data is 1 km. To make these two datasets comparable, the interpolation scheme and data selection are described below.

Instantaneous 2A23 data and ground radar data that are within a time lag with a maximum of 3 min are projected into a Cartesian coordinate with 5 km × 5 km horizontal resolution. Each ray of a PR swath is projected on the Cartesian grid by the status of the nearest pixel.

There are still steps needed to make the comparisons of two datasets meaningful. These steps are as follows: (1) a pixel is classified as stratiform by the 2A23 product if a BB is not detected and ref2km is greater than 40 dBz or if there is a BB detected and ref2km is greater than 42 dBz with a horizontal gradient greater than 3 dB/km; (2) a pixel is classified as convective by the 2A23 product if no BB is detected but ref2km is less than 40 dBz; and (3) a pixel is classified as convective by the 2A23 product if a BB is detected.

Table 1. Radar site information and data usage time.

Station	Date	Coordinate	Usage	Cases Number
Hefei	6 June 2010–10 June 2010	117.258°E, 31.867°N	Classification	2
Fuyang	25 June 2005–26 June 2005. 7 July 2007–9 July 2007	115.741°E, 32.879°N	Training and Classification	4
Lianyungang	1 July 2012–31 July 2012	119.294°E, 34.651°N	Training and Classification	7
Nanjing	1 July 2012–31 July 2012	118.698°E, 32.191°N	Classification	5
Guangzhou	4 June 2008–13 June 2008	120.976°E, 32.076°N	Training and Classification	4
Wenzhou	4 June 2008–13 June 2008	117.152°E, 34.293°N	Classification	2

3. Algorithm and Features

3.1. Overview of the K-Nearest Neighbor Method

KNN is a classification algorithm used to classify precipitation types in this paper. KNN does not have a display learning process. In the training phase, KNN simply saves the training samples and processes them after receiving the test samples [37]. Input samples with classification labels are used as KNN inputs for training samples. To achieve satisfactory classification results, a larger number of training samples are needed, and the proportion of each classification in the training samples should be as uniform as possible. In the actual precipitation process, the spatial and temporal extents of stratiform precipitation are usually larger than those of convective precipitation. In the interpolated and screened samples, the number of stratiform precipitation grid points is much larger than the number of convective precipitation grid points. If such data sets are used as KNN training samples, the classification results will be generally biased toward stratiform precipitation. The number of different types of precipitation samples in the training sample needs to be adjusted. Samples of different types of precipitation were randomly selected, and the training sample set was reconstructed according to stratiform cloud precipitation, convective precipitation and other precipitation with a ratio of 1:1:1.

When there are samples to be classified, to obtain the classification results, the distance between the sample to be classified and all the training samples is calculated. After calculating the distance, k training samples with the smallest distances from the sample to be classified are selected. The k training samples have the same influence factor, and the probability that the sample is classified as type j is as follows:

$$P_j = \frac{N_j}{\mathrm{k}}, \tag{1}$$

where P_j is the probability that the sample is classified as type j and N_j is the number of training samples with a classification label of type j among the k-nearest training samples. When the P_j value is a maximum, type j is the classification result.

3.2. Selection of Features

Using KNN to classify different precipitation types requires that the variables used in the classification have significant differences for different precipitation types, such as stratiform and convective precipitation, so that the precipitation types can be well distinguished. Using the 2A23 product as a reference, the variations in the frequency of the variables used for the classification of different precipitation types is determined and compared horizontally to validate the classification variable. However, if the bright band is present, the reflectivity will increase significantly, which will negatively influence the classification results. The bright band is not expected to appear at the time of classification. An altitude of 2 km is high enough to provide a sufficient amount of radar data out to a radius of approximately 150 km, and a 2-km altitude is low enough to avoid serious effects of the bright band, which usually appears at a height of 2.5 km to 4.5 km in tropical and sub-tropical areas [38].

Feature 1: Horizontal distribution characteristics of radar reflectivity at a height of 2 km (ref2km) [18]. ref2 km can often reflect the horizontal structural characteristics of convective systems. For stratiform systems, this height should be adjusted appropriately. In some cases, the temperature at 2 km in the vertical height layer is close to 0 °C, and there is a mixture of liquid and solid phase water and transitions between the two phases. During the conversion process of solid water to liquid water, a water-coating film is formed on the surface of melting water, and the difference between the negative refractive index values of the liquid phase particles and the solid phase particles will cause the back reflectance measured by the radar to increase, which may result in a flat and strong echo band. If such strong echo bands are not distinguished, it may cause an erroneous assessment of that type of precipitation in the area. Figure 1a is the frequency distribution diagram of ref2km. The frequencies of stratiform precipitation and convective precipitation increase below 30 dBz. In the range of 30–35 dBz, the frequency of stratiform precipitation reaches a maximum, and then the frequency decreases gradually with larger reflectivity values. At 40–45 dBz, the frequency drops to almost zero. The convective precipitation reaches a maximum frequency at 35–40 dBz, and the frequency decreases gradually with larger reflectivity values. However, when this frequency is above 50 dBz, there is still convective precipitation. The frequency graph of ref2km shows that the value of ref2km exhibits large differences and can be used to sufficiently discriminate among different precipitation types. It is reasonable to use ref2 km for the classification of precipitation types.

Feature 2: Vertically integrated liquid-water content (VIL) [39]. The liquid water content M and radar reflectivity Z can be defined as follows:

$$\mathrm{M} = \frac{\rho_w \pi}{6} \int_0^x n(a) a^3 da, \tag{2}$$

$$\mathrm{Z} = \int_0^x n(a) a^6 da, \tag{3}$$

where x is the maximum drop diameter, and ρ_w is the density of water. When the Marshall-Palmer drop size distribution is used in Equations (3) and (4), the error is small if the upper limit of integration, x, is replaced by ∞.

$$\mathrm{M} = \frac{N_0 \rho_w \pi}{6} \int_0^\infty \exp(-ba) a^3 da = \frac{N_0 \rho_w \pi}{6} \frac{\Gamma(4)}{b^4} = \frac{N_0 \rho_w \pi}{b^4}, \tag{4}$$

$$\mathrm{Z} = N_0 \int_0^\infty \exp(-ba) a^3 da = \frac{N_0 \Gamma(4)}{b^7} = \frac{720 N_0}{b^7}. \tag{5}$$

Eliminating the parameter b in Equations (5) and (6) yields

$$\mathrm{M} = \frac{N_0 \rho_w \pi}{[720 \times 10^{18} N_0]^{4/7}} Z^{4/7}. \tag{6}$$

For $N_0 = 8 \times 10^6\ \mathrm{m}^{-4}$ and $\rho_w = 10^6\ \mathrm{g/m}^3$,

$$\mathrm{M} = 3.44 \times 10^{-3} Z^{4/7}, \tag{7}$$

where the units of M are $\mathrm{g/m}^3$ and those of Z are $\mathrm{mm}^6/\mathrm{m}^3$.

$$M^* = \int_{h_{base}}^{h_{top}} \mathrm{M} dh' = 3.44 \times 10^{-6} \int_{h_{base}}^{h_{top}} Z^{\frac{4}{7}} dh', \tag{8}$$

Here, M^* is VIL, which is given in units of kg/m^2; Z is radar reflectivity, with units of mm^6/m^3; and h_{top} and h_{base} are the uppermost and lowermost layers of the radar echo, with units of meters. VIL reflects the overall vertical state of the echo area, and it is possible to filter the effects of false high echoes caused by bright bands and topographical factors. At the same time, changes in VIL are a good reflection of changes in a convective system. However, in nonconvective areas, VIL changes little, and the reference value decreases accordingly. Figure 1b shows the frequency distribution of VIL. The frequency of stratiform precipitation reaches a maximum for VIL of 2 kg/m^2 and then decreases rapidly. The VIL value of conditions with almost no stratiform precipitation could reach 4 kg/m^2. In contrast, the frequency of convective precipitation reaches a maximum near a VIL value of 4 kg/m^2 and then decreases, although convective precipitation exists even if the VIL value reaches 18 kg/m^2. Additionally, VIL considerably varies and sufficiently reflects different precipitation types. Thus, it is reasonable to use VIL for the classification of precipitation types.

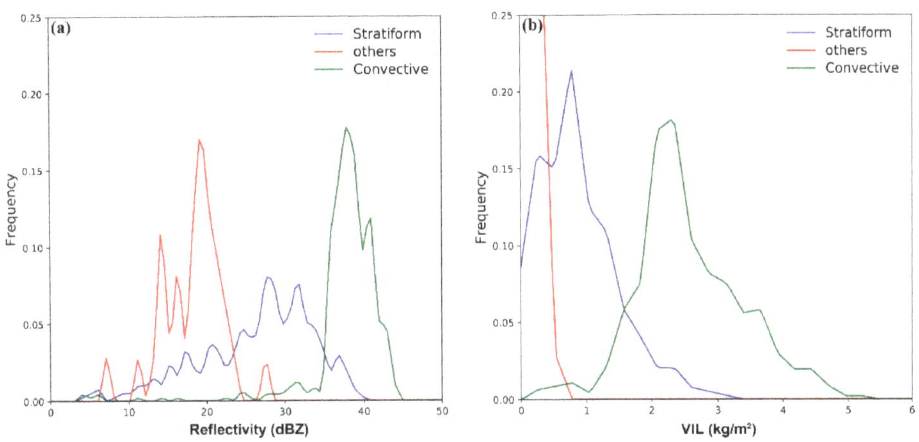

Figure 1. The variations in different types of precipitation frequency as a function of (**a**) Horizontal distribution characteristics of radar reflectivity at a height of 2 km (ref2km) and (**b**) Vertically integrated liquid water content (VIL).

Variables ref2km and VIL have different scales. During precipitation, ref2 km usually has a minimum of 16 dBz and maximum of 50 dBz, while VIL has a minimum of 0 kg/m^2 and maximum of 10 kg/m^2. When calculating the Euclidean distance, the effect of VIL on the distance can be significantly small due to the smaller scale. Thus, the data need to be normalized or standardized before calculating the distance, which could decease the influence of variables with different scales.

The standardized Euclidean distance can decrease the influence of variables with different scales by standardizing the data. The standardized Euclidean distance between sample x and sample y is calculated as follows:

$$\mathrm{d}(\mathrm{x},\mathrm{y}) = \sqrt{\sum_{i=1}^{n} \frac{(x_i - y_i)^2}{s_i^2}}, \tag{9}$$

where s_i is the standard deviation of x_i and y_i over the sample set.

The Euclidean distance, Manhattan distance, and standardized Euclidean distance are used to classify cases at the same time. Although the scales of ref2 km and VIL are not the same, the classification results of the standardized Euclidean distance, Euclidean distance and Manhattan distance do not differ substantially. To remove the possible occurrence of unstandardized adverse effects, the standardized Euclidean distance is used as the distance in the KNN in this study.

3.3. Training and Classification

In this paper, two variables, ref2km and VIL, are used as classification variables, and the corresponding precipitation classification results from the 2A23 product, as classification labels, are put into the KNN algorithm training process. An appropriate k value is selected, and ref2km and VIL of the sample to be classified are put into the KNN. The standardized Euclidean distance between the sample to be classified and each training sample stored in the KNN is calculated. Then, the class with the largest number of k training samples is taken as the classification result.

4. Results

In this section, typical independent individual cases are selected to determine whether the classifications are correct. In addition, an overall analysis is used to assess the performance of the KNN algorithm.

4.1. Evaluation Method

The KNN classification results were compared with the 2A23 product, and the results were evaluated based on the probability of detection (POD), false alarm rate (FAR), and cumulative success index (CSI):

$$\mathrm{POD} = \frac{n_s}{n_s + n_f}, \tag{10}$$

$$\mathrm{FAR} = \frac{n_{fa}}{n_s + n_{fa}}, \tag{11}$$

$$CSI = \frac{n_s}{n_s + n_f + n_{fa}}, \tag{12}$$

In the above three formulas, n_s, n_f and n_{fa} are the numbers of successful classifications, failed classifications and false alarm classifications, respectively. Success is counted when a method classification is similar to the PR 2A23 classification, failure is counted when a classification is not similar to the PR 2A23 class, and false alarm is counted when a pixel is classified opposite the PR 2A23 classification.

The POD can reflect the relationship between the number of successful classification points and the number of failed classification points; the higher the POD value is, the better the classification performance. The FAR can explain the proportion of false alarm points in the classification according to the number of correct points in the classification results. The lower the FAR value is, the better the classification performance. The CSI reflects the overall classification performance; it can explain the proportion of correctly classified points among all classified points, and when the CSI reaches a high value, the classification performance is satisfactory.

4.2. K Value

For a finite set classification, the classification error rate of the KNN tends to converge to a certain value as k increases [40]. When k is too large, the classification accuracy rate does not increase significantly, which results in wasted computational resources. When k is too small, the classification accuracy rate is low. Choosing the right k value helps improve the classification accuracy and reduce the calculation amount to improve the calculation speed.

Figure 2 shows the classification of an embedded convective process in the Guangzhou area at 05:28 (UTC) on 6 June, 2008, using the standardized Euclidean distance as the calculated distance. The effect of using different k values on the overall classification results is small. At the junction of different types of precipitation, the results of different k classifications are slightly different. When k is equal to 5, the boundary between stratiform and convective precipitation is rough, and when k is chosen to be 10 or more, the boundary is smooth.

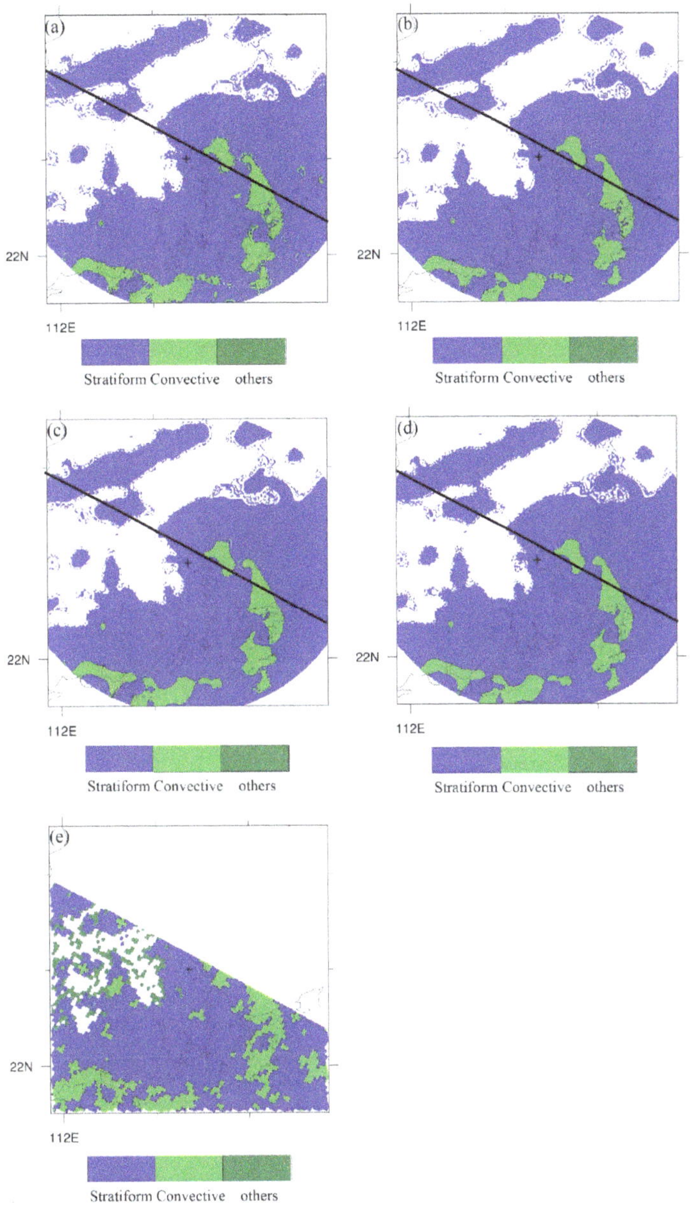

Figure 2. A case from Guangzhou at 05:28 (UTC) on 6 June, 2008, shows the classification results for different k values. (**a–d**) are the results for k = 5, 10, 15, and 20, and (**e**) is the classification of the 2A23 product. The bold black line represents the boundary of the PR scan range.

Other cases from Anhui, Jiangsu and Wenzhou were selected for analysis. The results are shown in Table 2. Although the classification result boundary is rough when k is equal to 5, this value yields the highest POD and CSI and a low FAR among several different k values. When k is equal to 10, the smallest FAR is observed, although the POD is not high and the CSI is low. When k is greater than

10, the POD, FAR and CSI differ, although the difference is not obvious. Thus, when k is equal to 5, the performance is obviously the best, therefore, the value of k is set to 5 in this paper.

Table 2. The POD, FAR and CSI for different k values in the same case.

K	POD	FAR	CSI
5	0.461	0.269	0.394
10	0.385	0.228	0.346
15	0.409	0.244	0.362
20	0.390	0.234	0.349

The classification of precipitation types for different types of weather processes can fully reflect the KNN classification performance. Squall line cases, embedded convective cases and stratiform cases are selected for KNN classification analyses.

4.3. Squall Line Case

Figure 3 shows a squall line case. Figure 3a shows the 2A23 product. Two northeast-southwest-oriented convective belts are classified within the scanning range. There are tiny gaps between the two band-shaped convective cells. Two northeast-southwest-oriented convective belts are classified within the scanning range. The cluster of convective cells is independent of the band-shaped cells. Outside the convective cells, stratiform precipitation covers large area. The southeastern part of Figure 3a is classified as an unknown type of precipitation. In this case, precipitation may occur, although the type of precipitation is unknown. Figure 3b shows the results of the KNN classification. There is a band-shaped northeast-southwest-trending convective cell, which is observed in the 2A23 product. However, the boundary between the two band-shaped convective cells is not obvious. In the northeast direction of the band-shaped convective cell, a cluster of convective cells is also classified, and the cluster shape is similar to that of the 2A23 product. There is also a massive convective cell in the northeast portion of the cluster of convective cells. In the 2A23 product, due to the sweep coverage, there are no corresponding data for this area. The northeast corner of the radar corresponds to the area classified as unknown in the 2A23 product. Because the KNN categorical variable data have no values in that area, no classification is provided. The southwest corner of Figure 3b is a void area due to the radar elevation angle.

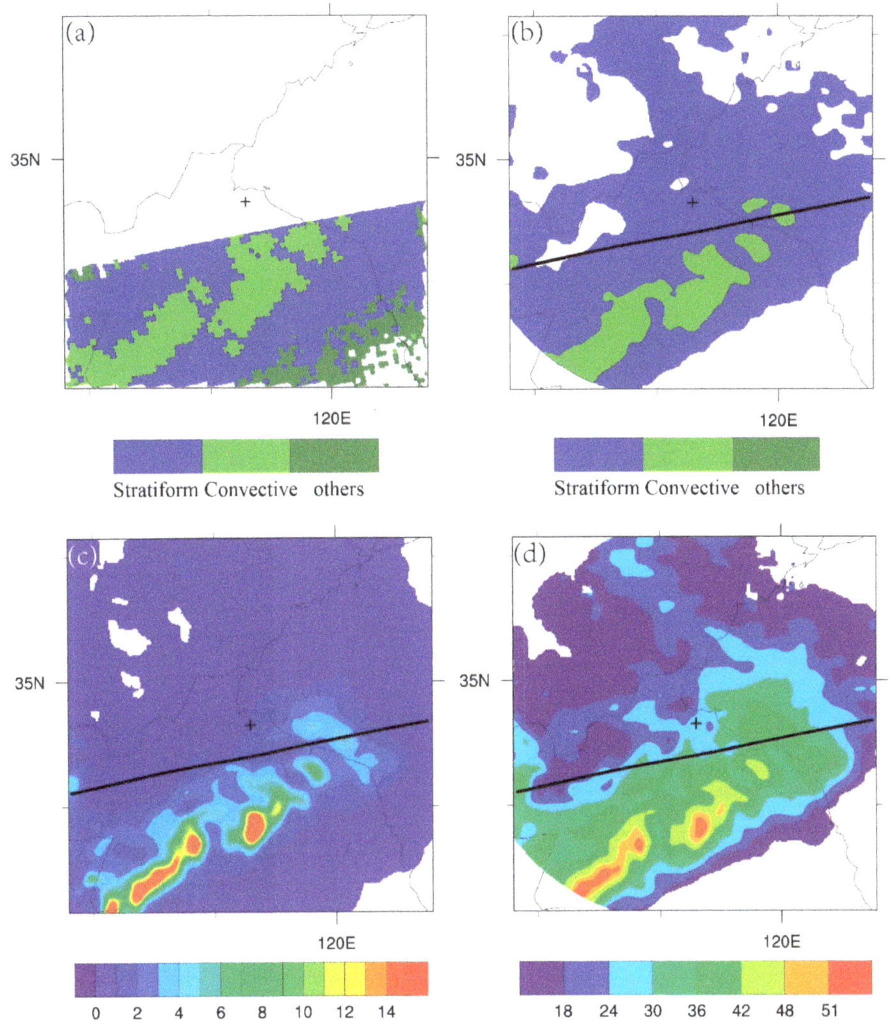

Figure 3. A squall line in Lianyungang at 13:25 (UTC) on July 4, 2012: (**a**) is the classification of the 2A23 product, (**b**) is the KNN classification, (**c**) is VIL, and (**d**) is ref2km. The bold black line represents the boundary of the PR scan range.

Figure 3c shows the VIL. In Figure 3c, there is a northeast-southwest band-shaped high-value area. There are multiple independent high-value centers in the high-value area. The values of all these centers exceed 14 kg/m^2. The value near the center also reaches or exceeds 4 kg/m^2, and there is a block-shaped high-value area in the northeast of the band-shaped high-value area, the value of which exceeds 6 kg/m^2. Additionally, in the northeastern part of the high-value area, there is an area with values exceeding 4 kg/m^2. The VIL values in the other areas are less than 2 kg/m^2. Figure 3d shows ref2km. The high-value area in the figure corresponds to the high-value area in Figure 3c, and the value of each high-value center exceeds 50 dBz. To the northeast of the band-shaped high-value area, there are also areas exceeding 40 dBz. The two high-value areas in the northeast direction of the band-shaped high-value area in Figure 3c,d are consistent with the area classified as convective by the KNN algorithm.

4.4. Embedded Convective Case

Figure 4 shows the classification results for an embedded convective scenario. In Figure 4a, an arched area in the center of the figure is classified as convective by the 2A23 product, and a large area on the west side of the arched convective area is classified as an unknown type of precipitation. There are small stratiform precipitation areas in the northwest corner and a large stratiform precipitation area on the east side of the arched convective cell. Due to the scope of the sweeping surface, there are no data available for the south side. The arched area in the center of Figure 4b is classified as convective precipitation. There is also a convective precipitation area outside the 2A23 product range, and there is a clear boundary between the two arched convective cells. There are large stratiform areas in the northeast portions of the two arched convective cells, and there are stratiform areas in the northwestern parts of the convective cells. The shape and location of the scattered stratiform areas are consistent with those in the 2A23 product. Most of the areas classified as unknown precipitation in the 2A23 product are due to missing values for the variables used for the classification. In Figure 4c, there are two arched high-value areas. The VIL values of the two high-value areas are greater than 4 kg/m^2, and there are obvious gaps between the two arched high-value areas. The VIL of the interval area is between 2 kg/m^2 and 3 kg/m^2. On the northeast side of the high-value area, the VIL is above 2 kg/m^2, and in some other areas, the value is more than 3 kg/m^2. These areas are classified as stratiform in the 2A23 product. There are scattered blocks with VIL values exceeding 1 kg/m^2 on the west side, the northwest side and the south side of the arched area, and the remaining VIL values are all below 0.5 kg/m^2. In Figure 4d, the radar reflectivity at the corresponding position of the high-value area in Figure 4c exceeds 36 dBz, and the reflectivity in the northeastern area of the arched high-value area exceeds 24 dBz. The reflectivity in some of this area exceeds 30 dBz, and in the scattered block area near this arched area, the reflectivity also reaches or exceeds 24 dBz.

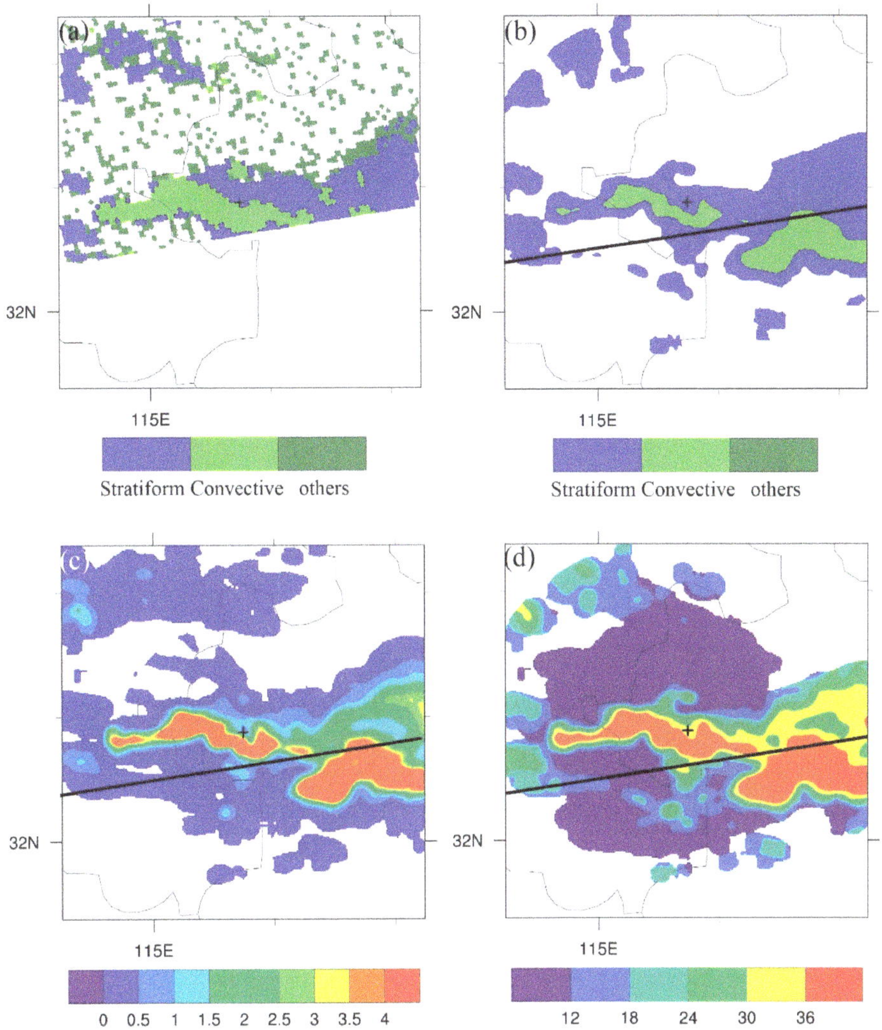

Figure 4. An embedded convective system in Fuyang at 01:41 (UTC) on 8 July 2007: (**a**) is the classification of the 2A23 product, (**b**) is the KNN classification, (**c**) is VIL, and (**d**) is ref2km. The bold black line represents the boundary of the PR scan range.

4.5. Stratiform Case

Figure 5 shows the classification result of a stratiform case. In Figure 5a, a large northwest-southeast-trending band-shaped area is classified as a stratiform area by the 2A23 product, and a small stratiform block is classified on the northwest side of the band-shaped area. In addition, parts of this area are classified as unknown or no precipitation areas. The southern part of the figure is beyond the PR scanning range; thus, there are no data in this area for the 2A23 product. The north side of the solid black line in Figure 5c is within the PR satellite scanning range, and the area and shape of the region classified as stratiform in this range are consistent with those of the 2A23 product. In Figure 5c, the VIL value of the northwest-southeast-trending area exceeds 0.5 kg/m^2, and the VIL value of the high-value area exceeds 1 kg/m^2. The VIL values of other areas are less than 0.5 kg/m^2. The reflectivity of the areas in Figure 5c,d is greater than 18 dBz, with some areas exceeding 24 dBz.

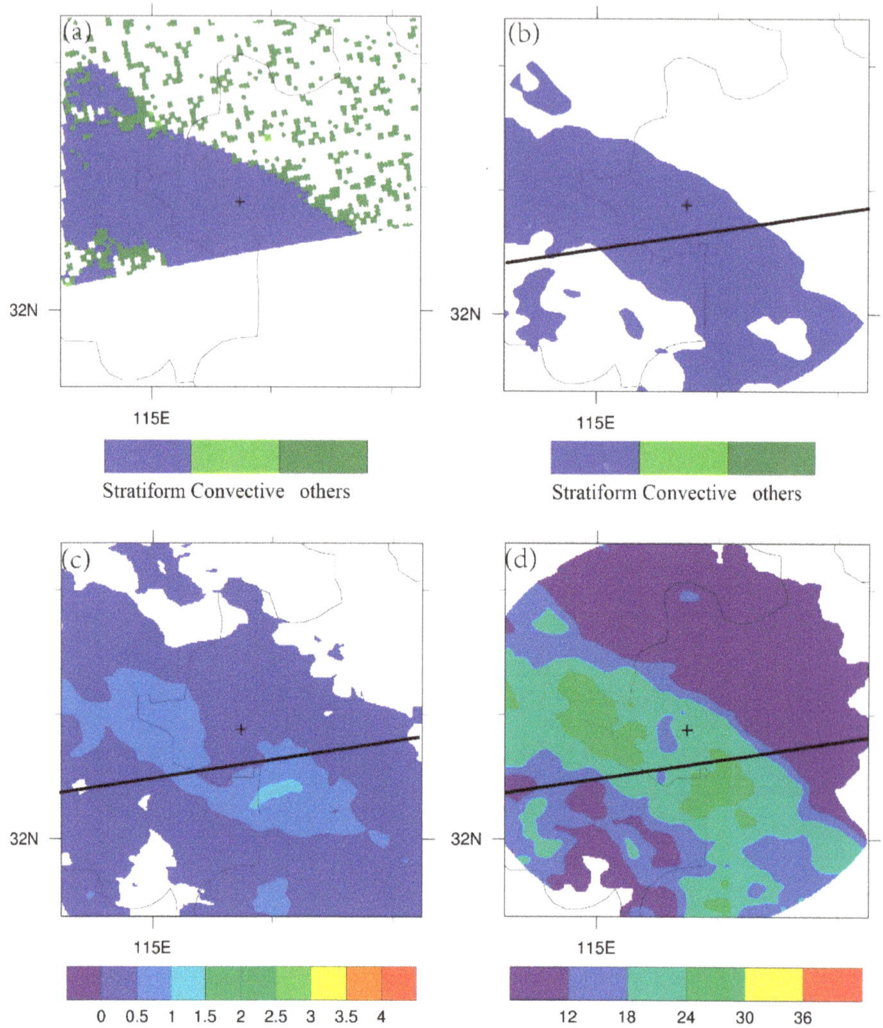

Figure 5. A stratiform precipitation system in Fuyang at 13:47 (UTC) on 7 June 2010: (**a**) is the classification of the 2A23 product, (**b**) is the KNN classification, (**c**) is VIL, and (**d**) is ref2km. The bold black line represents the boundary of the PR scan range.

4.6. Stability of the Algorithm

KNN can classify precipitation types well, but the effect of classifying continuous data is unknown. In fact, continuous data are more widely used and more meaningful. One-month continuous radar data from Lianyungang from 1 July 2012 to 31 July 2012 are used for continuous analysis, and Table 3 shows the result of the continuous analysis.

Table 3. Continuous analysis of Lianyungang in July 2012.

Time(UTC)	Precipitation	POD	FAR	CSI
0:00–12:00	Stratiform	0.978	0.123	0.860
	Convective	0.656	0.077	0.622
12:00–0:00	Stratiform	0.943	0.084	0.868
	Convective	0.784	0.153	0.687

Table 3 shows the result of continuous analysis of Lianyungang in July 2012. The time period of 0:00-12:00(UTC) is daytime in Lianyungang, and the time period of 12:00-0:00(UTC) is evening in Lianyungang. Table 3 shows that both the stratiform and convective classification results are better in evening than in daytime; however, the differences between the daytime and evening results are small, and the classification results are stable in different time periods.

Different geographical conditions may have an impact on the type of precipitation. The precipitation types of Lianyungang, Fuyang and Guangzhou stations were therefore classified to analyze the influence of geographical conditions on KNN. Lianyungang and Fuyang are both located in the subtropical zone, Lianyungang is located near the sea and Fuyang is located inland. The impact of coastal conditions on classification can also be analyzed. Both Guangzhou and Lianyungang are near the sea, Lianyungang is located in the subtropical zone, and Guangzhou is located in the tropical zone; consequently, the influence of latitude conditions on classification can be analyzed. The comparisons of the three sites are shown in Table 4.

Table 4. Comparison of the classification results of different geographical conditions.

Location	Precipitation	POD	FAR	CSI
Lianyungang	Stratiform	0.855	0.006	0.850
	Convective	0.986	0.270	0.722
Fuyang	Stratiform	0.869	0.012	0.859
	Convective	0.973	0.252	0.733
Guangzhou	Stratiform	0.900	0.004	0.896
	Convective	0.990	0.202	0.791

Table 4 shows the classification results under different geographical conditions. The classification results of Guangzhou have the best performance, and the classification results of Lianyungang have the worst performance. The POD values of the three sites are nearly the same for both precipitation types; Guangzhou has the lowest FAR and highest CSI. However, the CSI values of the three sites show few differences, and the results of classification are stable in different geographical conditions.

4.7. Overall Analysis

Table 5 shows the results of the evaluation of the KNN classification results by combining multiple cases of different processes for six Doppler radars in Jiangsu in July, 2012. The POD of KNN for stratiform classification reaches 0.950, the FAR is 0.085, and the CSI is 0.874. From a comprehensive perspective, it is possible to accurately classify more than 85% of the observed stratiform precipitation areas. The POD of the convective classification reaches 0.781, the FAR is 0.137, and the CSI is 0.695. Anagnostou [18] also uses the 2A23 product to classify precipitation types using neural networks, obtaining values of POD = 0.97, FAR = 0.07, and CSI = 0.90 for stratiform precipitation classification and POD = 0.52, FAR = 0.29 and CSI = 0.43 for the classification of convective precipitation. In that paper, the results of SHY95 were also evaluated by 2A23; the stratiform POD, FAR and CSI values were 0.85, 0.05 and 0.81, respectively, and the convective POD, FAR and CSI values were 0.72, 0.59 and 0.36. The cases used are not the same, but with the KNN classification, although the effectiveness for stratiform precipitation decreased, the classification accuracy of convective precipitation improved significantly.

Table 5. Comprehensive evaluation of the KNN classification results.

	POD	FAR	CSI
Stratiform	0.950	0.085	0.874
Convective	0.781	0.137	0.695

5. Conclusion

A KNN supervised machine learning algorithm is used in this paper to classify precipitation types with ground-based radar data. The ground-based radar data are from Anhui, Jiangsu, Guangdong and Zhejiang Provinces, and the classification results were evaluated using the 2A23 cloud classification product from the TRMM PR at the same time. The KNN algorithm is characterized by high precision, insensitivity to abnormal data, no data input assumptions, and a fast computational speed in the case of small data samples. The method performs well in the classification of precipitation types based on radar data. The radar reflectivity at a height of 2 km and VIL were selected as the classification variables. The values of these two variables in the cases of stratiform precipitation and convective precipitation were compared, and it was found that the two variables differ greatly for the different precipitation types. These two variables and corresponding precipitation types in the 2A23 product were input as training samples in the KNN algorithm. The algorithm calculates the distance between the input samples and the stored training samples (the standardized Euclidean distance was calculated in this paper). The maximum number of classification labels in the k samples closest to the input samples was taken as the classification result for the input samples. Samples can be classified into stratiform precipitation, convective precipitation and other types of precipitation.

Three different precipitation systems (stratiform precipitation, embedded convection, and squall lines) were analyzed. The KNN method is accurate in classifying the location and range of stratiform precipitation and can effectively describe the band arrangement pattern of multiple convective units in squall lines. Moreover, the position and shape of squall lines is well described, and the distribution of convective precipitation and stratiform precipitation is accurately described in the embedded convective systems.

The classification results and accuracy of all cases were analyzed, and the performance of the KNN algorithm in precipitation classification was evaluated. The statistical results confirm the results of the case analysis. Among the overall classification results of many processes and cases, the KNN algorithm is the most accurate in the classification of stratiform precipitation, with a POD of 0.950 and an FAR of only 0.085. The CSI, which reflects the overall classification, reaches 0.874. In all cases, the POD of convective classification is 0.781, the FAR is 0.137, and the CSI is 0.695. The evaluation results indicate that the KNN algorithm can accurately classify almost all stratiform precipitation, and most of the convective precipitation can also be classified accurately.

Because the duration of the radar data is insufficient, it is impossible to study the classification of precipitation types with the KNN algorithm in a certain area over a long period. Although the training and classification cases are limited, the results of the classification in different years and for different regional precipitation types could be important. If long-term radar data from a region were selected, more reliable and accurate classification results could be obtained, and the local climate characteristics and precipitation distribution could be better studied.

Author Contributions: Conceptualization, Z.Y. and Y.Y.; methodology, Z.Y., P.L. and Y.Y.; software, Z.Y. and P.L.; investigation, Z.Y. and P.L.; writing—original draft preparation, Z.Y.; writing—review and editing, Z.Y., P.L. and Y.Y.; visualization, Z.Y. and P.L.; supervision, Y.Y.

Funding: This research was supported by the National Key Research and Development Program of China (2017YFC1502102) and the National Nature Science Foundation of China (41675098).

Acknowledgments: We thank the National Aeronautics and Space Administration (NASA) and Japan Aerospace Exploration (JAXA) for providing the 2A23 TRMM precipitation radar rain characteristics product (https://disc.gsfc.nasa.gov/datasets/TRMM_2A23_V7/summary?keywords=2A23) as the training and evaluating data. We thank the Weather Service Forecast Office of Anhui Province and Jiangsu Province for providing radar data.

Conflicts of Interest: The authors declare no conflict of interest.

References

1. Houze, R.A. *Cloud Dynamics*; Academic Press: Cambridge, MA, USA, 1993.
2. Riehl, H. The heat balance of the equatorial trough zone, revisited. *Beitr. Phys. Atmos.* **1979**, *52*, 287–305.
3. Qi, Y.; Jian, Z.; Kingsmill, D.; Min, J.; Howard, K. Correction of radar qpe errors associated with low and partially observed brightband layers. *J. Hydrometeorol.* **2013**, *14*, 580–585. [CrossRef]
4. Lolli, S.; Girolamo, P.D.; Demoz, B.; Li, X.; Welton, E.J. Rain Evaporation rate estimates from dual-wavelength lidar measurements and intercomparison against a model analytical solution. *J. Atmos. Ocean. Technol.* **2017**, *34*, 829–838. [CrossRef]
5. Biggerstaff, M.I.; Listemaa, S.A. An improved scheme for convective/stratiform echo classification using radar reflectivity. *J. Appl. Meteorol.* **2000**, *39*, 2129–2150. [CrossRef]
6. Yanai, M.; Esbensen, S.; Chu, J.H. Determination of bulk properties of tropical cloud clusters from large-scale heat and moisture budgets. *J. Atmos. Sci.* **1973**, *30*, 611–627. [CrossRef]
7. Lolli, S.; D'Adderio, L.; Campbell, J.; Sicard, M.; Welton, E.; Binci, A.; Rea, A.; Tokay, A.; Comerón, A.; Barragan, R. Vertically resolved precipitation intensity retrieved through a synergy between the ground-based NASA MPLNET lidar network measurements, surface disdrometer datasets and an analytical model solution. *Remote Sens.* **2018**, *10*, 1102. [CrossRef]
8. Qi, Y.; Martinaitis, S.; Jian, Z.; Cocks, S. A real-time automated quality control of hourly rain gauge data based on multiple sensors in mrms system. *J Hydrometeorol.* **2016**, *17*, 1675–1691. [CrossRef]
9. Zhang, J.; Howard, K.; Langston, C.; Kaney, B.; Qi, Y.; Tang, L.; Grams, H.; Wang, Y.; Cocks, S.; Martinaitis, S. Multi-radar multi-sensor (mrms) quantitative precipitation estimation: Initial operating capabilities. *Bull. Am. Meteorol. Soc.* **2016**, *97*, 621–638. [CrossRef]
10. Rosenfeld, D.; Atlas, D.; Short, D.A. The estimation of convective rainfall by area integrals: 2. The Height-Area Rainfall Threshold (HART) method. *J. Geophys. Res. Atmos.* **1990**, *95*, 2161–2176. [CrossRef]
11. Qi, Y.; Jian, Z.; Kaney, B.; Langston, C.; Howard, K. Improving wsr-88d radar qpe for orographic precipitation using profiler observations. *J Hydrometeorol.* **2013**, *15*, 1135–1151. [CrossRef]
12. Austin, P.M.; Houze, R.A., Jr. Analysis of the structure of precipitation patterns in New England. *J. Appl. Meteorol.* **1972**, *11*, 926–935. [CrossRef]
13. Houze, R.A., Jr. A Climatological study of vertical transports by cumulus-scale convection. *J. Atmos. Sci.* **1973**, *30*, 1112–1123. [CrossRef]
14. Churchill, D.D.; Houze, R.A., Jr. Development and structure of winter monsoon cloud clusters on 10 December 1978. *J. Atmos. Sci.* **1984**, *41*, 933–960. [CrossRef]
15. Steiner, M.; Houze, R.A., Jr.; Yuter, S.E. Climatological characterization of three-dimensional storm structure from operational radar and rain gauge data. *J. Appl. Meteorol.* **1995**, *34*, 1978–2007. [CrossRef]
16. DeMott, C.A.; Cifelli, R.; Rutledge, S.A. An improved method for partitioning radar data into convective and stratiform components. In Proceedings of the 27th Conference on Radar Meteorology, Vail, CO, USA, 9–13 October 1995.
17. Bringi, V.N.; Chandrasekar, V.; Hubbert, J.; Gorgucci, E.; Randeu, W.L.; Schoenhuber, M. Raindrop size distribution in different climatic regimes from disdrometer and dual-polarized radar analysis. *J. Atmos. Sci.* **2003**, *60*, 354–365. [CrossRef]
18. Anagnostou, E.N. A convective/stratiform precipitation classification algorithm for volume scanning weather radar observations. *Meteorol. Appl.* **2004**, *11*, 291–300. [CrossRef]
19. Caracciolo, C.; Porcù, F.; Prodi, F. Precipitation classification at mid-latitudes in terms of drop size distribution parameters. *Adv. Geosci.* **2008**, *16*, 11–17. [CrossRef]
20. Zhang, J.; Qi, Y.C. A real-time algorithm for the correction of brightband effects in radar-derived qpe. *J Hydrometeorol.* **2010**, *11*, 1157–1171. [CrossRef]

21. Qi, Y.; Zhang, J.; Zhang, P. A real-time automated convective and stratiform precipitation segregation algorithm in native radar coordinates. *Q. J. R. Meteorol. Soc.* **2013**, *139*, 2233–2240. [CrossRef]
22. Qi, Y.; Jian, Z.; Zhang, P.; Cao, Q. Vpr correction of bright band effects in radar qpes using polarimetric radar observations. *J. Geophys. Res. Atmos.* **2013**, *118*, 3627–3633. [CrossRef]
23. Qi, Y.; Zhang, J.; Cao, Q.; Hong, Y.; Hu, X.M. Correction of radar qpe errors for nonuniform vprs in mesoscale convective systems using trmm observations. *J. Hydrometeorol.* **2013**, *14*, 1672–1682. [CrossRef]
24. Yang, Y.; Chen, X.; Qi, Y. Classification of convective/stratiform echoes in radar reflectivity observations using a fuzzy logic algorithm. *J. Geophys. Res. Atmos.* **2013**, *118*, 1896–1905. [CrossRef]
25. Yang, L.; Yang, Y.; Liu, P.; Wang, L. Radar-derived quantitative precipitation estimation based on precipitation classification. *Adv. Meteorol.* **2016**, *2016*, 2457489. [CrossRef]
26. Adler, R.F.; Negri, A.J. A satellite infrared technique to estimate tropical convective and stratiform rainfall. *J. Appl. Meteorol.* **1988**, *27*, 30–51. [CrossRef]
27. Adler, R.F.; Mack, R.A. Thunderstorm cloud height-rainfall rate relations for use with satellite rainfall estimation techniques. *J. Clim. Appl. Meteorol.* **1984**, *23*, 280–296. [CrossRef]
28. Goldenberg, S.B.; Houz, R.A., Jr.; Churchill, D.D. Convective and stratiform components of a winter monsoon cloud cluster determined from geosynchronous infrared satellite data. *J. Meteorol. Soc. Japan Ser II* **1990**, *68*, 37–63. [CrossRef]
29. Waka, J.; Iguchi, T.; Kumagai, H.; Okamoto, K. Rain type classification algorithm for TRMM precipitation radar. In Proceedings of the IEEE International Geoscience and Remote Sensing Symposium Proceedings, Remote Sensing-A Scientific Vision for Sustainable Development, Singapore, 3–8 August 1997.
30. Zhou, Z. *Machine Learning, 1st ed*; Tsinghua University Press: Beijing, China, 2016; pp. 1–2.
31. Fix, E.; Hodge, J.L., Jr. *Discriminatory Analysis-Nonparametric Discrimination: Consistency Properties*; California University Berkeley: Berkeley, CA, USA, 1951.
32. Cover, T.; Hart, P. Nearest neighbor pattern classification. *IEEE Trans. Inf. Theory* **1967**, *13*, 21–27. [CrossRef]
33. Barnes, S.L. A technique for maximizing details in numerical weather map analysis. *J. Appl. Meteorol.* **1964**, *3*, 396–409. [CrossRef]
34. Sorooshian, S.; Gao, X.; Hsu, K.; Maddox, R.A.; Hong, Y.; Gupta, H.V.; Imam, B. Diurnal variability of tropical rainfall retrieved from combined GOES and TRMM satellite information. *J. Clim.* **2002**, *15*, 983–1001. [CrossRef]
35. Hou, A.Y.; Zhang, S.Q.; da Silva, A.M.; Olson, W.S.; Kummerow, C.D.; Simpson, J. Improving global analysis and short-range forecast using rainfall and moisture observations derived from TRMM and SSM/I passive microwave sensors. *Bull. Am. Meteorol. Soc.* **2001**, *660*, 659–679. [CrossRef]
36. Yao, Z.; Li, W.; Zhu, Y.; Zhao, B.; Yong, C. Remote Sensing of Precipitation on the Tibetan Plateau Using the TRMM Microwave Imager. *J. Appl. Meteorol.* **2001**, *40*, 1381–1392. [CrossRef]
37. Zhang, M.L.; Zhou, Z.H. ML-KNN: A lazy learning approach to multi-label learning. *Pattern Recognit.* **2007**, *40*, 2038–2048. [CrossRef]
38. Gao, J.; Tang, G.; Hong, Y. Similarities and improvements of GPM Dual-frequency Precipitation Radar (DPR) upon TRMM Precipitation Radar (PR) in global precipitation rate estimation, type classification and vertical profiling. *Remote Sens.* **2017**, *9*, 1142. [CrossRef]
39. Greene, D.R.; Clark, R.A. Vertically integrated liquid water—A new analysis tool. *Mon. Weather Rev.* **1972**, *100*, 548–552. [CrossRef]
40. Keller, J.M.; Gray, M.R.; Givens, J.A. A fuzzy k-nearest neighbor algorithm. *IEEE Trans. Syst. Man Cybern.* **1985**, *4*, 580–585. [CrossRef]

remote sensing

MDPI

Article

Using the Digital Elevation Model (DEM) to Improve the Spatial Coverage of the MODIS Based Reservoir Monitoring Network in South Asia

Shuai Zhang [1] **and Huilin Gao** [2,*]

[1] Department of Geological Sciences, University of North Carolina at Chapel Hill, Chapel Hill, NC 27599, USA; zshuai@email.unc.edu

[2] Zachry Department of Civil Engineering, Texas A&M University, College Station, TX 77843, USA

* Correspondence: hgao@civil.tamu.edu; Tel.: +1-979-845-2875

Received: 31 December 2019; Accepted: 19 February 2020; Published: 25 February 2020

Abstract: Satellite remote sensing of near real-time reservoir storage variations has important implications for flood monitoring and water resources management. However, satellite altimetry data, which are essential for estimating storage variations, are only available for a limited number of reservoirs. This lack of high-density spatial coverage directly hinders the potential use of remotely sensed reservoir information for improving the skills of hydrological modeling over highly regulated river basins. To solve this problem, a reservoir storage dataset with high-density spatial coverage was developed by combining the water surface area estimated from Moderate Resolution Imaging Spectroradiometer (MODIS) imageries with the Digital Elevation Model (DEM) data collected by the Shuttle Radar Topography Mission (SRTM). By including more reservoirs, this reservoir dataset represents 46.6% of the overall storage capacity in South Asia. The results were validated over five reservoirs where gauge observations are accessible. The storage estimates agree well with observations, with coefficients of determination ranging from 0.47 to 0.91 and normalized root mean square errors (NRMSE) ranging from 15.46% to 37.69%. Given the general availability of MODIS and SRTM data, this algorithm can be potentially applied for monitoring global reservoirs at a high density.

Keywords: reservoir storage; MODIS; SRTM

1. Introduction

Human-made reservoirs, which are managed by storing and releasing water under predetermined operation rules, play an important role in mitigating floods and improving the efficiency of the water supply for municipal, industrial, and agricultural demands [1–4]. Although most (if not all) human operated reservoirs are monitored in real-time, reservoir storage information is not commonly available to the public. Indeed, this directly limits the effectiveness of reservoir flow regulation with regard to flood control, water supply, and other purposes—especially for those reservoirs located within transboundary river basins. For instance, the lack of reservoir information for the Mekong River delta has created challenges with regard to flood forecasting in this region [5,6]. In addition, when assessing and predicting the impacts of droughts, the lack of reservoir storage information reduces the reliability of drought analysis systems [7,8].

Due to the limited availability of gauge observations—especially with regard to remote locations, restricted locations, and/or observations over large geographical areas—remote sensing technology provides a promising alternative by monitoring reservoirs from space [4,9–12]. With remotely sensed water surface area and elevation data, reservoir storage information can be inferred. Reservoir surface area is commonly estimated by classifying optical satellite imageries [13,14] and surface elevation values are typically obtained from satellite radar altimetry [15,16]. The Geoscience Laser Altimeter System (GLAS) onboard the Ice, Cloud, and Land Elevation Satellite (ICESat) and the Advanced Topographic Laser Altimeter System (ATLAS) onboard ICESat-2 were used to measure the elevation values of relatively small lakes and reservoirs [4,17–20].

Even though a variety of remote sensing approaches were developed to monitor reservoir storage from space [21–23], they are still insufficient in terms of spatial and temporal coverage—which hinders their applications when high-density reservoir network information is required. For radar altimetry, the restrictions are mainly due to the coarse spatial resolution. With about 3–20 km footprints, it is difficult to capture water surface level values using radar altimetry over reservoirs that are either not large enough or do not overlap with the satellite tracks [24]. Even for lakes that are detectable by radar altimeters, the data may not be accurate enough for applications if the surrounding topography is complex. Consequently, as of 2015 less than 200 large lakes and reservoirs have been observed using the past and current set of radar altimeters [24]. Compared with radar altimeters, the ICESat/GLAS instrument has a distinct advantage with its small footprint (70 m)—but this comes at the cost of a very long return period (91 days). By combining ICESat elevation values and Moderate Resolution Imaging Spectroradiometer (MODIS) area estimations, Zhang et al. [25] developed an algorithm which is partially capable of monitoring South Asian reservoirs at 16-day intervals, with 28% of the total capacity of in the region covered. Despite such progress, the reservoir observation network is still too sparse due to the large spaces between satellites tracks. Water surface area from Landsat and the area-elevation relationship provided by the Shuttle Radar Topography Mission (SRTM) were combined to infer the water level and reservoir storage variations [26–28]. Landsat can be used to estimate water surface area for smaller reservoirs and lakes due to its high spatial resolution (30 m). However, its repeat period of 16 days limits its ability to monitor reservoir storage at high temporal resolution—especially when cloud coverage is too thick. Therefore, the lack of dense spatial and temporal representation from satellite altimeters remains a major challenge for collecting reservoir storage information on a large scale.

South Asia, which contains one of the largest and densest populations, suffers the most from the dearth of reservoir storage data sharing. The deficient communications with regard to reservoir storage (and management decisions) further exacerbate the casualties and economic losses from flood events. According to past statistical records, South Asia experiences one of the highest fatality rates in the world caused by floods [29]. The available remotely sensed reservoir storage datasets only sparsely cover the region. For instance, radar altimetry data are only available for six reservoirs in this region, which accounts for 10.70% of the total capacity in South Asia (according to Hydrology by altimetry data from Laboratoire d'Etudes en Géophysique et Océanographie Spatiales (LEGOS) [30] and the Global Reservoir Lake Monitor [31]). Although the use of ICESat elevation data improved the coverage to around 28% of South Asian reservoirs [25], it still does not meet the strong societal need. Therefore, acquiring reservoir storage information with large spatial coverage is critical for minimizing the vulnerabilities and maximizing the benefits to communities in this region through good reservoir management practices.

To extend the spatial coverage where remote sensing reservoir storage data are available, a reservoir storage dataset was developed by leveraging the global coverage capability of the Digital Elevation Model (DEM) collected by SRTM. Although DEMs have been most commonly used for generating river routing networks [32,33], they have also been adopted in studies to estimate glacier variations [34,35] and surface water storage change [36]. Due to its high consistency, accuracy, and global coverage [35,37], the SRTM DEM was used to extract the area-elevation (*A-H*) relationship for calculating reservoir storage in this study.

Our overarching goal was to improve the spatial coverage of the remotely sensed reservoir storage dataset in the South Asia region. To this end, the *A-H* relationship of a given reservoir was first derived from MODIS water surface area values and SRTM DEM surface heights, and then combined with the area time series to estimate storage variations. The results were validated with gauge observations. The performance of the generated reservoir dataset was compared with the ICESat based algorithm reported by Zhang et al. [25]. In addition to the data analysis and the results validation, storage estimation uncertainties due to reservoir surface area retrieval algorithm parameterization and elevation measurement errors were also quantified.

2. Data

2.1. Remote Sensing Data

In this study, the two primary remote sensing datasets were the STRM DEM and the MODIS imageries. The DEM was used for inferring the *A-H* relationship. The DEM data were collected by SRTM during an 11-day mission in February 2000, covering a near-global domain from 56° S to 60° N [38]. The relative vertical accuracy was ~6m, and the absolute accuracy was ~16 m [37]. The NASA SRTM V3.0 dataset provides land surface elevation values at a 30-m spatial resolution globally. Here, the global SRTM DEM dataset was obtained from the U.S. Geological Survey's Long Term Archive [39].

For each given reservoir, MODIS imageries were used to derive surface area estimations, which were then applied to the *A-H* relationship to generate a long-term time series of reservoir storage. The reservoir surface area was calculated from the MODIS/Terra 16-day, 250-m resolution vegetation indices product (MOD13Q1). Specifically, an image classification algorithm (Section 3.2.1) was applied to the Normalized Difference Vegetation Index (NDVI) imageries to extract the reservoir area. From 2000 to 2015, a total of 365 imageries were processed for each reservoir.

2.2. Data for Validations

Gauge observations released by the Indian Central Electricity Authority (CEA, [40]) were used to validate the remotely sensed reservoir storage dataset. This gauge data contained daily reservoir water level and storage information for 30 hydropower reservoirs. We downloaded the record from 2008 to 2011 and from 2013 to 2016 in May 2016.

Additionally, the reservoir storage results derived from MODIS and SRTM were compared against the previous results from MODIS and ICESat [25]. Because the Zhang dataset contains results from 21 South Asian reservoirs, this cross-validation helped us to better understand the overall performance of this new dataset on a regional scale.

3. Reservoir Selection and Methodology

3.1. Reservoir Selection

Two criteria were used to identify the reservoirs included in this study: First, the reservoir maximum area at capacity needed to be larger than 55 km^2. The threshold of 55 km^2 was based on a comprehensive consideration of both estimation accuracy and spatial coverage. This would guarantee that the surface area could be estimated with high accuracy using medium-resolution MODIS imageries. Reservoirs larger than 55 km^2 account for ~46.6% of the total South Asian reservoir capacity. Second, the surface area according to the SRTM DEM for a reservoir of interest should not reach its maximum surface area (estimated from MODIS). Otherwise, the respective ranges of area and elevation detected by SRTM DEM would have been too small to infer the *A-H* relationship accurately. Following the above criteria, a total of 28 reservoirs were chosen from the Global Reservoir and Dam (GRanD) database [41]. Figure 1 shows the locations of these reservoirs, and compares the reservoirs from this study with those in Zhang et al. [25], with details shown in Table 1.

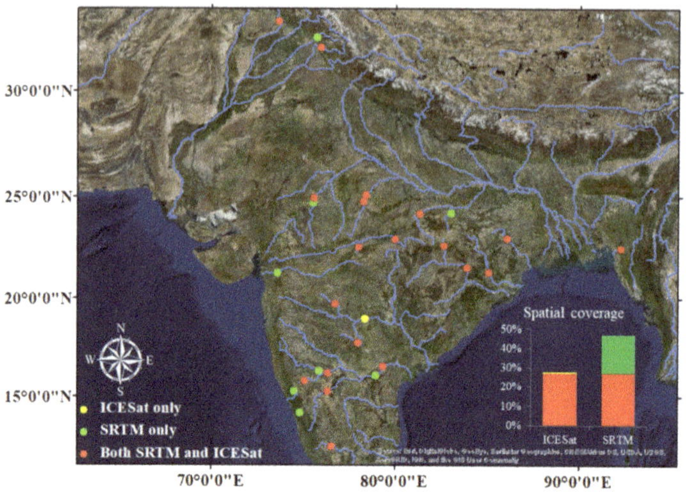

Figure 1. Locations of 28 reservoirs that can be monitored using a remote sensing approach. Yellow dots represent reservoirs that can only be monitored by the Moderate Resolution Imaging Spectroradiometer-Ice, Cloud, and Land Elevation Satellite (MODIS-ICESat). Green dots are reservoirs that can only be monitored through the MODIS-Shuttle Radar Topography Mission (STRM). Red points are reservoirs that can be monitored by both approaches. For each reservoir, detailed information is provided in Table 1.

Table 1. Detailed information for the 28 reservoirs.

I.D.	Reservoir	Country	Location (°N, °E)	Area at Capacity (km^2)	Capacity (km^3)	Purpose [a]	*A-H* Relationship [b]
01	Almatti	India	16.33, 75.89	424	2.63	E	$y = 0.026\ x + 507.17$
02	Bango	India	22.61, 82.60	104	3.41	I,E	$y = 0.201\ x + 332.57$
03	Bansagar	India	24.19, 81.29	384	5.41	I,E	$y = 0.713\ x + 315.71$
04	Bargi	India	22.95, 79.93	268	3.92	I,E	$y = 0.104\ x + 400.28$
05	Chandil	India	22.98, 86.02	139	1.96	I,E	$y = 0.166\ x + 170.15$
06	Gandhi Sagar	India	24.71, 75.55	578	5.60	E	$y= 0.034\ x + 378.24$
07	Hirakud	India	21.52, 83.85	603	4.08	I,E	$y = 0.270\ x + 174.48$
08	Karnafuli	Bangladesh	22.5, 92.23	777	6.48	I,E,F	$y = 0.024\ x + 23.375$
09	Krisharaja Sagar	India	12.42, 76.57	100	1.37	I,E,W	$y = 0.134\ x + 736.91$
10	Linganamakki	India	14.18, 74.85	316	4.18	E	$y = 0.079\ x + 542.95$
11	Mangla	Pakistan	33.13, 73.64	251	7.30	I,E,F	$y = 0.166\ x + 319.61$
12	Malaprabha	India	15.82, 75.09	130	1.07	I,E	$y = 0.136\ x + 619.53$
13	Matatila	India	25.10, 78.37	139	1.13	I,E	$y = 0.095\ x + 292.84$
14	N. J. Sagar	India	16.57, 79.31	240	6.54	I,E	$y = 0.270\ x + 118.8$
15	Narayanapura	India	16.22, 76.35	102	1.07	I	$y = 0.105\ x + 482.91$
16	Pong	India	31.97, 75.95	260	6.95	I,E	$y = 0.212\ x + 366.98$
17	Rajghat	India	24.76, 78.23	224	2.17	I,E	$y = 0.070\ x + 350.35$
18	Ranjit Sagar	India	32.44, 75.73	56	2.20	E	$y = 1.284\ x + 441.10$
19	Rengali	India	21.28,85.03	392	3.17	I	$y = 0.070\ x + 100.88$
20	Rihand	India	24.20, 83.01	485	5.85	I,E	$y = 0.083\ x + 232.99$
21	R. P. Sagar	India	24.92, 75.58	210	1.57	I,E	$y = 0.123\ x + 325.49$
22	Singur	India	17.75, 77.93	129	0.85	W	$y = 0.053\ x + 517.21$
23	Srisailam	India	16.09, 78.90	560	7.11	I,E	$y = 0.042\ x + 254.05$
24	Supa	India	15.28, 74.53	120	4.18	E	$y = 0.460\ x + 506.89$
25	Tawa	India	22.56, 77.98	200	2.31	I	$y = 0.117\ x + 338.36$
26	Tungabhadra	India	15.27, 76.33	390	3.76	I,E	$y = 0.052\ x + 483.92$
27	Ukai	India	21.25, 73.59	512	6.20	I,E,F	$y = 0.042\ x + 81.364$
28	Yeldari	India	19.72, 76.73	82	0.93	I,E	$y = 0.223\ x + 443.45$

[a] I, irrigation; E, electricity generation; W, water supply; F, flood control; [b] y, water surface height; x, area.

3.2. Methodology for Reservoir Storage Estimation

The MODIS-SRTM-based reservoir storage estimation algorithm—referred to as the "MODIS-SRTM algorithm" hereafter—is illustrated using the flowchart in Figure 2. It mainly contains three step: First, the water surface area was estimated from MODIS NDVI imageries via an enhanced classification procedure; second, the *A–H* relationship was generated from the DEM information by regressing the cumulative area values against their corresponding elevation values (within the delineated reservoir maximum domain; and third, by applying the water surface area estimations to the *A–H* relationship, the reservoir storage variations were calculated. Further details of these steps are provided as below.

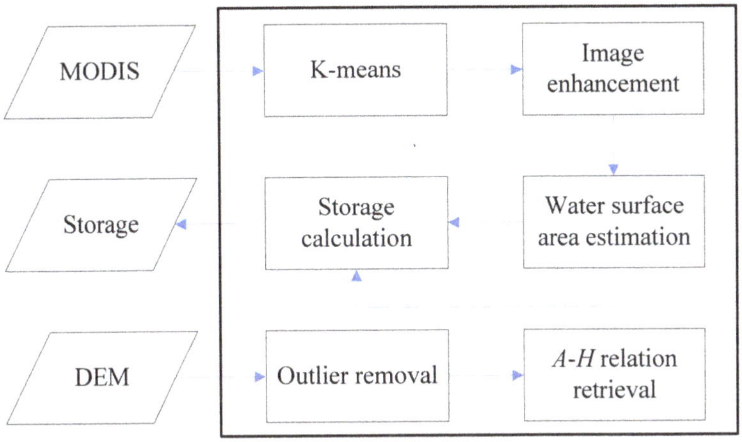

Figure 2. Flowchart of the MODIS-SRTM based reservoir storage estimation algorithm.

3.2.1. Surface Area Estimation

For each given reservoir, the water surface area was estimated using the enhanced K-means classification approach developed by Zhang et al. [25]. First, a threshold of 0.1 was applied to each 16-day MODIS NDVI image from 2000 to 2015, where pixels with NDVI values less than 0.1 were considered water. Based on these simplified classifications, a mask image was created to represent the water coverage percentile and to delineate the domain of the reservoir. Then, the K-means clustering algorithm [42] was used to identify all water pixels within the masked area of the MODIS NDVI images. Finally, a classification enhancement procedure was applied to finetune the results from the K-means clustering. The main purpose of the enhancement was to use the water occurrence map as a reference to correct misclassified pixels and/or to assign an appropriate class to the unclassified pixels [25].

3.2.2. Area-Elevation (*A-H*) Relationship Development

The SRTM DEM data were used to develop the *A-H* relationship for each reservoir. As a valid approximation, the relationships for all reservoirs were assumed to be linear ($H = kA + b$, where k is the slope of the *A-H* relationship, and b is the intercept) [43]. To capture the relationship, we first delineated the water surface area from the DEM for each reservoir of interest. For a given reservoir, the water surface area during the SRTM acquisition time was expanded to include its surrounding pixels by gradually increasing the surface elevation threshold, with the water surface elevation corresponding to the DEM area as the initial value. During this process, all pixels that were not directly connected to the increasing water area were discarded as noise. This expansion continued until the new area on this DEM reached the maximum reservoir area estimated from the MODIS images (from 2000 to 2015). This maximum reservoir area was then delineated from the SRTM DEM. A simplified example of a delineated reservoir is shown in Figure 3a. After delineating the maximum coverage of the reservoir from the DEM, the cumulative area (e.g., A_3) at any given elevation value (e g., H_3) could be estimated by counting the number of pixels with elevations equal to or smaller than that value (i.e., H_3). By regressing the cumulative area values against the elevation values, the *A-H* relationship for the reservoir of interest was established (Figure 3b). A real example of the *A-H* relationship development for the Pong reservoir is provided in Figure 3c,d.

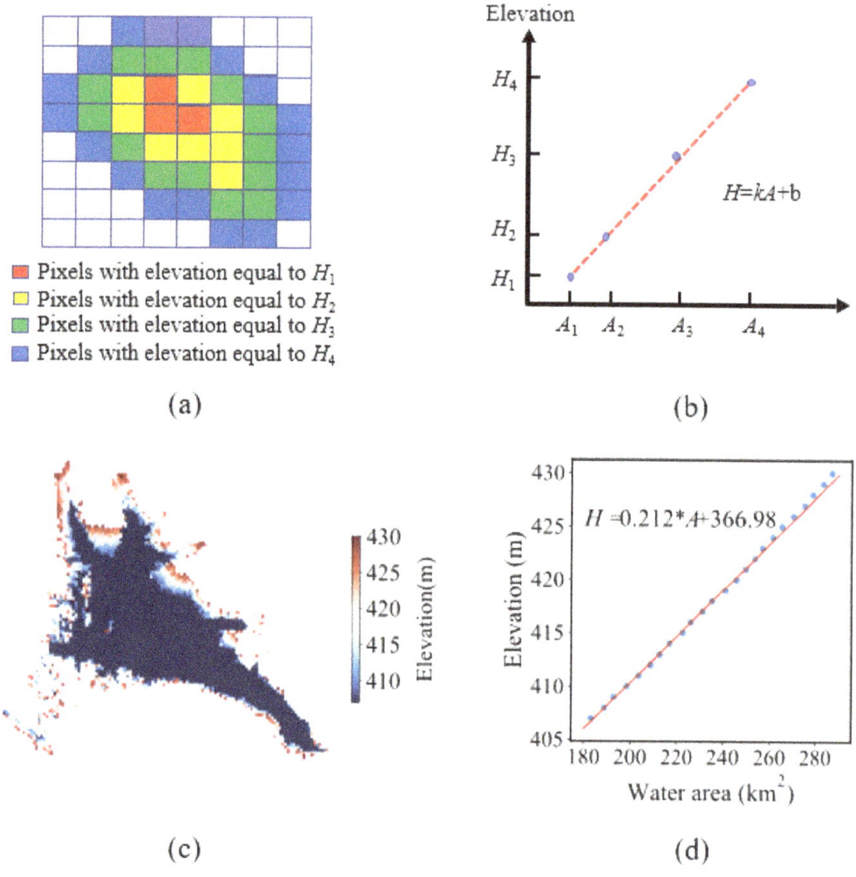

Figure 3. (**a**) A simplified example of a delineated reservoir from the SRTM Digital Elevation Model (DEM), where $H_1 > H_2 > H_3 > H_4$; (**b**) the corresponding *A-H* relationship inferred from a simplified example; (**c**) real example of a delineated reservoir from the SRTM DEM over the Pong reservoir; (**d**) the corresponding *A-H* relationship inferred from the Pong reservoir.

An example of the *A-H* relationship over the Hirakud reservoir is shown in Figure 4a. This *A-H* relationship was also compared with that derived from MODIS area values and ICESat elevations for cross-validation purposes. The MODIS-ICESat-based *A-H* relationship was adopted from Zhang et al. [2014]. The *A-H* relationship from the MODIS-ICESat algorithm is capable of capturing a larger range of water surface elevation values due to its longer temporal coverage period (seven years). The range of elevation values associated with the SRTM based *A-H* relationship depends on how full the reservoir was during the SRTM flight time—the fuller the reservoir at the overpass time, the smaller the elevation range above the water. The slopes for the two relationships are fairly similar with only with a small bias.

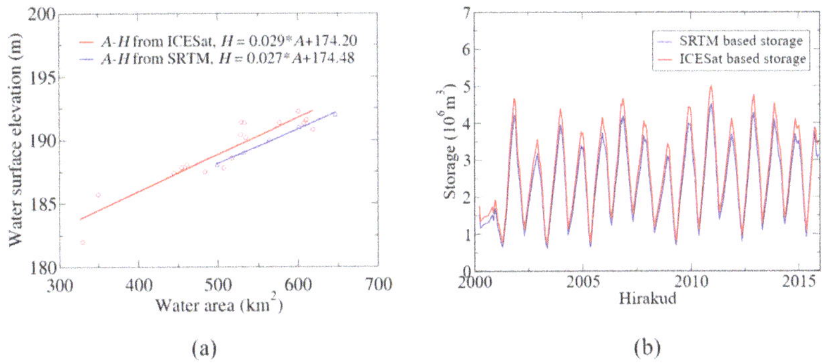

Figure 4. (**a**) The *A-H* relationship developed from SRTM compared with the relationship derived from ICESat, (**b**) time series of the storage estimation values for the Hirakud reservoir from both the SRTM- and the ICESat-based approaches.

3.2.3. Storage Estimation

Reservoir storage can be estimated based on the remotely sensed water surface area and elevation using Equation (1):

$$V_{RS} = V_C - (A_C + A_{RS})(H_C - H_{RS})/2 \quad (1)$$

where V_C, A_C, and H_C represent the storage, area, and water elevation at capacity, respectively. V_{RS}, A_{RS}, and H_{RS} are the remotely sensed storage, area, and water height at the monitoring time.

In this MODIS-SRTM algorithm, since H_{RS} can be calculated by applying the *A-H* relationship to the MODIS area estimation (i.e., A_{RS}), the reservoir storage is calculated through Equation (2) (which was transformed from Equation (1)).

$$V_{RS} = V_C - (A_C + A_{RS})(A_C - A_{RS})k/2 \quad (2)$$

Using the methods explained in this section, the reservoir storage was calculated for the 28 selected reservoirs in South Asia from 2000 to 2015. Using the Hirakud reservoir as an example, Figure 4b compares the time series of reservoir storage from this MODIS-SRTM algorithm with that from the MODIS-ICESat algorithm by Zhang et al. [25]. Results suggest that these two sets of storage estimations are in good agreement. However, compared with the MODIS-ICESat-based algorithm, the storage values from this study tend to be underestimated (due to the different *A-H* relationships). To better understand the error statistics of these two approaches, validations using gauge data were conducted and are reported on in Section 4.1.

4. Results

The MODIS-SRTM-based reservoir storage dataset was examined comprehensively from three perspectives: First, the reliability of the dataset was tested by validating the MODIS-SRTM based reservoir storage results with both in situ gauge data and the MODIS-ICESat based results. Second, the enhanced spatial coverage from this new dataset was compared with the existing reservoir storage dataset in South Asia. Third, the uncertainties associated with the algorithm and dataset were analyzed.

4.1. Validation Results

The MODIS-SRTM-based reservoir storage was validated over 11 reservoirs (Table 2) where gauge observation data were available. The performance of the results was evaluated using Equations (3)–(5), which represent three statistical criteria: the coefficient of determination (R^2), the relative bias (B), and the normalized root mean square error (NRMSE):

$$R^2 = \frac{\sum_{i=1}^{n}\left(RS_i - \overline{RS_i}\right)\left(OBS_i - \overline{Obs}\right)}{\sqrt{\sum_{i=1}^{n}\left(RS_i - \overline{RS_i}\right)^2}\sqrt{\sum_{i=1}^{n}\left(Obs_i - \overline{Obs}\right)^2}} \tag{3}$$

$$B = \frac{\overline{RS} - \overline{Obs}}{\overline{Obs}} \times 100\% \tag{4}$$

$$NRMSE = \frac{\sqrt{\sum_{i=1}^{n}\frac{(RS_i - Obs_i)^2}{n}}}{\overline{Obs}} \times 100\% \tag{5}$$

where RS represents the remotely sensed results, Obs is the gauge data, i denotes the ith record, n is the total number of data points, and $\overline{RS}$ and $\overline{Obs}$ are the average values of the remote sensing results and the gauge data, respectively.

Table 2. Statistical validation results for the remotely sensed reservoir storage data obtained from the MODIS-SRTM approach.

ID	Reservoir Name	R^2	Bias (%)	NRMSE (%)
01	Almatti	0.84	12.40	35.87
05	Gabdhi Sagar	0.69	6.25	15.46
06	Hirakud	0.88	−11.07	18.44
14	N. J. Sagar	0.82	2.80	27.95
15	Pong	0.88	19.25	24.52
17	Ranjit Sagar	0.47	17.77	37.69
18	Rengali	0.79	−13.43	23.81
19	Rihand	0.84	−16.22	28.69
20	R. P. Sagar	0.91	−1.79	15.00
22	Srisailam	0.90	−31.7	32.75
26	Ukai	0.81	−14.76	15.93

As shown in Table 2, most of these results were highly correlated with CEA gauge observations. The R^2 values ranged from 0.47 to 0.91, with a mean of 0.8. The lowest R^2 was found over the Ranjit Sagar reservoir. This reservoir has a relatively small area (56 km^2 at capacity) and is very meandering with a high shoreline to area ratio, complicating the accurate estimation of the surface area from the medium spatial resolution MODIS data [9]. This multicriteria evaluation provided a comprehensive understanding of the results. Using the Srisailam reservoir as an example, its R^2 value was the second highest among all of the validated reservoirs, but its NRMSE was relatively large. Because the slope of the A-H relationship (k, in Equation (2)) is constant, a high R^2 value suggests that the area estimations are accurate. Thus, the large NRMSE was mainly caused by errors associated with the slope of the A-H relationship. Because the area error was proven to be small as indicated by the large R^2, the SRTM DEM was thus the primary error source for the storage results for this reservoir. Another example is the Ranjit Sagar reservoir. Although it had an extremely low R^2 value due to the large amount of error in the surface area estimations, the storage bias was close to those of the Pong and Rihand reservoirs, which indicates a relatively more accurate A-H relationship over this reservoir.

The performance of this algorithm was also compared with the MODIS-ICESat algorithm by Zhang et al. [25] (Table 3). The remotely sensed reservoir storage data from these two algorithms were validated over five reservoirs (Hirakud, N. J. Sagar, Pong, Rengali, and R. P. Sagar) where gauge observations and *A-H* values were available (from both MODIS-ICESat and SRTM).

Table 3. Comparison of the validation results between the MODIS-SRTM and MODIS-ICESat approaches.

		Hirakud	N.J. Sagar	Pong	Rengali	R.P. Sagar
NRSME (%)	ICESat	14.58	26.50	15.21	19.69	18.18
	SRTM	18.44	27.95	24.52	23.81	15.00
Relative Bias (%)	ICESat	−1.88	4.13	0.41	−2.63	−8.97
	SRTM	−11.07	2.80	19.25	−13.43	−1.79
R^2	ICESat	0.94	0.85	0.98	0.85	0.92
	SRTM	0.88	0.82	0.88	0.79	0.91

As shown in Figure 5a, both the MODIS-SRTM and MODIS-ICESat-based approaches performed well overall. The time series from these two algorithms closely matched the gauge values for reservoir storage. To highlight the differences between the DEM and ICESat based algorithms, Figure 5b compares the storage errors against the gauge observations from these two datasets. The error statistics are provided in Table 3. Among each of the five reservoirs, the NRMSE of the MODIS-SRTM algorithm ranged from 18.14% to 27.95%, with a mean value of 21.94%. The relative bias values ranged from −11.07% to 19.25%. The NRMSE of the MODIS-ICESat algorithm ranged from 14.58% to 26.50%, with a mean value of 18.83%. The bias values ranged from −8.97% to 4.41%. In terms of accuracy, the two approaches performed relatively similarly, with the MODIS-ICESat algorithm slightly better than the DEM based algorithm. For the N. J. Sagar reservoir, the NRMSE was 27.95% for the MODIS-SRTM algorithm and 26.50% for the ICESat-based algorithm. For this reservoir, the DEM results were more accurate than the ICESat results. The NRMSE was 15.00%, which was 3.18% better than the ICESat based algorithm. For the Hirakud, Pong, and R. P. Sagar reservoirs, the MODIS-ICESat algorithm showed a superior accuracy when validated against the gauge data. The higher accuracy of the MODIS-ICESat algorithm at these three reservoirs may be attributed to the higher vertical accuracy of the ICESat elevation values, and/or the longer observation period of ICESat (than the DEM, which results in a more representative *A-H* relationship). Because the ICESat and SRTM approaches both use the same MODIS water area values, the larger bias of storage from the SRTM DEM implies that the lower accuracy of SRTM could reduce the quality of the reservoir storage product. As stated by the authors of [44], the components of the SRTM error include baseline roll error, phase error, beam differential errors, and timing and position errors. However, the SRTM DEM errors related to terrain, timing, and position—along with the low vertical resolution (1-m intervals)—still influenced the accuracy of the *A-H* relationship, which led to a higher bias of the storage calculation. Overall, the MODIS-SRTM algorithm performed reasonably well.

4.2. Spatial Coverage of the Reservoir Storage Dataset

With full-coverage of two-dimensional elevation data at a fine spatial resolution (30 m), the MODIS-DEM algorithm generated storage time series for the 28 reservoirs in South Asia from 2000 to 2015 (Figure 6). These reservoirs had an integrated capacity of 124.17 km^3 (46.6% of the region's total capacity). Compared with the MODIS-ICESat algorithm, the MODIS-SRTM algorithm enabled the monitoring of eight additional reservoirs (Figure 1), which represented a 18.6% increase of the overall storage capacity. Sriram Sagar, which was almost at its maximum level during the SRTM flight time, was the only reservoir for which the *A-H* relationship could be generated by MODIS-ICESat but not by the DEM.

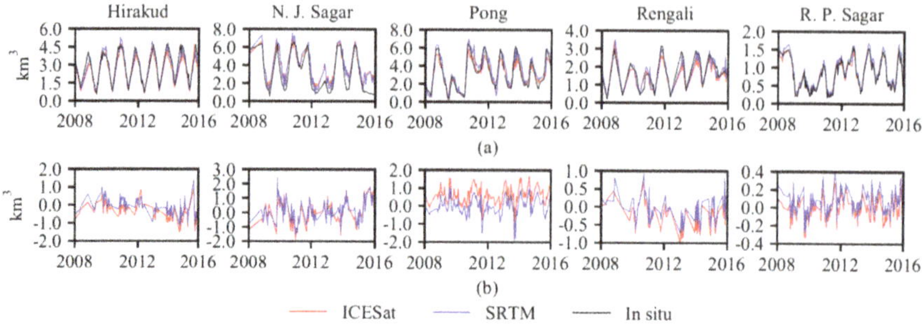

Figure 5. Validation results by comparing the remotely sensed storage values with gauge observations: (**a**) Comparison among absolute storage values; (**b**) comparison of storage difference (remotely sensed storage minus gauge data).

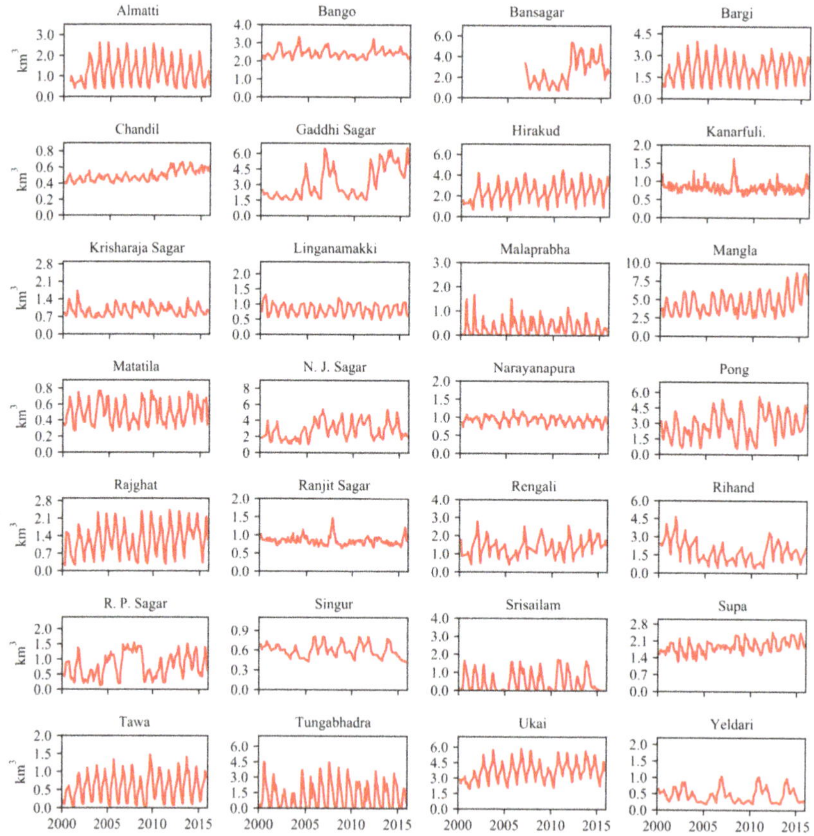

Figure 6. Combined remotely sensed storage time series of the South Asian reservoirs analyzed in this study.

The new dataset contains the storage variation information over multiple reservoirs at the basin scale, which is essential for regional water management purposes. For instance, with two additional reservoirs included in the dataset, the total storage of the monitored reservoirs in the Krishna river basin (KRB) increased from 33.4% to 67.0% (i.e., from 9.70 to 19.44 km^3) of the basin's capacity (29 km^3). The Krishna River is the fourth largest river in India, with its basin extending over an area of 259,948 km^2 (about 8% of India). Most of the KRB is relatively flat, with about 76% of the basin covered by agricultural land. Many hydroelectric power stations are distributed along the Krishna River, providing clean energy to a large area of India. Therefore, the improved spatial representativeness of reservoirs in this river basin is essential for hydrologic modeling and water management. The Ukai Dam across the Tapti River was constructed for the purposes of irrigation, hydropower generation, and flood control. The Tapti River basin accounts for nearly two percent of the total area of India. However, before this study, no reservoir in this basin had remotely sensed elevation or storage data from space. In 2000, a severe drought occurred in the Tapti basin, causing drinking water scarcity in some villages [45]. In 2009, many districts in this basin were declared to be under drought conditions due to the deficiency of rainfall from June to September [46]. The low storage values around 2000 and 2009 (Figure 6) reflect this water scarcity. Figure 6 also shows an increase of maximum storage in the Mangla Reservoir after 2012. This is attributed to the enhanced storage capacity, that was used to increase the reservoir's irrigation capability [47]. Another example is the Yeldari reservoir. According to media reports, two severe drought events occurred in the region in 2004 and from 2012 to 2015—and, in both cases, the Yeldari reservoir almost dried up [48,49].

4.3. Uncertainty Analysis

The storage uncertainty associated with the *A-H* relationship is primarily attributed to the use of partial bathymetry information to represent the *A-H* relationship for the entire reservoir. Because the DEM dataset only represents the part of the bathymetry that was above the water surface when the SRTM measurements were collected, it assumes that the unmeasured part below the water surface shares the same *A-H* relationship. To quantify the uncertainty associated with this assumption, we compared the storage estimations from three different scenarios (Figure 7). In each case, a simplified cross-sectional view of the reservoir was used—with the water surface area collected by the SRTM (in 2000) indicated as A_1, and the area of the reservoir bottom indicated as A_2. Under all scenarios, the storage volume below the DEM water surface was preserved. The first scenario (Figure 7a) follows the algorithm used in this study, which assumes that the *A-H* relationship remains the same across the entire profile. The second scenario (Figure 7b) assumes that the area of the reservoir bottom is zero, and thus the *A-H* relationship of the unknown part below the water surface has the smallest possible slope of k_{min}. The third scenario (Figure 7c) assumes that the minimum area from the MODIS estimations is the area of the reservoir bottom, and thus the *A-H* relationship of the unknown part below the water surface has the largest possible slope of k_{max}.

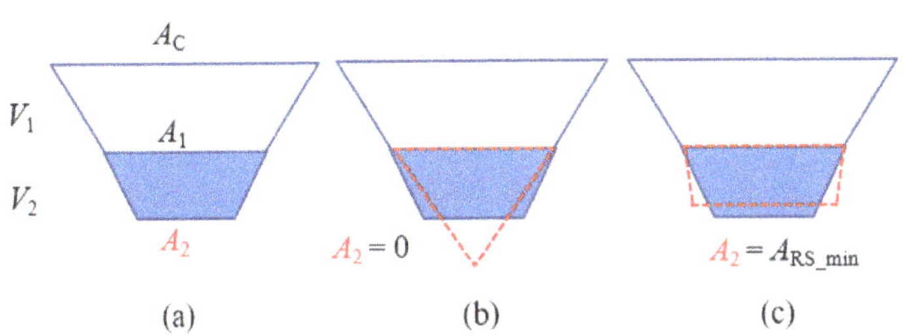

Figure 7. Illustration of the process for quantifying the uncertainty associated with the extrapolation of the *A-H* relationship: (**a**) an example of a simplified reservoir cross section, with a bottom area of A_2 identified by assuming the unmeasured portion shares the same *A-H* relationship, (**b**) the reservoir cross section by assuming the bottom area A_2 is 0, (**c**) the reservoir cross section by assuming the reservoir bottom area A_2 equals $A_{\text{RS\_min}}$.

Using the slope of the upper portion (i.e., k) as estimated from the DEM, the reservoir storage value when the DEM was constructed (i.e., V_2) can be calculated using Equation (6):

$$V_2 = V_c - V_1 = V_c - (A_c + A_1)(A_c - A_1)k/2 \tag{6}$$

As shown in Figure 7b, the minimum value of A_2—which is 0—can be used to estimate $k_{\min}$ via Equation (7):

$$k_{min} = \frac{2V_2}{(A_1 + A_2)(A_1 - A_2)} = \frac{2V_2}{A_1^2} \tag{7}$$

Similarly, the maximum value of A_2—which is equal to the minimum water surface area from MODIS during the research period (Figure 7c)—can be used to estimate $k_{\max}$ after Equation (8):

$$k_{max} = \frac{2V_2}{(A_1 + A_2)(A_1 - A_2)} = \frac{2V_2}{\left(A_1 + A_{RS}^{min}\right)\left(A_1 - A_{RS}^{min}\right)} \tag{8}$$

Thus, for any MODIS remotely sensed area (A_{RS}) that is less than A_1, the storage can range between a minimum possible value of ${V_{\mathrm{RS}}}^{\min}$ (Equation (9)) and a maximum value of ${V_{RS}}^{\max}$ (Equation (10)):

$$V_{RS}^{\min} = V_2 - (A_1 + A_{RS})(A_1 - A_{RS})k_{\min}/2 \tag{9}$$

$$V_{RS}^{\max} = V_2 - (A_1 + A_{RS})(A_1 - A_{RS})k_{\max}/2 \tag{10}$$

Therefore, the uncertainties associated with the constant slope assumption can be represented by the difference between the two storage estimates described below using Equation (11):

$$\Delta V = (A_1 + A_{RS})(A_1 - A_{RS})(k_{\max} - k_{\min})/2 \tag{11}$$

The uncertainties associated with this source are illustrated in Figure 8. For all 28 reservoirs, the absolute uncertainty due to the unmeasured *A-H* relationship ranged from 0 km^3 to 0.54 km^3, with an average of 0.23 km^3. Among these reservoirs, the Rihand reservoir had the largest absolute uncertainty (0.54 km^3), primarily because this large reservoir was at a relatively high level when the DEM data were collected. The surface area of the Rihand reservoir—as measured by DEM—was 388.96 km^2, whereas its surface area at full capacity is 485 km^2. Considering all of the reservoirs, we found a significant increasing trend of the absolute uncertainty as the reservoir capacity increased. For every 1 km^3 increase in reservoir capacity, the uncertainty increased by 0.034 km^3 ($p < 0.01$). The averaged relative uncertainty caused by the unmeasured *A-H* relationship was 4.68%. However, we observed no significant relationship between the relative uncertainty and the capacity.

The uncertainties from the area estimation algorithm were quantified thoroughly by Zhang et al. [2014]. From this source, the absolute uncertainties were also found to be highly correlated with the storage at capacity, where the absolute uncertainties had an average value of 3.90%. This is a similar uncertainty range but lightly larger than the unmeasured *A-H* relationship.

The vertical error of SRTM DEM could be another source of uncertainty. This was not analyzed in this study because the storage calculation (in this study) was based on the slope of the *A-H* relationship and area estimations, rather than using the absolute elevation value from the SRTM DEM directly. Since the slope of the A-H relationship is determined by the elevation difference of reservoir pixels, the absolute vertical DEM error can be offset during the process, reducing its influence on the storage estimation.

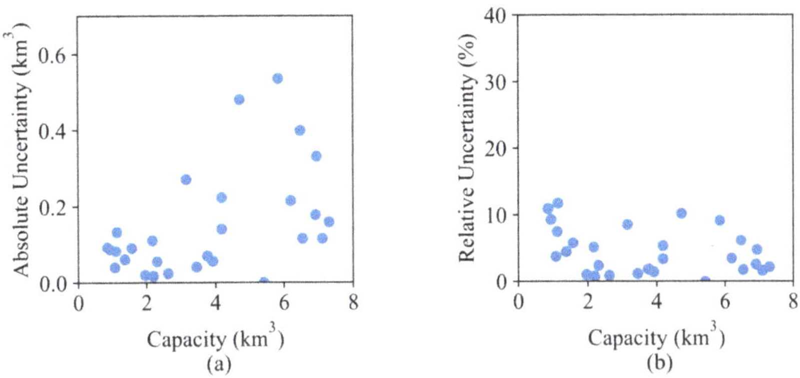

Figure 8. Uncertainty analysis results: (**a**) Absolute uncertainty; (**b**) relative uncertainty due to the unmeasured *A-H* relationship of SRTM DEM.

5. Conclusions

In this study, an algorithm that leverages the SRTM DEM data was developed to improve the spatial coverage of the reservoir monitoring network in South Asia. By combining water surface area from MODIS for reservoir storage estimations, we were able to take the advantage of high temporal resolution of MODIS and large spatial coverage of SRTM. Furthermore, validation results against gauge observations over 11 reservoirs in South Asia suggested that the storage estimations had a good level of accuracy (with R^2 values ranging from 0.47 to 0.91). The integrated storage capacity of these reservoirs was 118.76 km^3, which represents 46.6% of the overall storage in the region.

This algorithm still has some limitations that need to be noted. First, the accuracy of the proposed algorithm depends on the water level at the time the DEM data were collected. For certain reservoirs that were almost full during the SRTM acquisition time, this approach did not work. Due to the assumption that the *A-H* relationship derived from the DEM above the water surface represented the full bathymetry, uncertainties in storage estimations were introduced in addition to those from the area retrieval algorithm. Second, the low vertical resolution of SRTM DEM and the errors from different sources may reduce the accuracy of the storage estimation [44]. Therefore, examining the DEM errors with respect to the terrain of the reservoirs could help us to better understand the error characteristics of the storage estimation bias. Third, due to the medium resolution of MODIS, the accuracy of reservoir storage estimation decreased for the reservoir with the smallest surface area (56 km^2). Nonetheless, the benefits of the extended number of reservoirs outweigh the constraints.

The algorithm proposed in this study can provide reservoir storage products that support water management on a large scale. For instance, given the long-term availability of high spatial resolution sensors, this approach could be used to monitor much smaller sized reservoirs than possible using existing techniques. This algorithm may also contribute to future satellite missions such as the Surface Water Ocean Topography (SWOT) mission, which will provide a direct water surface measurement for about two-thirds of global lakes and reservoirs, including those with an individual water area > 0.06 km^2.

Author Contributions: Conceptualization, S.Z. and H.G.; Methodology, S.Z. and H.G.; Writing—Original Draft Preparation, S.Z.; Writing—Review & Editing, S.Z. and H.G.; Funding Acquisition, H.G. All authors have read and agreed to the published version of the manuscript.

Funding: This work was supported by the NASA Science of Terra, Aqua, and Suomi NPP (TASNPP) Program (80NSSC18K0939) provided to Texas A&M University. It has benefitted from the computing resources of the Texas A&M Supercomputing Facility (http://sc.tamu.edu). The authors would also like to thank the Central Electricity Authority of India for providing the reservoir gauge observations.

Conflicts of Interest: The authors declare no conflict of interest.

References

1. Bai, T.; Wu, L.; Chang, J.-X.; Huang, Q. Multi-objective optimal operation model of cascade reservoirs and its application on water and sediment regulation. *Water Resour. Manag.* **2015**, *29*, 2751–2770. [CrossRef]
2. Haddeland, I.; Heinke, J.; Biemans, H.; Eisner, S.; Flörke, M.; Hanasaki, N.; Konzmann, M.; Ludwig, F.; Masaki, Y.; Schewe, J. Global water resources affected by human interventions and climate change. *Proc. Natl. Acad. Sci. USA* **2014**, *111*, 3251–3256. [CrossRef] [PubMed]
3. Zhao, G.; Gao, H.; Naz, B.S.; Kao, S.-C.; Voisin, N. Integrating a reservoir regulation scheme into a spatially distributed hydrological model. *Adv. Water Resour.* **2016**, *98*, 16–31. [CrossRef]
4. Li, Y.; Gao, H.; Jasinski, M.F.; Zhang, S.; Stoll, J.D. Deriving High-Resolution Reservoir Bathymetry From ICESat-2 Prototype Photon-Counting Lidar and Landsat Imagery. *IEEE Trans. Geosci. Remote Sens.* **2019**, *57*, 7883–7893. [CrossRef]
5. Lauri, H.; de Moel, H.; Ward, P.; Räsänen, T.; Keskinen, M.; Kummu, M. Future changes in Mekong River hydrology: Impact of climate change and reservoir operation on discharge. *Hydrol. Earth Syst. Sci.* **2012**, *16*, 4603–4619. [CrossRef]
6. Le, T.V.H.; Nguyen, H.N.; Wolanski, E.; Tran, T.C.; Haruyama, S. The combined impact on the flooding in Vietnam's Mekong River delta of local man-made structures, sea level rise, and dams upstream in the river catchment. *Estuar. Coast. Shelf Sci.* **2007**, *71*, 110–116. [CrossRef]
7. Pulwarty, R.S.; Sivakumar, M.V. Information systems in a changing climate: Early warnings and drought risk management. *Weather Clim. Extrem.* **2014**, *3*, 14–21. [CrossRef]
8. Vicente-Serrano, S.M.; Beguería, S.; Gimeno, L.; Eklundh, L.; Giuliani, G.; Weston, D.; El Kenawy, A.; López-Moreno, J.I.; Nieto, R.; Ayenew, T. Challenges for drought mitigation in Africa: The potential use of geospatial data and drought information systems. *Appl. Geogr.* **2012**, *34*, 471–486. [CrossRef]
9. Gao, H.; Birkett, C.; Lettenmaier, D.P. Global monitoring of large reservoir storage from satellite remote sensing. *Water Resour. Res.* **2012**, *48*, W09504. [CrossRef]
10. Lettenmaier, D.P.; Alsdorf, D.; Dozier, J.; Huffman, G.J.; Pan, M.; Wood, E.F. Inroads of remote sensing into hydrologic science during the WRR era. *Water Resour. Res.* **2015**, *51*, 7309–7342. [CrossRef]
11. Rodrigues, L.N.; Sano, E.E.; Steenhuis, T.S.; Passo, D.P. Estimation of small reservoir storage capacities with remote sensing in the Brazilian Savannah Region. *Water Resour. Manag.* **2012**, *26*, 873–882. [CrossRef]
12. Pereira, B.; Medeiros, P.; Francke, T.; Ramalho, G.; Foerster, S.; De Araújo, J.C. Assessment of the geometry and volumes of small surface water reservoirs by remote sensing in a semi-arid region with high reservoir density. *Hydrol. Sci. J.* **2019**, *64*, 66–79. [CrossRef]
13. Zhao, G.; Gao, H. Automatic correction of contaminated images for assessment of reservoir surface area dynamics. *Geophys. Res. Lett.* **2018**, *45*, 6092–6099. [CrossRef]
14. Yao, F.; Wang, J.; Wang, C.; Crétaux, J.-F. Constructing long-term high-frequency time series of global lake and reservoir areas using Landsat imagery. *Remote Sens. Environ.* **2019**, *232*, 111210. [CrossRef]
15. Berry, P.; Garlick, J.; Freeman, J.; Mathers, E. Global inland water monitoring from multi-mission altimetry. *Geophys. Res. Lett.* **2005**, *32*. [CrossRef]
16. Birkett, C.M. Contribution of the TOPEX NASA radar altimeter to the global monitoring of large rivers and wetlands. *Water Resour. Res.* **1998**, *34*, 1223–1239. [CrossRef]
17. Wang, X.; Gong, P.; Zhao, Y.; Xu, Y.; Cheng, X.; Niu, Z.; Luo, Z.; Huang, H.; Sun, F.; Li, X. Water-level changes in China's large lakes determined from ICESat/GLAS data. *Remote Sens. Environ.* **2013**, *132*, 131–144. [CrossRef]
18. Zhang, G.; Xie, H.; Kang, S.; Yi, D.; Ackley, S.F. Monitoring lake level changes on the Tibetan Plateau using ICESat altimetry data (2003–2009). *Remote Sens. Environ.* **2011**, *115*, 1733–1742. [CrossRef]
19. Zhang, S.; Gao, H. A novel algorithm for monitoring reservoirs under all-weather conditions at a high temporal resolution through passive microwave remote sensing. *Geophys. Res. Lett.* **2016**, *43*, 8052–8059. [CrossRef]
20. Zhang, G.; Chen, W.; Xie, H. Tibetan Plateau's lake level and volume changes from NASA's ICESat/ICESat-2 and Landsat missions. *Geophys. Res. Lett.* **2019**, *46*, 13107–13118. [CrossRef]
21. Liebe, J.; Van De Giesen, N.; Andreini, M. Estimation of small reservoir storage capacities in a semi-arid environment: A case study in the Upper East Region of Ghana. *Phys. Chem. Earth Parts A/B/C* **2005**, *30*, 448–454. [CrossRef]

22. Smith, L.C.; Pavelsky, T.M. Remote sensing of volumetric storage changes in lakes. *Earth Surf. Processes Landf.* **2009**, *34*, 1353–1358. [CrossRef]
23. Busker, T.; de Roo, A.; Gelati, E.; Schwatke, C.; Adamovic, M.; Bisselink, B.; Pekel, J.-F.; Cottam, A. A global lake and reservoir volume analysis using a surface water dataset and satellite altimetry. *Hydrol. Earth Syst. Sci.* **2019**, *23*, 669–690. [CrossRef]
24. Gao, H. Satellite remote sensing of large lakes and reservoirs: From elevation and area to storage. *Wiley Interdiscip. Rev. Water* **2015**, *2*, 147–157. [CrossRef]
25. Zhang, S.; Gao, H.; Naz, B.S. Monitoring reservoir storage in South Asia from multisatellite remote sensing. *Water Resour. Res.* **2014**, *50*, 8927–8943. [CrossRef]
26. Bonnema, M.; Hossain, F. Inferring reservoir operating patterns across the M ekong B asin using only space observations. *Water Resour. Res.* **2017**, *53*, 3791–3810. [CrossRef]
27. Tseng, K.-H.; Shum, C.; Kim, J.-W.; Wang, X.; Zhu, K.; Cheng, X. Integrating Landsat imageries and digital elevation models to infer water level change in Hoover Dam. *IEEE J. Sel. Top. Appl. Earth Obs. Remote Sens.* **2016**, *9*, 1696–1709. [CrossRef]
28. Getirana, A.; Jung, H.C.; Tseng, K.-H. Deriving three dimensional reservoir bathymetry from multi-satellite datasets. *Remote Sens. Environ.* **2018**, *217*, 366–374. [CrossRef]
29. Adhikari, P.; Hong, Y.; Douglas, K.R.; Kirschbaum, D.B.; Gourley, J.; Adler, R.; Brakenridge, G.R. A digitized global flood inventory (1998–2008): Compilation and preliminary results. *Natl. Hazards* **2010**, *55*, 405–422. [CrossRef]
30. Hydrology by Altimetry. Available online: http://www.legos.obs-mip.fr/soa/hydrologie/hydroweb/Page_2.html (accessed on 31 December 2019).
31. Global Reservoirs and Lakes Monitor. Available online: https://ipad.fas.usda.gov/cropexplorer/global_reservoir/ (accessed on 31 December 2019).
32. Goteti, G.; Famiglietti, J.S.; Asante, K. A catchment-based hydrologic and routing modeling system with explicit river channels. *J. Geophys. Res. Atmos.* **2008**, *113*. [CrossRef]
33. Lehner, B.; Grill, G. Global river hydrography and network routing: Baseline data and new approaches to study the world's large river systems. *Hydrol. Processes* **2013**, *27*, 2171–2186. [CrossRef]
34. Berthier, E.; Arnaud, Y.; Vincent, C.; Remy, F. Biases of SRTM in high-mountain areas: Implications for the monitoring of glacier volume changes. *Geophys. Res. Lett.* **2006**, *33*. [CrossRef]
35. Surazakov, A.B.; Aizen, V.B. Estimating volume change of mountain glaciers using SRTM and map-based topographic data. *IEEE Trans. Geosci. Remote Sens.* **2006**, *44*, 2991–2995. [CrossRef]
36. Papa, F.; Frappart, F.; Güntner, A.; Prigent, C.; Aires, F.; Getirana, A.C.; Maurer, R. Surface freshwater storage and variability in the Amazon basin from multi-satellite observations, 1993–2007. *J. Geophys. Res. Atmos.* **2013**, *118*, 11951–11965. [CrossRef]
37. Rabus, B.; Eineder, M.; Roth, A.; Bamler, R. The shuttle radar topography mission—A new class of digital elevation models acquired by spaceborne radar. *ISPRS J. Photogramm. Remote Sens.* **2003**, *57*, 241–262. [CrossRef]
38. Farr, T.G.; Kobrick, M. Shuttle Radar Topography Mission produces a wealth of data. *Eos Trans. Am. Geophys. Union* **2000**, *81*, 583–585. [CrossRef]
39. U.S. Geological Survey's Long Term Archive. Available online: https://lta.cr.usgs.gov/SRTM1Arc (accessed on 31 December 2019).
40. Indian Central Electricity Authority. Available online: http://www.cea.nic.in (accessed on 30 May 2016).
41. Lehner, B.; Liermann, C.R.; Revenga, C.; Vörösmarty, C.; Fekete, B.; Crouzet, P.; Döll, P.; Endejan, M.; Frenken, K.; Magome, J. High-resolution mapping of the world's reservoirs and dams for sustainable river-flow management. *Front. Ecol. Environ.* **2011**, *9*, 494–502. [CrossRef]
42. Jain, A.K. Data clustering: 50 years beyond K-means. *Pattern Recognit. Lett.* **2010**, *31*, 651–666. [CrossRef]
43. Gao, H.; Zhang, S.; Durand, M.; Lee, H. Satellite remote sensing of lakes and wetlands. In *Hydrologic Remote Sensing*; CRC Press: Boca Raton, FL, USA, 2016; pp. 57–72.
44. Rodriguez, E.; Morris, C.S.; Belz, J.E. A global assessment of the SRTM performance. *Photogramm. Eng. Remote Sens.* **2006**, *72*, 249–260. [CrossRef]
45. Jairath, J. *Droughts and Integrated Water Resource Management in South Asia: Issues, Alternatives and Futures*; SAGE Publications: Southend Oaks, CA, USA, 2008.

46. India Speed. This Year's Drought Is Severe, But Not Unprecedented. 2016. Available online: https://everylifecounts.ndtv.com/this-years-drought-is-severe-but-not-unprecedented-2230 (accessed on 31 December 2019).
47. Kayani, S.-A. *Mangla Dam Raising Project (Pakistan): General Review and Socio-Spatial Impact Assessment*; Hal-00719226: Islamabad, Pakistan, 2012.
48. Sud, S. 38 Reservoirs Down to 30 per Cent Storage. Rediff Business. 2004. Available online: https://www.rediff.com/money/report/water/20040728.htm (accessed on 31 December 2019).
49. Bhosale, J. You Don't Get Water Even If You Are Ready to Pay for It. The Economic Times. 2019. Available online: https://economictimes.indiatimes.com/news/politics-and-nation/you-dont-get-water-even-if-you-are-ready-to-pay-for-it/articleshow/69066949.cms?from=mdr (accessed on 31 December 2019).

remote sensing

MDPI

Article

Addressing Challenges for Mapping Irrigated Fields in Subhumid Temperate Regions by Integrating Remote Sensing and Hydroclimatic Data

Tianfang Xu [1,2,*], Jillian M. Deines [2,3], Anthony D. Kendall [2], Bruno Basso [2,4] and David W. Hyndman [2,*]

1 Utah Water Research Laboratory, Civil and Environmental Engineering, Utah State University, Logan, UT 84321, USA

2 Department of Earth and Environmental Sciences, Michigan State University, East Lansing, MI 48824, USA; jillian.deines@gmail.com (J.M.D.); kendal30@msu.edu (A.D.K.); basso@msu.edu (B.B.)

3 Department of Earth Systems Science and Center for Food Security and the Environment, Stanford University, Stanford, CA 94305, USA

4 W.K. Kellogg Biological Station, Michigan State University, East Lansing, MI 48824, USA

* Correspondence: tianfang.xu@usu.edu or xutianfangnju@gmail.com (T.X.); hyndman@msu.edu (D.W.H.)

Received: 23 December 2018; Accepted: 3 February 2019; Published: 12 February 2019

Abstract: High-resolution mapping of irrigated fields is needed to better estimate water and nutrient fluxes in the landscape, food production, and local to regional climate. However, this remains a challenge in humid to subhumid regions, where irrigation has been expanding into what was largely rainfed agriculture due to trends in climate, crop prices, technologies and practices. One such region is southwestern Michigan, USA, where groundwater is the main source of irrigation water for row crops (primarily corn and soybeans). Remote sensing of irrigated areas can be difficult in these regions as rainfed areas have similar characteristics. We present methods to address this challenge and enhance the contrast between neighboring rainfed and irrigated areas, including weather-sensitive scene selection, applying recently developed composite indices and calculating spatial anomalies. We create annual, 30m-resolution maps of irrigated corn and soybeans for southwestern Michigan from 2001 to 2016 using a machine learning method (random forest). The irrigation maps reasonably capture the spatial and temporal pattern of irrigation, with accuracies that exceed available products. Analysis of the irrigation maps showed that the irrigated area in southwestern Michigan tripled in the last 16 years. We also discuss the remaining challenges for irrigation mapping in humid to subhumid areas.

Keywords: irrigation mapping; remote sensing; random forest; subhumid region

1. Introduction

Agriculture is the sector with the largest consumptive use of water across the globe. While crop water demand is largely met by irrigation in arid to semiarid regions, farmers in humid regions traditionally rely on rainfall. However, irrigation has become more common in humid to subhumid regions [1], driven by the growth of demand for corn grain bioethanol, the need to increase yield given current low prices of corns and soybeans [2], the ready availability of more water and energy efficient irrigation technologies, and increasing climate variability.

The rapid expansion of irrigation has important implications for terrestrial water balances, food production, and local to regional climate [3–6]. Land surface models have been increasingly used as quantitative tools to estimate the effects of land use change and other human activities on terrestrial water and energy cycles. However, these models require high-resolution observations at the model scale to fully vet the irrigated area [7,8]. Thus, detailed depiction of spatiotemporal patterns of irrigation is needed for modelers and decision makers [9]. However, accurate monitoring of irrigated area can be difficult in humid to subhumid regions (hereafter humid regions), primarily because of the similarity of signals from rainfed and irrigated areas in such regions [10].

Remote sensing provides valuable information to delineate irrigated areas. Within the U.S., the Moderate Resolution Imaging Spectroradiometer (MODIS) Irrigated Agriculture Dataset for the United States (MIrAD-US) national irrigation dataset was developed by the U.S. Geological Survey (USGS) by integrating U.S. Department of Agriculture (USDA) county statistics, MODIS satellite imagery and a national land cover map [11]. The MIrAD-US product has 250-m resolution and is available in 2002, 2007 and 2012. MIrAD-US revealed significant temporal variability and suggests the need for regular periodic mapping of irrigated areas [12]. Later studies have used higher resolution imagery (10–30 m) from Landsat and Sentinel-1 satellites to develop more detailed irrigation maps for local to regional studies [13–15]. In particular, annual irrigation maps were developed for the Republican River Basin from 1999 to 2016 (AIM-RRB), leveraging recent advances in cloud computing, machine learning, and increasingly accessible Landsat data [13].

In southwestern Michigan, a subhumid region in the midwestern U.S., water consumption by agriculture has rapidly increased over the past two decades. Irrigation of row crops (primarily corn and soybean) was once practiced only on a small fraction of the total crop land across the upper Midwest. However, in the last two decades there has been a dramatic expansion in irrigation use [2], mostly from groundwater pumping [16]. Large acreages of fields in southwestern Michigan are devoted to producing seed corn, commercial corn, and soybeans [16]. The prevailing sandy soils [17] and shallow depths to groundwater [18] in this region allows adoption of central pivot irrigation systems with limited operation costs. Given the strong connection between groundwater and surface water, irrigation in southwestern Michigan has the potential to reduce the health of some local surface water ecosystems [19].

Remote sensing methods are able to map irrigated fields in arid and semi-arid environments with satisfactory accuracy, however the accuracy of satellite-based irrigation mapping techniques in more humid regions is still unknown [11]. The objective of this study was to create high-resolution, annual maps of irrigated fields in a sub-humid region by integrating remote sensing imagery with climate and land surface modeling data. We identified three methods to increase remote sensing accuracy: (1) use weather-sensitive selection of imagery timing, (2) test the transferability of recently-developed composite indices for detecting irrigation in arid areas [13] to humid regions, and (3) calculate spatial anomaly indices. We demonstrated this approach in southwestern Michigan (SW MI) where corn and soybeans are the two principal irrigated crops. We also evaluated the accuracy of irrigation mapping under various climate conditions in this region, which provided insights into the applicability to other humid regions.

2. Materials and Methods

2.1. Study Area

In this study, we considered ten counties (Allegan, Barry, Eaton, van Buren, Kalamazoo, Calhoun, Berrien, Cass, St. Joseph and Branch) covering 28,281 km^2 in southwestern Michigan (Figure 1a). The study area is part of the US Corn Belt with a subhumid climate [20], with annual precipitation ranging from 590 mm (in the 2012 drought) to 858 mm during the study period 2001–2016 (Figure 1b, [21]). Short term droughts, common in this region, induce plant water stress and reduce grain yields of corn and soybeans, which account for 45% and 32% of total agricultural area, respectively [22]. During extended drought periods, the sandy soils prevalent in this region cannot store sufficient soil water to allow crops to reach full yield potential.

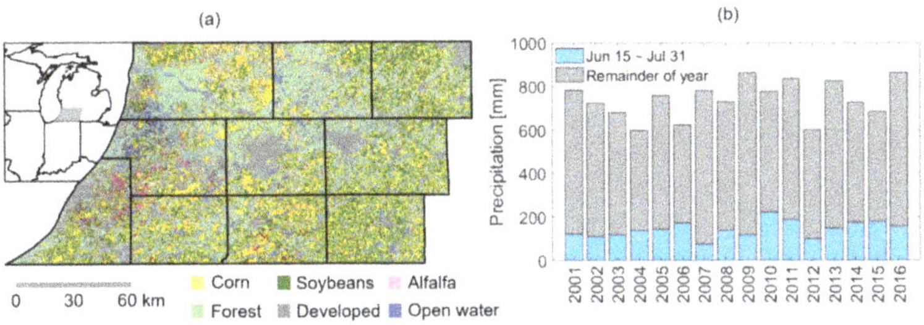

Figure 1. (**a**). The location (inset) and remotely sensed crop types (CDL, [22]) of the study area, including ten southwestern Michigan counties. (**b**) Precipitation from 15 June to 31 July (blue), and the remainder of year (gray) in the study area.

2.2. Basic Remotely Sensed, Land Surface Model, and Climate Input Data

We used a variety of time-varying and static input data for the random forest (RF) classification model (summarized in Tables 1 and 2). The static input variables describe terrain, soil, and geographic location. Dynamic inputs are derived from remote sensing and climate data, as well as land surface model output. For most time varying data, we focus on the June 15th to July 31st period, which is the time before canopy closure occurs in corn and soybeans to avoid reflectance saturation. Data were either obtained from, or uploaded to, the Google Earth Engine (GEE) cloud computing platform [23] for classification.

Table 1. Basic input variables and indices used to calculate derived input variables.

Variable	Description	Source
EVI	Enhanced Vegetation Index	Landsat
GI	Green Index	Landsat
NDWI	Normalized Difference Water Index	Landsat
NDVI	Normalized Difference Vegetation Index	Landsat
Thermal	Landsat 5 & 7: 10.40–12.50 μm band Landsat 8: 11.50–12.51 μm band	Landsat
Dryspell	*See text*	Derived: PRISM
P	Precipitation	PRISM
VPD	Mean daily max. vapor pressure deficit	PRISM
GDD	Growing degree-day	PRISM
Aridity	Total precipitation/PET, May–Aug	Derived: GRIDMET
PDSI	Palmer Drought Severity Index	GRIDMET
Soil moisture	Root zone soil moisture	NLDAS-2 Noah
AWC	Available water capacity	SSURGO
K_{sat}	Vertical saturated hydraulic conductivity	SSURGO

Table 2. Weather-sensitive remote sensing, spatial anomaly and composite indices.

Group	Variable Code or Suffix	Description
Weather-sensitive remote sensing indices	VDPMaxGI	3-day average VPD before maximum Landsat GI day
	dryspellMaxGI	Number of consecutive days with rainfall ≤ 5 mm before maximum GI day
		NDVI, EVI, GI and NDWI calculated using the Landsat scene after a dry period identified using three criteria
	_SM	Descending soil moisture
	_pdsi	Lowest PDSI
	_ppt	Longest dryspell
Spatial anomaly indices		NDVI, EVI, GI and NDWI statistics subtracted by neighborhood %
	_p40	40%
	_p90	90%

Table 2. *Cont.*

Group	Variable Code or Suffix	Description
Composite indices	WGI	Maximum GI × mean NDWI (water-adjusted GI, [13]
	AGI	Maximum GI/aridity (aridity normalized GI, [13]
	WGI_ppt, AGI_ppt	WGI and AGI calculated using GI from scenes that immediately follows a dry period

We included five temporally static input variables to account for possible relations between yield and terrain attributes, soil properties, crop characteristics, and geographic information. These variables are: (1) slope calculated from 30-m National Elevation Dataset (NED) Digital Elevation Model (DEM) [24], (2) soil available water capacity (AWC, field capacity minus wilting point), (3) vertical saturated hydraulic conductivity (k_{Sat}), (4) latitude (lat), and (5) longitude (long). The AWC and k_{Sat} are based on the top 25 cm soil properties provided by the USDA Soil Survey Geographic Dataset (SSURGO) Web Soil Survey [17]. For each SSURGO map unit polygon, we calculated depth- and component-fraction weighted averages of all soil horizon textures (%sand, %clay) within the top 25 cm. We then used the ROSETTA database [25] to relate these textures to estimates of field capacity, wilting point, and soil hydraulic conductivity. Vertical hydraulic conductivity was calculated for each component as the harmonic mean of individual horizontal saturated conductivities.

Climate inputs were derived from daily 4-km resolution PRISM [21] and Gridded Surface Meteorological dataset (GRIDMET) [26]: (1) precipitation, (2) aridity calculated as the ratio of growing season rainfall to potential evapotranspiration (PET), (3) average Palmer Drought Severity Index (PDSI), (4) dryspell (maximum consecutive days with less than 5 mm precipitation), (5) average daily maximum VPD (vapor pressure deficit), (6) daily mean temperature, (7) heatwave (maximum consecutive days with daily mean temperature above 25 °C, (8) GDD (growing degree days = cumulative degree obtained from the difference between air temperature and base temperature for corn and soybeans, 25 °C in this study), and (9) as a measure of pre- and in season-wetness, we calculated the total precipitation before June 15th (p_early) and from June 15th to Jul. 31st (p_sum), respectively.

Irrigation decisions are often based on soil water content [8]. Here, we use the root zone soil water content at noon from NLDAS-Noah with 1/8° spatial resolution and hourly time step [27], which is currently the best readily available product at regional scale that has sufficiently fine temporal resolution for our application. The NLDAS-Noah product does not implement irrigation, thus its soil water content data serves as a reference that represents wetness under rainfed conditions.

We used remote sensing data from Landsat Surface Reflectance Products at an 8-day interval in all years except 2012, when there was a 16-day interval since only Landsat 7 ETM+ was operational. We included all scenes within 5 days of the June 15th to July 31st key growing season period. The actual number of available scenes during this period varies spatially and inter-annually as the 8-day (16-day in 2012) return interval is simultaneously reduced by cloud coverage and augmented by overlapping scene edges. From 2001 to 2016, the average number of valid observations among pixels in the study domain varied from 1.67 (2012) to 6.35 (2001) (Table S1, Supplementary Material). All Landsat 7 images collected after May 31, 2003 have data gaps due to the Scan Line Corrector (SLC) failure, which leads to significant data shortage in 2012. We used a moving window average method to fill in the gaps caused by the SLC failure. For every pixel within a gap, we set its value as the average within a five pixel by one pixel rectangle, oriented perpendicular to the scanline.

After filling the data gaps, we extracted the thermal, near-infrared, short-wave infrared, red, green, and blue bands from Landsat images between June 10th and August 5th and calculated NDVI, EVI, GI and NDWI. We then created statistical composites from the available imagery following a best pixel approach [28] to generate mean, maximum, minimum and range (i.e., maximum subtracted by minimum) composites for all four indices and the thermal band.

2.3. Weather-Sensitive Scene Selection, Spatial Anomaly Calculation, and Novel Composite Indices

Initially we applied the methods from previous studies (i.e., [12–14]) to construct a RF classifier for this region, but found the accuracies were inadequate. We thus developed and tested a variety of approaches, including several that we deemed unsuccessful because they did not increase the accuracy of the RF classifier. Ultimately, we implemented three methods to create layers beyond the basic remotely sensed, land surface model (LSM), and climate indices described in Section 2.2.

First, we calculated two composite indices recently proposed in [13] that adjust GI in an attempt to account for regional variations in available water. The first composite variable is the product of maximum GI and growing-season mean NDWI (i.e., water-adjusted green index, WGI); we anticipate that for irrigated fields GI and NDWI will both be high. A pixel with high GI but low NDWI may be rainfed (low NDWI) but treated with abundant nutrients (high GI). The other composite variable is maximum GI divided by seasonal aridity (i.e., aridity-normalized green index, AGI). The rationale behind AGI is that irrigated fields should have high GI even during relatively dry growing conditions.

Second, we developed weather-sensitive remote sensing indices calculated from Landsat scenes at the time most favorable for distinguishing irrigated from rainfed fields. We assumed this time immediately follows a dry period identified based on three separate criteria: (1) maximum consecutive days with monotonically descending root zone (up to 1-m depth) soil water content, (2) lowest PDSI of the season, and (3) greatest number of consecutive days with daily precipitation less than 5 mm. The criteria were calculated using climate and LSM model outputs listed in Table 1. The resulting input variables are denoted with suffix _SM, _pdsi, _ppt, respectively. Irrigated crops generally exhibit higher vegetation indices and NDWI than rainfed crops [13,14,27,29,30]; we expect that this difference is amplified under water stress conditions during dry periods. Further, the three-day average vapor pressure deficit (VPD) before the day of maximum Landsat GI (VPDMaxGI) and number of consecutive days with rainfall not exceeding 5 mm before maximum GI day (dryspellMaxGI), were calculated as rainfed crops are unlikely to exhibit maximum GI when VPD is high or after a dry period.

Third, we calculated spatial anomaly remote sensing indices to better distinguish irrigated from rainfed fields. We first calculated the neighborhood percentiles (40% and 90%) of the vegetation indices using a circular kernel with a 10-km radius for every year. This radius was selected based on the range of a spherical fit to the empirical variogram of the climate and LSM model outputs. Two percentiles were selected to provide anomalies that would be useful in areas where irrigation is relatively sparse (where an anomaly relative to the 90%would be more appropriate), and where irrigation is predominant (similarly, where the 40% might indicate irrigated fields). We then subtract the neighborhood percentiles from the vegetation indices to produce annual anomaly maps; resulting input variables are denoted with suffix _p40 and _p90, respectively. A positive value points to a higher-than-neighbor vegetation index under similar climate conditions, which we expected to be related to irrigation activity.

All together, the basic remote sensing, climate, and LSM simulated indices (Section 2.2) and the weather-sensitive remote sensing, spatial anomaly, and composite indices comprise 98 input variables of the RF classifier. A complete list of these variables is provided in Table S2, Supplementary Material.

Several "failed" attempts to define improved indices were made, and then abandoned based on lack of improvement in classification accuracy. We provide these here as information for others seeking to further the work of humid region irrigation remote sensing. A full list of these variables is included in Table S3, Supplementary Material. Many of these variables were extracted from MODIS products. Due to short overpass time, MODIS products are less subject to cloud coverage than Landsat products. We expected that MODIS thermal bands, terrestrial evapotranspiration (ET), and potential evapotranspiration (PET) estimates [31] would provide valuable information to identify irrigation activity [32]. We thus used monthly statistics of these products as well as calculated composite indices, including the difference between precipitation and MODIS ET, and ET divided by VPD. Another climate index that we tested is temporal anomaly of precipitation (annual precipitation subtracted by the multi-year average precipitation). We also derived the monthly ratio of vegetation indices such as maximum GI in July divided by maximum GI in June, ratio among vegetation indices such as GI divided by EVI, and GI divided by NLDAS-Noah soil water content. Our preliminary results suggested that these indices did not improve classification accuracy on our validation points and were thus not used to generate the final results. Likely reasons include the coarse spatial resolution of MODIS and climate products as well as possible errors embedded in these products. These may be useful in areas with larger fields than the pilot study area.

2.4. Random Forest Classifier

We use a random forest (RF) machine learning algorithm to inductively build a classifier of irrigated versus rainfed areas. We selected RF because the algorithm was successful in various hydrologic and remote sensing applications (e.g., [13,14,33–35] is robust with relatively large number of inputs, provides input variable importance measures and probabilistic outputs [36], and is supported in the GEE cloud computing platform.

A random forest is comprised of an ensemble of decision trees. Given a set of training data $\{x_i, y_i\}, i = 1, \ldots, n$, where x_i denotes input variables, and y_i is the corresponding output. In this study, y_i is a categorical variable with two classes: irrigated and rainfed. The algorithm randomly draws n samples with replacements from the training dataset to train a single tree. The process is repeated N times, resulting in a forest of N trees. Once trained, each tree predicts the class of a new data point, and the N trees may predict M classes. The RF algorithm outputs the percentage of trees that provide a prediction of the M classes. The class that receives the highest probability is the final prediction.

In the mapping process, a composite of images is created for each year as input data layers (Tables 1 and 2, Table S2, Supplementary Material). For each year since 2007, the composite is masked using the Cropland Data Layers (CDL, [22]) to keep only corn and soybean fields for this region. For years before 2007, the composite is masked using the National Land Cover Dataset (NLCD, [37]), the primary available product for the study region. The NLCD-based crop mask includes all row crop fields because NLCD does not distinguish among row crops. The trained RF classifier is then applied to input composites and labels each pixel as either irrigated or rainfed. In this way, we develop irrigation probability maps for every year from 2001 to 2016. The probability value ranges from 0 to 1, with higher values suggesting larger likelihood of irrigation activity in the pixel. A pixel is classified as irrigated if it receives a probability greater than 0.5, and as rainfed otherwise. The resulting binary maps are postprocessed in two steps. Due to cloud coverage, 2014 and 2015 have 8.2% and 5.7% pixels, respectively, with no Landsat scene from June 10th to August 5th. The gap pixels in 2014 are labeled irrigated if it was classified as irrigated in both 2012 and 2013. Similarly, the gap pixels in 2015 are labeled irrigated if it is classified as irrigated in the 2013 map and gap-filled 2014 map. In the second step, all pixels that are classified as irrigated only once during 2001–2015 are labeled as rainfed as it is unlikely that farmers will irrigate only one time due to high infrastructure costs. We then examine the final irrigation maps to track the spatial extent and the changing irrigation dynamics.

During training, the RF algorithm also calculates a variable importance score based on the total decrease in node impurities by splitting on the variable, averaged over all trees [38]. The variable importance scores provide a measure of the relative importance of each input variable for capturing the spatiotemporal irrigation pattern. In this study, we used all 98 input variables described in Sections 2.2 and 2.3, and used the RF calculated variable importance measure to draw insights into the data worth of various indices in similar irrigation mapping applications. We note that RF algorithms are robust with the presence of a large number of inputs. Depending on specific applications, and especially when using other machine learning algorithms that are less robust to high input number, more sophisticated feature selection techniques (e.g., [34,39]) can be used to constrain the input space dimension.

2.5. Manually Labeled Dataset

The RF classifier is built based on a training dataset compiled from high-resolution aerial photography that was acquired during growing seasons (National Agriculture Imagery Program, NAIP [40]) from the study area. Approximately half of the training points are sampled in two representative years, 2012 and 2014. 2012 is known as a dry year in which limited rainfall induced water stress during the crop growth period. We expect that the irrigated crops are more productive than rainfed crops in this year, and the difference should be reflected in our selected indices. Therefore, 2012 represents an "easy" irrigation mapping case for the classifier. On the other hand, the 2014 growing season receives plenty of precipitation and thus represents a challenging case for the algorithm. In addition, we sampled 100 locations across multiple years (2005, 2006, 2009, 2010) to track shifts between rainfed and irrigated fields. The locations of the data points were randomly sampled after applying an agriculture land cover mask. From 2001 to 2006 the mask is derived from the NLCD and included pixels categorized as cultivated crops. Since 2007 when CDL was first available in the study area, the masks include pixels labeled as corn and soybean fields in CDL.

Through the GEE cloud computing platform, we manually labeled each point as either irrigated or rainfed based on multiple lines of evidence, including the presence of visible irrigation infrastructure, high vegetation indices, and limited water supply from remote sensing and climate data. The presence of irrigation infrastructure, primarily central pivot irrigation systems, is identified from NAIP images. When such infrastructure is identified, we examine the time series of vegetation indices, NDWI, precipitation, and NLDAS-Noah root zone soil water content to estimate whether a particular location is irrigated. As described previously, the vegetation indices and NDWI are derived from available Landsat scenes, with precipitation data from PRISM. A data point is discarded when a decision cannot be made. In total, the manually labeled dataset include 1536 data points (locations in Figure 2).

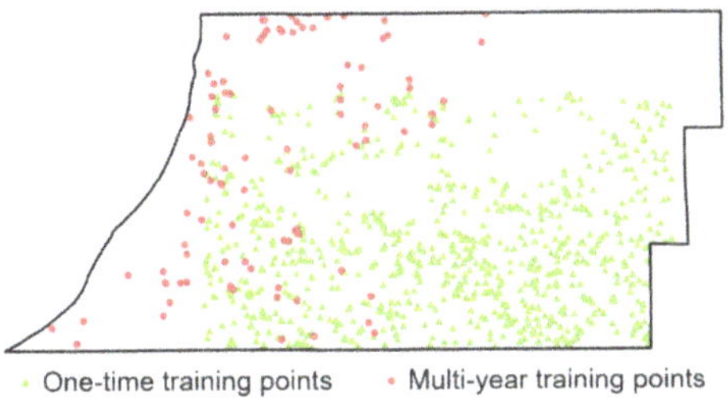

Figure 2. Training points are randomly generated, scattered in crop areas. One-time training points (green triangles) are generated for 2012 (dry) and 2014 (wet) years. Additional points (red dots) are generated for 2005, 2006, 2009, 2010, 2012 and 2014.

2.6. Classification Accuracy Assessment

We evaluated the accuracy of irrigation mapping using two validation data sources. First, we randomly divided the manually labeled dataset into training (80%) and validation (20%). We trained a random forest on the training dataset, and then tested its performance on the validation data points. To reduce the effects of random sampling, we repeated this sample-and-train process 20 times. We note that some of the remote sensing and climate information are used both in the manual labeling of the validation dataset as well as inputs to the random forest classifier. Such overlap may favorably influence classifier performance evaluation on the manually labeled validation dataset. However, the manually labeled reference dataset primarily relies upon visual cues in the NAIP high-resolution imagery, which was not included in the RF classification. Overlapping datasets provided only supporting evidence for manual labeling. Therefore, the validation points still provide valuable insights into irrigation mapping accuracy, especially given the lack of ground truth data. As a second, independent assessment, we calculated the total irrigated area of corn and soybean for each county in the study domain. The results are compared with county-wise statistics of the two crops available through NASS Agricultural Census [2] in the available years (2002, 2007, 2012).

For comparison, we also evaluate the accuracy of MIrAD-US, a national, 250-m resolution irrigation dataset available in 2002, 2007 and 2012 [12], using the manually labeled dataset. The MIrAD-US product identifies irrigated pixels when MODIS annual peak NDVI exceeds a threshold, which is adjusted for each county such that the resulting total irrigated acreage agrees with the USDA NASS statistics. We compare MIrAD-US label (irrigated versus rainfed) with the manually labeled dataset (Section 2.5), with error rates reported in Section 3.1. The MIrAD-US error rate is then compared with RF classification validation error for data points in 2012 (no training data is generated in 2002 and 2007 due to lack of NAIP images), averaged over 20 repeated experiments.

3. Results and Discussions

We developed annual irrigation probability maps for 2001 to 2016 (Figure 3) by integrating Landsat remote sensing imagery and hydroclimatic variables in a RF analysis as discussed above. Figure 3 shows the irrigation probability maps for a 2.7 by 3.2 km area in 2012. Comparing the irrigation maps with NAIP imagery (Figure 3b,c), the RF classifier can identify irrigation with detailed sub-field spatial pattern, which national products such as MIrAD-US cannot capture due to its coarse resolution (Figure 3e). This is important, as small fragmented fields are common in the study area. It is important to note that NAIP imagery was only used to label training points and not included in the input data to produce the irrigation maps. Figure 3c shows the green indices calculated from the Landsat scene that immediately follows a dry period (GI_ppt), which is identified as the longest consecutive days with daily precipitation less than 5 mm (Section 2.4). This variable is the most important input variable according to RF important score (see Section 3.2 for more details).

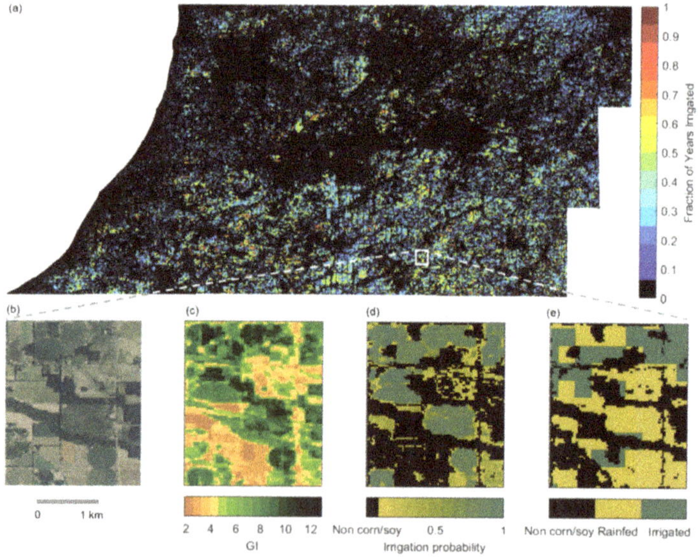

Figure 3. (**a**) Map of fraction of years classified as irrigated since earliest year irrigated according to the fandom forest (RF)-based annual irrigation maps spanning 2001–2016. For example, a pixel that is irrigated every year since the start of irrigation in 2012 will receive a fraction of 1.0. 2012 insets of (**b**) NAIP aerial image showing irrigated farms with varying sizes, (**c**) GI calculated from the Landsat scene that immediately follows the largest dryspell (GI_ppt), (**d**) random forest-based irrigation probability map with 30-m resolution and (**e**) MIrAD-US irrigation map with 250-m resolution. Images (**b**–**e**) are for 2012. Areas not classified as corn or soybeans (USDA-NASS, 2016) are shown in dark.

3.1. Classification Accuracy

First, we examined the error rate of the RF classifier on the validation points we reserved before training the classifier, as described in Section 2.5. Accuracies for all years, dry years (6/15 to 7/31 precip. < 2001–2016 mean), and wet years (6/15 to 7/31 precip. > 2001–2016 mean) are 82%, 85% and 78%, respectively (Table 3). It is not surprising that the classification accuracy is lower in wet years that received plenty of precipitation. In wet years, the input indices describing the crop status may not be significantly different between rainfed and irrigated fields, inducing higher commission error and lower omission error than in dry years (Table 3). It is notable that the difference between the accuracies in dry and wet years is small. This suggests the utility of spatial anomaly and weather-sensitive remote sensing indices in distinguishing irrigated fields from rainfed even under humid conditions.

Table 3. Irrigation mapping accuracy evaluated using manually labeled data points. Omission error describes the percentage of irrigated training points that are classified as rainfed (false negative), while commission error describes the percentage of rainfed training points that were classified as irrigated (false positive). The accuracies of RF classifier and MIrAD-US [12] are compared for 2012 when both the manually labeled data points and MIrAD-US map are available.

Year	Omission Error	Commission Error	Overall Accuracy
Dry (2009, 2012)	40%	9%	85%
Wet (2005, 2006, 2010, 2014)	38%	14%	78%
All years	39%	13%	82%
2012 RF (This study)	39%	6%	84%
2012 MIrAD-US [12]	49%	16%	74%

We then compared the county irrigated area classified by RF with NASS Agricultural Census statistics [2] in 2002, 2007, 2012, as shown in Figure 4. For county statistics, there is a good overall agreement ($r^2 = 0.69$). Figure 5 reports the annual total irrigated area in the study domain. While spread is noticeable in the county data (Figure 4), the total irrigated area agrees well with NASS statistics.

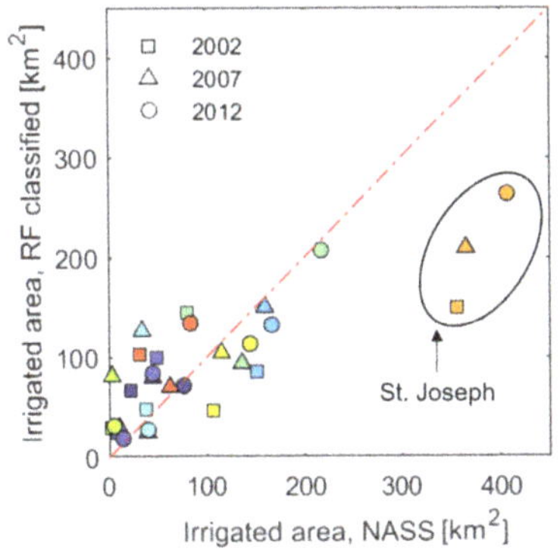

Figure 4. County irrigated area for 2002, 2007, 2012 according to the RF-based irrigation maps. Color encodes different counties. As explained in Section 3.1, RF significantly underestimates irrigated area in St. Joseph county due to widespread seed corn production in that area.

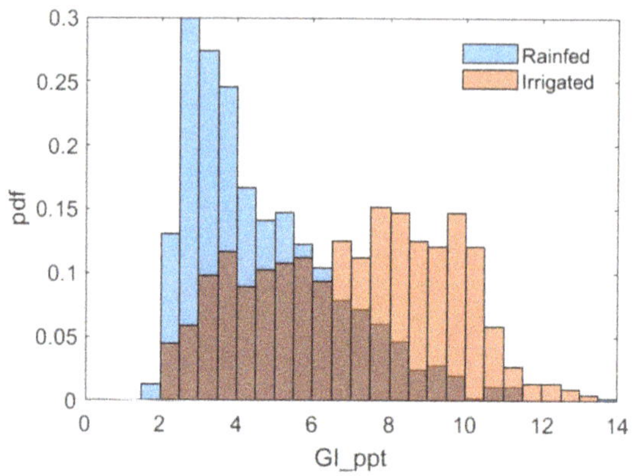

Figure 5. Empirical probability density function of green index following maximum dryspell (GI_ppt) estimated from manually labelled dataset.

Next, we compared the error rate of RF classifier and MIrAD-US using the manually labeled data (Sections 2.5 and 2.6) in 2012 when both the manually labeled data points and MIrAD-US map are available. As shown in Table 3, the MIrAD-US error rate (defined as 1—accuracy) is 26%, and the RF classifier error rate is 16%. In addition, RF irrigation maps have lower omission and commission errors than MIrAD-US, suggesting a higher accuracy of RF-derived irrigation maps.

The high omission error of the irrigation classification (Table 3) may be due to the agricultural management practice of deficit irrigation in the study area. Notably from Figure 4, the RF classifier significantly underestimated irrigated area in St. Joseph county where seed corn is the dominant crop [18], and deficit irrigation in late season (August) is commonly applied to dry up corn in plots. These locations would thus exhibit lower vegetation indices. The irrigation maps may underrepresent locations where a deficit irrigation strategy is applied in the rest part of the study domain.

The accuracy assessed on the validation points is lower than previous study that used a similar method to map irrigated area in a semi-arid to arid region [13]. In a more arid climate, the vegetation indices of rainfed crops are distinctively lower than those of irrigated crops. This is not the case in humid to subhumid climates. As shown in Figure 5, while the mean value of GI_ppt is higher in irrigated fields, the distributions of the two classes largely overlap. Such mixing also occurs for other input variables, making separation of the two classes challenging in humid regions.

The accuracy of our irrigation maps is also subject to the uncertainties of the input data. As described in Section 2.2, the irrigation maps are developed based on crop masks derived from NLCD and CDL. Thus, misclassification of either product affects the validity of training points and the accuracy of irrigation maps. For years before 2007, the crop mask based on NLCD includes all row crop fields, thus the classified irrigated fields likely include irrigated fields other than corn and soybean. Furthermore, cloud coverage inevitably leads to missing scenes during the critical crop development phase. For locations with few available Landsat scenes, important information regarding the crop status may be missing, and resulting classification may be misled. The issue of cloud coverage may be alleviated using fusion of remote sensing products across recent platforms [41,42] such as radar imagery [15]. Finally, fields smaller than the 30-m resolution may not be well captured by the Landsat-based mapping method.

3.2. Important Input Variables

During the training process, the RF classifier calculates the variable importance scores as how much impurity (i.e., irrigated versus rainfed) can be explained by each input variable. Figure 6 depicts the 30 variables receiving the highest scores. Most of these higher-ranking variables are weather-sensitive remote sensing indices from Landsat scenes immediately following a dry period identified by low soil water content, negative PDSI, or limited precipitation. This finding highlights the importance of selecting Landsat scenes that provide the most information for separating irrigated fields from rainfed ones. These critical scenes capture crop performance under a water stress condition; irrigated crops are expected to exhibit higher vegetation indices than rainfed crops. Such differences are not captured by peak vegetation indices, as rainfed crops may exhibit vegetation indices as high as irrigated crops during periods with sufficient precipitation. Simple remote sensing indices such as peak vegetation indices are not among the variables that explain most of irrigation spatiotemporal variability. In humid regions, maximum vegetation indices can be biased due to extensive cloud coverage.

Composites and some spatial anomaly indices receive high importance score, suggesting the utility of these variables to identify irrigated fields. Besides climate and remote sensing data, latitude is among the most important variables, likely due to the gradient of increasing fraction of seed corn from north to south. It is common to regularly irrigate seed corn as required in contracts. In addition, water supply indicators such as PDSI and pre-season precipitation (p_early) receive high ranks because they can explain the interannual climate variability. Other climate variables received lower importance scores.

We found that soil properties and slope are not important factors for simulating the spatial distribution of irrigation in this region. This is not surprising because sandy soil with low AWC and mild terrain are prevalent in the agricultural lands of the study area. Other variables that do not appear to be important include soil water content, precipitation, aridity and extreme weather condition indices such as GDD, dryspell and heatwave, likely because the resolution of the meteorological data used to calculate these indices is too coarse to capture the fine-scale heterogeneity of irrigation. However, these variables portray large-scale water supply and demand, and we have shown that they can be used to select Landsat scenes that provide the best information for separating irrigated fields from rainfed ones. In particular, soil water content is simulated by NLDAS-Noah, which does not account for irrigation and estimates wetness under rainfed conditions.

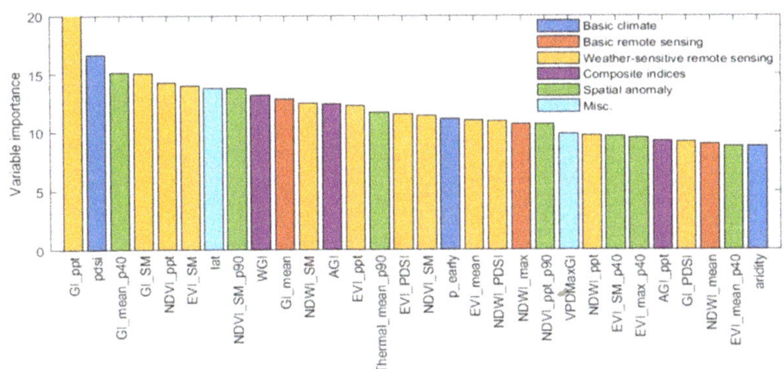

Figure 6. Top 30 (of 98) important variables as identified by RF; variables are grouped into six categories as indicated by different colors (Sections 2.2 and 2.3, Tables 1 and 2, Table S2, Supplementary Material).

3.3. *Expansion of Irrigation*

From the RF classified irrigation maps we calculated the total irrigated area for the study region for 2001–2016 (Figure 7) and compared this to NASS statistics. Temporal fluctuation is noticeable, with limited irrigated area in 2009–2011 and high irrigated area in 2013–2014. The peak in 2014 is likely a combination of three factors. First, the critical crop development phase in 2014 had 21% higher than average (Figure 7) and more frequent precipitation (the dryspell of the study area is 13 days in 2014 and 17 averaged over 2001–2016), leading to robust rainfed crops and correspondingly high vegetation index values across the region. Thus, the RF classifier may overestimate the irrigated area in this year. Second, as described in Section 2.5, 2014 has 8.2% pixels with no Landsat scenes during the critical crop development phase. To fill in the gaps, we labeled pixels as irrigated in 2014 if they were classified as irrigated in both 2012 and 2013. This may result in commission errors (i.e., classifying rainfed fields as irrigated) in those pixels. Third, farmers may have switched to irrigated agriculture after the crop losses in the 2012 drought.

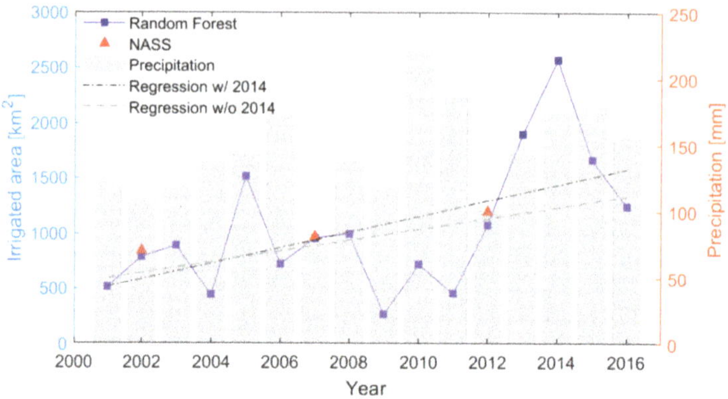

Figure 7. Annual total irrigated area in SW MI according to the random forest-based irrigation maps (line with squares), NASS Consensus (triangles, [2]). Gray bars show precipitation during the 15 June–31 July period, averaged over the study domain. Linear regression with and without 2014 irrigated area is also shown.

There is a statistically significant ($p < 0.05$) increasing trend in irrigation in this study region despite notable interannual variability. Over the 16-year period, irrigated area tripled (increased by 204%), according to the linear regression shown in Figure 7. The estimated slope of 70.8 km^2/year is approximately twice the estimate from NASS statistics in 2002, 2007 and 2012 (slope = 35.6 km^2/year). In order to isolate the likely skewness due to high irrigated area estimated in 2014, we performed another linear regression excluding 2014 (Figure 7). An increasing trend is statistically significant ($p < 0.05$) with estimated slope of 49.1 km^2/year.

We also calculated the change in irrigated area for corn and soybeans, respectively, for 2007–2016, when CDL is available for the study area. The irrigated area fractions for corn and soybeans increased from 19.1% in 2007 to 24.9% in 2016, and from 9.2% in 2007 to 17.9% in 2016, respectively.

To examine the spatial pattern, irrigation trends are calculated based on linear regression (Figure 8). To do this, we aggregated the 30-m irrigation maps to a larger grid to perform linear regression on the irrigated area through time. We chose the 9-km^2 grid to examine relatively fine-scale spatial patterns of irrigation trends across the region (vs. county level, for instance). Slight decreasing trends in lakeshore area suggests discontinuation of irrigation. The highest increase rate (up to 0.25 km^2/year per 9-km^2 cell) was found in southern part of the study area. The irrigation expansion is believed to be associated with the promotion of seed corn in this area [16,18]. Seed corn is usually irrigated because irrigation is typically required by the contracts between farmers and seed corn companies.

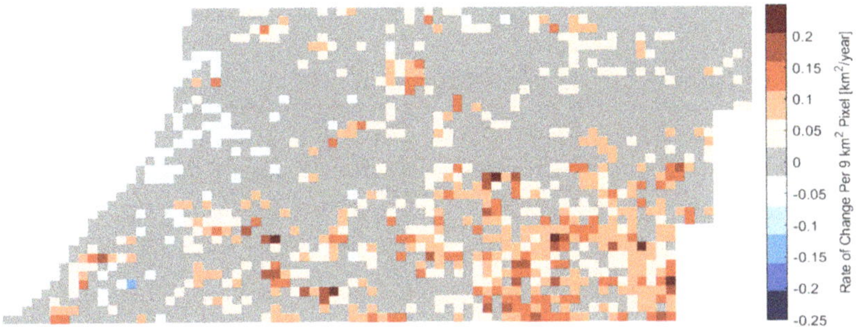

Figure 8. Rate of change of irrigated area over time based on linear regression calculated for an aggregated 9 km^2 grid. Gray indicates non-significant trend ($\alpha > 0.1$).

In addition, correlation analyses suggest crop commodity price is another factor affecting irrigation decisions. The annual irrigated area of the study region is found to be correlated with previous year's corn price [2] (Figure 9, $r = 0.66$, $p = 0.009$). Irrigation may double corn yields and increase soybean yields by more than 66% in SW MI [16]. Given the easy accessibility to irrigation water, adoption of irrigation will likely increase farmers' revenue. Irrigation expansion may be further encouraged by devastating crop losses in the 2012 drought in fields without irrigation [43,44].

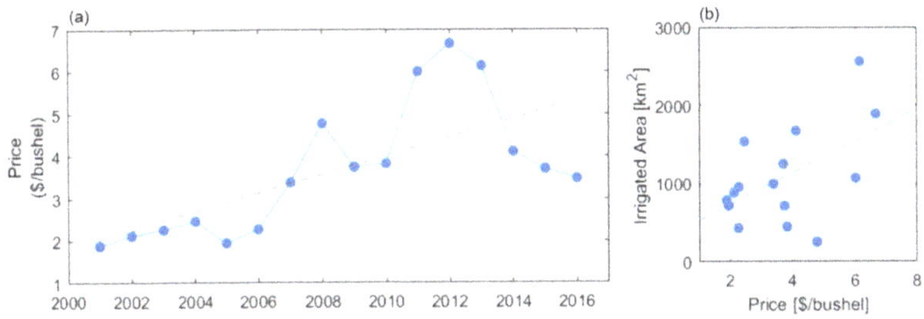

Figure 9. Commodity price of corn (**a**) and its correlation with irrigated area (**b**) 2001–2016.

4. Conclusions

By integrating satellite imagery and hydroclimatic information using a machine learning algorithm, we created annual irrigation maps for a subhumid area in southwestern Michigan for 2001–2016. The maps capture the spatiotemporal pattern of irrigation at a high spatial resolution (30 m) and indicate that irrigated area in southwestern Michigan roughly tripled in the last 16 years according to linear regression.

We demonstrated the utility of novel input variables including weather-sensitive remote sensing, spatial anomalies, and recently-developed composite indices. In particular, we found that those vegetation indices following dry periods are the most important to distinguish irrigated fields from rainfed. This not only reduces the number of scenes (thus memory and computational expense) to process, but also avoids possible confounding effects of high vegetation indices captured during a wet period.

The annual irrigation maps are validated using multiple data sources. Reasonable accuracy is achieved despite the difficulties involved with estimating irrigated area in a region with a subhumid climate and heterogeneous agricultural management practices (e.g., deficit irrigation strategy for seed corn). We found that the mapping accuracy in dry years is higher than in wet years with a narrow margin. The small difference between accuracies may be attributed to the use of spatial anomaly and weather-sensitive remote sensing indices, which were able to distinguish irrigated from rainfed fields even under subhumid conditions.

We identified several challenges and limitations for mapping irrigated areas in subhumid to humid regions, including the dependency on the quality of input data (e.g., land cover) and cloud coverage, which is more frequent in such regions. The substantial efforts and difficulty involved in generating training data are also noteworthy and call for in season high-resolution imagery. Nevertheless, the promising results underscore the potential of using remote sensing and cloud computing to provide valuable information for water resources decision makers and hydrologic studies at regional scales.

Supplementary Materials: The annual irrigation maps 2001–2016 can be downloaded at https://doi.org/10.4211/hs.3766845be72d45969fca21530a67bb2d. In addition, the following are available online at http://www.mdpi.com/2072-4292/11/3/370/s1, Table S1: The mean and quantiles for the cumulative probabilities 0.025 and 0.975 of number of available scenes for all pixels in the study domain between June 10th and August 5th for each year in the study period (2001–2016), Table S2: All input variables of the random forest classifier grouped into seven categories, Table S3. Unsuccessful input variables that were not used in the final random forest classifier.

Author Contributions: T.X., J.M.D., A.D.K., and D.W.H. conceived and planned the experiments. T.X. carried out the experiments. T.X. and J.M.D. developed codes. T.X., J.M.D., and A.D.K contributed to preparation of data. T.X. drafted the manuscript. All authors contributed to analytical methods, interpretation of results, and editing of the manuscript.

Funding: This work was supported by USDA NIFA Grants 2015-68007-23133 and 2018-67003-27406, and NSF Grants 1027253 and 1637653. We thank Jeremy Rapp for his contribution to training dataset preparation. We are grateful for the thoughtful review and constructive comments by the four anonymous reviewers.

Conflicts of Interest: The authors declare no conflicts of interest.

References

1. Food and Agriculture Organization of the United Nations (FAO) AQUASTAT Main Database. Available online: http://www.fao.org/nr/aquastat (accessed on 1 January 2016).
2. U.S Department of Agriculture. National Agricultural Statistics Service (USDA NASS) QuickStats Ad-hoc Query Tool. Available online: https://quickstats.nass.usda.gov/ (accessed on 1 November 2017).
3. Cotterman, K.A.; Kendall, A.D.; Basso, B.; Hyndman, D.W. Groundwater depletion and climate change: Future prospects of crop production in the Central High Plains Aquifer. *Clim. Change* **2018**, *146*, 187–200. [CrossRef]
4. Pei, L.; Moore, N.; Zhong, S.; Luo, L.; Hyndman, D.W.; Heilman, W.E.; Gao, Z. WRF Model Sensitivity to Land Surface Model and Cumulus Parameterization under Short-Term Climate Extremes over the Southern Great Plains of the United States. *J. Clim.* **2014**, *27*, 7703–7724. [CrossRef]
5. Pei, L.; Moore, N.; Zhong, S.; Kendall, A.D.; Gao, Z.; Hyndman, D.W. Effects of Irrigation on Summer Precipitation over the United States. *J. Clim.* **2016**, *29*, 3541–3558. [CrossRef]
6. Smidt, S.J.; Haacker, E.M.K.; Kendall, A.D.; Deines, J.M.; Pei, L.; Cotterman, K.A.; Li, H.; Liu, X.; Basso, B.; Hyndman, D.W. Complex water management in modern agriculture: Trends in the water-energy-food nexus over the High Plains Aquifer. *Sci. Total Environ.* **2016**, *566–567*, 988–1001. [CrossRef] [PubMed]
7. Kumar, S.V.; Peters-Lidard, C.D.; Santanello, J.A.; Reichle, R.H.; Draper, C.S.; Koster, R.D.; Nearing, G.; Jasinski, M.F. Evaluating the utility of satellite soil moisture retrievals over irrigated areas and the ability of land data assimilation methods to correct for unmodeled processes. *Hydrol. Earth Syst. Sci.* **2015**, *12*, 5967–6009. [CrossRef]
8. Lawston, P.M.; Santanello, J.A.; Franz, T.E.; Rodell, M. Assessment of irrigation physics in a land surface modeling framework using non-traditional and human-practice datasets. *Hydrol. Earth Syst. Sci.* **2017**, *21*, 2953–2966. [CrossRef] [PubMed]
9. McInerney, D.; Thyer, M.; Kavetski, D.; Githui, F.; Thayalakumaran, T.; Liu, M.; Kuczera, G. The Importance of Spatiotemporal Variability in Irrigation Inputs for Hydrological Modeling of Irrigated Catchments. *Water Resour. Res.* **2018**, *54*, 6792–6821. [CrossRef]
10. Levin, S.B.; Zarriello, P.J. *USGS Scientific Investigations Report 2013–5066: Estimating Irrigation Water Use in the Humid Eastern United States*; USGS: Reston, VA, USA, 2013.
11. Pervez, M.S.; Brown, J.F. Mapping irrigated lands at 250-m scale by merging MODIS data and National Agricultural Statistics. *Remote Sens.* **2010**, *2*, 2388–2412. [CrossRef]

12. Brown, J.F.; Pervez, M.S. Variability and trends in irrigated and non-irrigated croplands in the central U.S. In Proceedings of the 2013 Second International Conference on Agro-Geoinformatics (Agro-Geoinformatics); IEEE, Fairfax, VA, USA, 12–16 August 2013; pp. 102–105.
13. Deines, J.M.; Kendall, A.D.; Hyndman, D.W. Annual Irrigation Dynamics in the U.S. Northern High Plains Derived from Landsat Satellite Data. *Geophys. Res. Lett.* **2017**, *44*, 9350–9360. [CrossRef]
14. Deines, J.M.; Kendall, A.D.; Butler, J.J.; Hyndman, D.W. Quantifying water use and farmer adaptation strategies in response to novel stakeholder-driven groundwater management in teh US High Plains Aquifer. *Environ. Res. Lett.* 2019. [CrossRef]
15. Gao, Q.; Zribi, M.; Escorihuela, M.; Baghdadi, N.; Segui, P.; Gao, Q.; Zribi, M.; Escorihuela, M.J.; Baghdadi, N.; Segui, P.Q. Irrigation Mapping Using Sentinel-1 Time Series at Field Scale. *Remote Sens.* **2018**, *10*, 1495. [CrossRef]
16. Michigan State University (MSU) Extension. *Value of Irrigation to the Southwest Michigan Economy*; MSU: Lansing, MI, USA, 2014.
17. USDA Natural Resources Conservation Service (NRCS) Web Soil Survey. Available online: https://websoilsurvey.nrcs.usda.gov/ (accessed on 1 November 2017).
18. Kaercher, M.; Neumann, B. *St. Joseph County Agriculture: Past, Present and Future*; MSU: Centreville, MI, USA, 2006.
19. Kraft, G.J.; Clancy, K.; Mechenich, D.J.; Haucke, J. Irrigation Effects in the Northern Lake States: Wisconsin Central Sands Revisited. *Ground Water* **2012**, *50*, 308–318. [CrossRef] [PubMed]
20. Wolock, D.M.; Winter, T.C.; McMahon, G. Delineation and Evaluation of Hydrologic-Landscape Regions in the United States Using Geographic Information System Tools and Multivariate Statistical Analyses. *Environ. Manag.* **2004**, *34*, S71–S88. [CrossRef] [PubMed]
21. Daly, C.; Halbleib, M.; Smith, J.I.; Gibson, W.P.; Doggett, M.K.; Taylor, G.H.; Curtis, J.; Pasteris, P.P. Physiographically sensitive mapping of climatological temperature and precipitation across the conterminous United States. *Int. J. Climatol.* **2008**, *28*, 2031–2064. [CrossRef]
22. U.S. Department of Agriculture. National Agricultural Statistics Service (USDA NASS) Cropland Data Layer. Available online: https://www.nass.usda.gov/Research_and_Science/Cropland/Release/index.php (accessed on 1 April 2018).
23. Gorelick, N.; Hancher, M.; Dixon, M.; Ilyushchenko, S.; Thau, D.; Moore, R. Google Earth Engine: Planetary-scale geospatial analysis for everyone. *Remote Sens. Environ.* **2017**, *202*, 18–27. [CrossRef]
24. USGS 1 Arc-second Digital Elevation Models (DEMs)—USGS National Map 3DEP Downloadable Data Collection. Available online: https://www.sciencebase.gov/catalog/item/4f70aa71e4b058caae3f8de1 (accessed on 1 November 2017).
25. Schaap, M.G.; Leij, F.J.; van Genuchten, M.T. ROSETTA: A computer program for estimating soil hydraulic parameters with hierarchical pedotransfer functions. *J. Hydrol.* **2001**, *251*, 163–176. [CrossRef]
26. Abatzoglou, J.T. Development of gridded surface meteorological data for ecological applications and modelling. *Int. J. Climatol.* **2013**, *33*, 121–131. [CrossRef]
27. Xia, Y.; Mitchell, K.; Ek, M.; Sheffield, J.; Cosgrove, B.; Wood, E.; Luo, L.; Alonge, C.; Wei, H.; Meng, J.; et al. Continental-scale water and energy flux analysis and validation for the North American Land Data Assimilation System project phase 2 (NLDAS-2): 1. Intercomparison and application of model products. *J. Geophys. Res. Atmos.* **2012**, *117*. [CrossRef]
28. White, J.C.; Wulder, M.A.; Hobart, G.W.; Luther, J.E.; Hermosilla, T.; Griffiths, P.; Coops, N.C.; Hall, R.J.; Hostert, P.; Dyk, A.; et al. Pixel-Based Image Compositing for Large-Area Dense Time Series Applications and Science. *Can. J. Remote Sens.* **2014**, *40*, 192–212. [CrossRef]
29. Gao, B. NDWI—A normalized difference water index for remote sensing of vegetation liquid water from space. *Remote Sens. Environ.* **1996**, *58*, 257–266. [CrossRef]
30. Ozdogan, M.; Gutman, G. A new methodology to map irrigated areas using multi-temporal MODIS and ancillary data: An application example in the continental US. *Remote Sens. Environ.* **2008**, *112*, 3520–3537. [CrossRef]
31. Running, S.; Mu, Q.; Zhao, M.; Moreno, A. *MODIS Global Terrestrial Evapotranspiration (ET) Product (NASA MOD16A2/A3) Collection 5*; NASA Headquarters: Washington, DC, USA, 2013.

32. McAllister, A.; Whitfield, D.; Abuzar, M. Mapping Irrigated Farmlands Using Vegetation and Thermal Thresholds Derived from Landsat and ASTER Data in an Irrigation District of Australia. *Photogramm. Eng. Remote Sens.* **2015**, *81*, 229–238. [CrossRef]
33. Naghibi, S.A.; Pourghasemi, H.R.; Dixon, B. GIS-based groundwater potential mapping using boosted regression tree, classification and regression tree, and random forest machine learning models in Iran. *Environ. Monit. Assess.* **2016**, *188*, 44. [CrossRef] [PubMed]
34. Xu, T.; Valocchi, A.J. Data-driven methods to improve baseflow prediction of a regional groundwater model. *Comput. Geosci.* **2015**, *85*, 124–136. [CrossRef]
35. Xu, T.; Valocchi, A.J.; Ye, M.; Liang, F.; Lin, Y.F. Bayesian calibration of groundwater models with input data uncertainty. *Water Resour. Res.* **2017**, *53*, 3224–3245. [CrossRef]
36. Breiman, L. Random Forests. *Mach. Learn.* **2001**, *45*, 5–32. [CrossRef]
37. Fry, J.; Xian, G.Z.; Jin, S.; Dewitz, J.; Homer, C.G.; Yang, L.; Barnes, C.A.; Herold, N.D.; Wickham, J.D. Completion of the 2006 national land cover database for the conterminous united states. *Photogramm. Eng. Remote Sens.* **2011**, *77*, 858–864.
38. Hastie, T.; Friedman, J.; Tibshirani, R. *The Elements of Statistical Learning*; Springer: New York, NY, USA, 2001.
39. Abe, S. *Feature Selection and Extraction*; Springer: London, UK, 2010; pp. 331–341.
40. National Agriculture Imagery Program (NAIP) USDA Farm Service Agency National Agriculture Imagery Program. Available online: https://www.fsa.usda.gov/programs-and-services/aerial-photography/imagery-programs/naip-imagery/ (accessed on 1 November 2017).
41. Boschetti, L.; Roy, D.P.; Justice, C.O.; Humber, M.L. MODIS–Landsat fusion for large area 30 m burned area mapping. *Remote Sens. Environ.* **2015**, *161*, 27–42. [CrossRef]
42. Wang, Q.; Blackburn, G.A.; Onojeghuo, A.O.; Dash, J.; Zhou, L.; Zhang, Y.; Atkinson, P.M. Fusion of Landsat 8 OLI and Sentinel-2 MSI Data. *IEEE Trans. Geosci. Remote Sens.* **2017**, *55*, 3885–3899. [CrossRef]
43. Boyer, J.S.; Byrne, P.; Cassman, K.G.; Cooper, M.; Delmer, D.; Greene, T.; Gruis, F.; Habben, J.; Hausmann, N.; Kenny, N.; et al. The U.S. drought of 2012 in perspective: A call to action. *Glob. Food Sec.* **2013**, *2*, 139–143. [CrossRef]
44. Zhang, T.; Lin, X.; Sassenrath, G.F. Current irrigation practices in the central United States reduce drought and extreme heat impacts for maize and soybean, but not for wheat. *Sci. Total Environ.* **2015**, *508*, 331–342. [CrossRef] [PubMed]

remote sensing

MDPI

Article

Detecting Winter Wheat Irrigation Signals Using SMAP Gridded Soil Moisture Data

Zhen Hao [1], Hongli Zhao [2,*], Chi Zhang [1], Hao Wang [2] and Yunzhong Jiang [2]

1 Faculty of Infrastructure Engineering, Dalian University of Technology, Dalian 116034, China; haozhen@mail.dlut.edu.cn (Z.H.); czhang@dlut.edu.cn (C.Z.)

2 Department of Water Resources Research, China Institute of Water Resources and Hydropower Research, Beijing 100038, China; Wanghao@iwhr.com (H.W.); lark@iwhr.com (Y.J.)

* Correspondence: zhaohl@iwhr.com; Tel.: +86-136-8121-8530

Received: 9 August 2019; Accepted: 14 October 2019; Published: 15 October 2019

Abstract: The southern part of the Hebei Province is one of China's major crop-producing regions. Due to the continuous decline in groundwater level, agricultural water use is facing significant challenges. Precision agricultural irrigation management is undoubtedly an effective way to solve this problem. Based on multisource data (time series soil moisture active passive (SMAP) data, Moderate Resolution Imaging Spectroradiometer (MODIS) normalized difference vegetation index (NDVI) and evapotranspiration (ET), and meteorological station precipitation), the irrigation signal (frequency, timing and area) is detected in the southern part of the Hebei Province. The SMAP data was processed by the 5-point moving average method to reduce the error caused by the uncertainty of the microwave data derived SM. Irrigation signals can be detected by removing the precipitation effect and setting the SM change threshold. Based on the validation results, the overall accuracy of the irrigation signal detection is 77.08%. Simultaneously, considering the spatial resolution limitation of SMAP pixels, the SMAP irrigation area was downscaled using the winter wheat area extracted from MODIS NDVI. The analytical results of 55 winter wheat samples (5 samples in a group) showed that winter wheat covered by one SMAP pixel had an 82.72% growth consistency in surface water irrigation period, which can indicate a downscaling effectiveness. According to the above statistical analysis, this paper considers that although the spatial resolution of SMAP data is insufficient, it can reflect the change of SM more sensitively. In areas where the crop pattern is relatively uniform, the introduction of high-resolution crop pattern distribution can be used not only to detect irrigation signals but also to validate the effectiveness of irrigation signal detection by analyzing crop growth consistency. Therefore, the downscaling results can indicate the true winter wheat irrigation timing, area and frequency in the study area.

Keywords: irrigation signal; SMAP; irrigation intensity; winter wheat

1. Introduction

Winter wheat is the main crop in the North China Plain (NCP). Due to the high irrigation demand of winter wheat, more than 70% of the irrigated water resources are used for winter wheat irrigation every year [1]. The increasing population has led to a corresponding increase in the demand for agricultural, industrial and domestic water in the NCP. The surface water resources are insufficient, and groundwater has become the main source of water for the NCP [2]. In recent decades, the overexploitation of groundwater has led to a significant decline in groundwater levels, which increases not only environmental problems but also the pressure on agricultural food production [3,4]. Groundwater is the main source of water for NCP agriculture irrigation. Long-term dependence on groundwater for agricultural irrigation has resulted in groundwater over-exploitation, and agricultural water irrigation needs to be reduced; however, the sustainable of food crop production must also be ensured [5,6]. Timely and effective monitoring of irrigation water is of great significance for agricultural water management and water resources protection. The irrigation signal includes the time, frequency and area of irrigation. Irrigation time can be used to dynamically correct irrigation schedules, while irrigation frequency and area can be used for the estimation and dynamic monitoring of agricultural irrigation water use [7–10]. This study prepares to establish a model that can be used to detect irrigation signals and dynamically acquire irrigation information. The results of the irrigation signal will be used for the dynamic monitoring of agricultural irrigation water to achieve refined management of agricultural irrigation.

With the continuous development of remote sensing technology, more remote sensing data can be used for irrigation information detection [11–15]. Compared with traditional agricultural statistical methods, remote sensing has a wide range of multifrequency, high spatial and temporal resolution advantages and has been widely used in agricultural management [16–18]. Representative data sources include Moderate Resolution Imaging Spectroradiometer (MODIS), which provides 250 m, 500 m and 1 km resolution daily surface reflectance data. The richness of time series and improvement in remote sensing data spatial resolution has greatly improved the accuracy of irrigated area identification [19]. In recent research, the Normalized Difference Vegetation Index (NDVI) has been extensively used as an effective indicator for irrigated area recognition based on optical remote sensing data [19–21]. An analysis of the time-varying pattern of NDVI is the primary method for identifying irrigated and non-irrigated areas. In particular, wheat and maize are affected by irrigation, and their NDVIs will appear to be higher than other vegetation [20,22]. Although the identification method for irrigated areas has been comprehensive, this irrigated area extraction method based on optical remote sensing data is mostly used for long-term irrigated area monitoring to analyze trends in irrigated areas over multiple years. Chen et al. [23] proposed a method for detecting irrigation extent, timing and frequency based on MODIS and Landsat remote sensing data, which is an important irrigation property for understanding the sustainability of water resources in arid and semiarid regions. The irrigation signal detection method based on the visible vegetation index must model the daily scale data, and this method is more suitable for irrigation signal detection in regions with less cloud cover. Remote sensing images of areas with more clouds are likely to miss the critical period of irrigation signal detection due to cloud pollution. Moreover, in addition to the influence of image quality, the response of vegetation to irrigation is lagged, which increases the uncertainty of irrigation timing detection.

In addition to the method of identifying the irrigated area by using vegetation index information, the change in the wetness index can also be used to identify the irrigation signal [24]. Based on the SM being higher in the irrigated area than in the non-irrigated area, some researchers have identified irrigated areas based on different principles. Based on the MODIS enhanced vegetation index (EVI) and land surface water index (LWSI) ratio method, Peng et al. [25] introduced the variable EVI/LWSI threshold function to improve the detection ability of this method in different rice crops under mixed rice crop patterns (single-season rice, early-season rice, and late-season rice). Abuzar et al. [26] used vegetation and thermal thresholds derived from Landsat and Advanced Spaceborne Thermal Emission and Reflection Radiometer (ASTER) data to detect the irrigated area in an Australian irrigation district based on the soil temperature in the irrigated area being lower than that in the non-irrigated area. Although different researchers use SM information to detect irrigated area information from different aspects, they do not use SM indicators because optical and thermal infrared remote sensing data cannot directly obtain SM information.

Active and passive microwave satellites have proven to be effective tools for retrieving soil water variations at regional and global scales [27–29]. NASA's Soil Moisture Active Passive (SMAP) satellite, launched on 31 January 2015, provides a new source of data for near-surface (0–5 cm) soil water monitoring on a global scale. Colliander et al. [30] validated the SMAP surface SM product through the core validation site. The results indicate that the SMAP radiometer-based SM data product meets the expected performance of 0.04 m^3/m^3 volumetric SM (unbiased root mean square error) and that the combined radar-radiometer product is close to its expected performance of 0.04 m^3/m^3. Chan et al. and

Zhang et al. [11,31] evaluated the results of different SMAP products in different regions and obtained similar conclusions to those of Colliander. SMAP has more information improvements than previous SM satellites, which has raised interest in whether SMAP can improve irrigation monitoring [32]. Subsequently, Lawston et al. [33] explored the use of SMAP data in identifying irrigation areas and timing in the Sacramento Valley, San Luis Valley and Columbia River Valley. However, the study did not identify the irrigation timing in the irrigated area. Since the detection of the irrigated area is a combination of changes in SM over a period of time, the time scale is the entire period of the crop. Compared with optical/thermal infrared methods, SMAP's method of detecting irrigated areas has unique advantages in terms of temporal resolution and ability to directly acquire SM [34]. The SMAP data spatial resolution is a major limiting factor that affects its use.

Obtaining irrigation time, area and frequency will help estimate irrigation water volume and provide data support for agricultural irrigation management. Despite having the low spatial resolution, SMAP provides high temporal resolution SM products. To address the spatial resolution issues, this paper will be studied in the following three aspects: 1) Based on SMAP and meteorological data, the irrigation signal in the study area was detected, which solved the problem of optical data not being applicable in cloudy regions; 2) MODIS remote sensing data were used to downscale the detection results to solve the low spatial resolution problem of SMAP data; and 3) through an analysis of the consistency of winter wheat growth covered by SMAP pixels, the SMAP data effectiveness in downscaling the winter wheat irrigation results in this study area was evaluated.

2. Study Area

The region of interest in this paper is located in the southern part of the Hebei Province and belongs to the NCP. The boundaries of the study area are city administrative boundaries, including Shi Jiazhuang, Baoding, Langfang, Hengshui, Cangzhou, Xingtai and Handan, with a total area of 8.9×10^4 km^2 (as shown in Figure 1). Although precipitation in the study area is not scarce, the distribution of precipitation during the year is extremely uneven. The study area is dominated by a temperate monsoon climate with mean annual precipitation of 472.7–889.2 mm, and 70% of the annual precipitation occurs between June and September [35]. Under the irrigation conditions of the study area in recent years, the main crop pattern is the winter wheat-summer maize double crop rotation. Winter wheat and summer maize are also the main irrigated crops in this region [36]. The lower amount of precipitation in spring is not enough to provide sufficient water for winter wheat growth, and groundwater irrigation has been the main irrigation method for winter wheat and summer maize for a long time. Winter wheat is generally irrigated 4–5 times, and precipitation has little effect on the number of irrigations due to the severe shortage of precipitation during the winter wheat growing period. Summer maize is usually irrigated before planting, and if effective precipitation has occurred before planting and the soil moisture meets the sowing requirements, the crop will not be irrigated during the growing period. The Middle Route of the South-to-North Water Transfer Project (MSWTP) was launched at the end of 2014, and this project provided a new source of water for agricultural irrigation in the NCP [37].

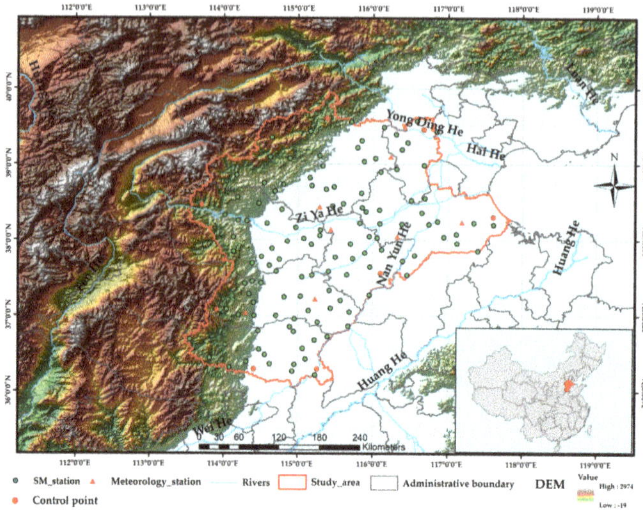

Figure 1. Study area and meteorological sites locations and the spatial distribution of SM stations.

3. Materials and Methods

The flow chart (shown in Figure 2) of this paper includes the processing of collected data (Section 3.1), selection of samples (Section 3.2.1), the application of algorithms (Section 3.2.2) and validation of accuracy (Section 3.2.3).

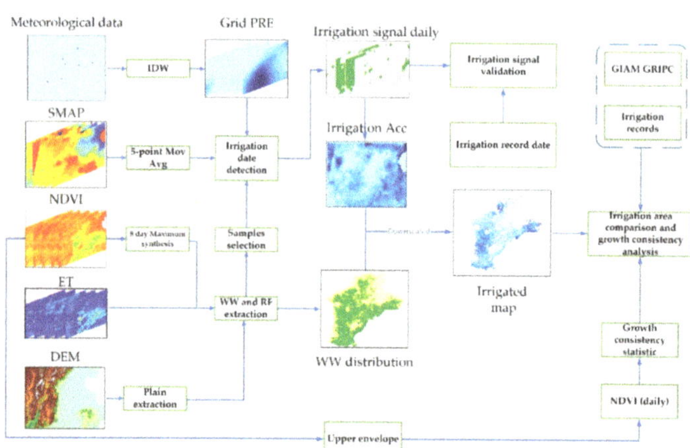

Figure 2. Flow chart for this study. Here, 5-point Mov Avg represents the 5-point moving average and Avg and Std represent the average and standard deviation, respectively. The irrigation Acc accumulates as a result of the irrigation signal.

3.1. Data Collection and Pre-Processing

3.1.1. SMAP

SMAP is an orbiting observatory capable of measuring the amount of water in the top 5 cm of soil at global scales. To meet the various needs of soil moisture monitoring, the SMAP mission uses an L-band radar and an L-band radiometer for concurrent, coincident measurements integrated into a single observational system and ultimately produce a variety of spatial and temporal resolution SM products [38]. Since the successful launch of SMAP in January 2015, it has provided SM products of many levels worldwide. After validating the accuracy of SMAP products, the SMAP products meet the mission requirements and can also be used to assess hydrologic processes [30,31,39]. In this study, considering the spatial and temporal resolution of the SMAP products required for the study area, SMAP enhanced L3 radiometer global daily 9 km EASE-grid soil moisture version 1 was used as the final data source [40]. The study used the SMAP SM datasets from the end of Feb to the end of May (2015–2018) corresponding to the winter wheat irrigation period in study area. Although the SMAP dataset can provide daily SM products, due to satellite orbits, SM products do not cover the spatial extent of the study area every day due to satellite orbits. To select the SMAP data that can completely cover the study area, 8 control points are used to filter the data that meet the requirements. The eight control points are evenly distributed at the vertices of the study area boundary, and the judgement equation is as follows:

$$X_i = \begin{cases} 1, \ 0 < V_i < 1 \\ 0, \ V_i = null \end{cases} \quad (1)$$

$$J = \begin{cases} R, \ \sum_i^8 X \geq 6 \\ D, \ \sum_i^8 X < 6 \end{cases} . \quad (2)$$

where X is the judgement result of the SMAP pixel value (V) and null is no-data in this pixel, i is the number of the control point. If the value of the SMAP pixel is between 0 and 1, X = 1; and if the SMAP pixel value is null, then X = 0. J is the judgement result of whether the SMAP data are retained, and R and D represent the retention and deletion of SMAP data, respectively. If the sum of the 8-control point X ≥ 6, it indicates that SMAP data can cover a large area (more than 75% of the study area is covered) of the study area and this SMAP data is retained; if it less than 6, the data are deleted. The programming language for batch filtering, processing and extracting of SMAP data is python 2.7, and the arcpy function provided by ArcGIS 10.4 (Environmental Systems Research Institute in California) is also used. Regarding the extraction of pixel values in this paper, the "ExtractValuesToPoints" function in arcpy is used.

3.1.2. MODIS

MODIS provides researchers with stable, long time series global remote sensing data. Some global land use/land cover (LULC) datasets based on MODIS data have been generated [41,42]. MOD09GA and MOD16A2 provide daily surface reflectance with a spatial resolution of 500 m and evapotranspiration of 500 m every 8 days [43,44]. MOD09GA and MOD16A2 were used in this study for irrigated area downscaling, while the latter is based on 8-day synthetic data and does not require further processing. MOD09GA is daily surface reflectance data, and cloud pollution has a large impact on the use of data. First, the NDVI is calculated based on the MOD09GA dataset.

$$NDVI = \frac{NIR - RED}{NIR + RED} \quad (3)$$

where NIR and RED are the surface reflectance factors for the presented wavelengths. Second, the 8-day maximum value composite (MVC) method was used for the daily NDVI time series dataset, and the method is to composite a new NDVI image by using the daily maximum value of NDVI within 8 days of each pixel in the image as a valid pixel value [45]. This processing method reduces the impact of clouds on the dataset and keeps the time resolution of the two MODIS products consistent. The batch redefinition projection and raster attribute conversion of MODIS data are based on MRT (MODIS Reprojection Tool supported by NASA, referenced by Dwyer et al. [46]), and the maximum synthesis of the NDVI is based on MATLAB 2018b.

3.1.3. Precipitation

The National Meteorological Information Centre of China provides daily precipitation data (meteorology station) from 1961 to present [47]. There are 2472 meteorology stations in China, and there are 7 meteorological stations in this study area. Daily precipitation data were collected from March 2015 to December 2018. Since precipitation data must be coordinated with SMAP data for irrigation information monitoring, the spatiotemporal resolution of the precipitation data must be processed. The daily precipitation data include statistical results for two periods (20:00–8:00 and 8:00–20:00) in Beijing time. The current method for the L3_SM product is to use the nearest 6:00 AM local solar time criterion to perform Level 3 compositing [38]. The precipitation from 8:00–8:00 is summed as the daily precipitation, and the station data are interpolated into the grid data using the inverse distance weighting (IDW) method based on python 2.7.

3.1.4. Irrigated Map

A global irrigated area map (GIAM) and global rainfed, irrigated, and paddy cropland (GRIPC) were also collected in this study for irrigated area validation. Based on the unsupervised classification method, GIAM provides irrigated area recognition results with a spatial resolution of 1 km in the year 2000 [34]. GRIPC is the result of the decision tree method for the classification of MODIS data and the spatial resolution is 500 m in year 2005 [17].

3.1.5. In Situ SM Measurement Data and Irrigation Records

There are 135 SM stations in the study area, and SM data are provided every 10 days (1st, 11th and 21st). The SM data measurement (oven-drying method) depths include 10 cm, 20 cm and 40 cm, and the measurement time is concentrated at 8:00 AM Beijing time. These sites also provide information on precipitation and irrigation times between measurements. The recorded irrigation data include areas of agricultural irrigation, irrigation crops, timing and volume. Although the recorded irrigation information is relatively abundant, the spatial scale is the agricultural irrigation region. The data collected in this study are shown in Table 1. Since this study only collected information on irrigation records in 2018, only the SM changes in 2018 were plotted during sample training and validation.

Table 1. Datasets collected in this study.

Data Source	Temporal Resolution	Spatial Resolution	Time Period	Data Access
SMAP	daily	9 km	March 2015 to December 2018	https://nsidc.org/data/SPL3SMP_E/versions/2
PRE	daily	site	March 2015 to December 2018	http://data.cma.cn/
MOD09GA	daily	500 m	March 2015 to December 2018	https://ladsweb.modaps.eosdis.nasa.gov/
MOD16A2	8-day	500 m	March 2015 to December 2018	https://ladsweb.modaps.eosdis.nasa.gov/
Irrigated Map	year	1 km and 500 m		http://www.iwmi.cgiar.org/ https//dl.dropboxusercontent.com/u/12683052/GRIPCmap.zip
Irrigation Records	10-day	site	January 2018 to December 2018	

PRE: precipitation.

3.2. Methods

3.2.1. Established SMAP Training Samples for Winter Wheat and Rainfed Crops

The selection of training samples is important before establishing a model of irrigation signal detection. Since the SMAP data have a low spatial resolution, the training samples should be selected to ensure that the surrounding crops are consistent. In this paper, samples were selected using a combination of MODIS NDVI and MODIS ET. Since the winter wheat (WW) NDVI in the NCP was significantly higher than other crop in March, the spatial distribution of WW can be extracted based on the March NDVI data. However, this spatial distribution may include other vegetation with a higher NDVI (such as landscape forest), and ET is needed to improve the extraction accuracy of WW. Since March to April is the main irrigation period for WW, the cumulative ET value of WW is significantly higher than that of other vegetation during this period [48]. This indicates that the extraction accuracy of WW can be improved by adding ET as a limiting condition. Using these two features (NDVI and ET), the WW pixels can be extracted more accurately. WW and rainfed crop pixels were extracted by the decision tree model in Figure 3.

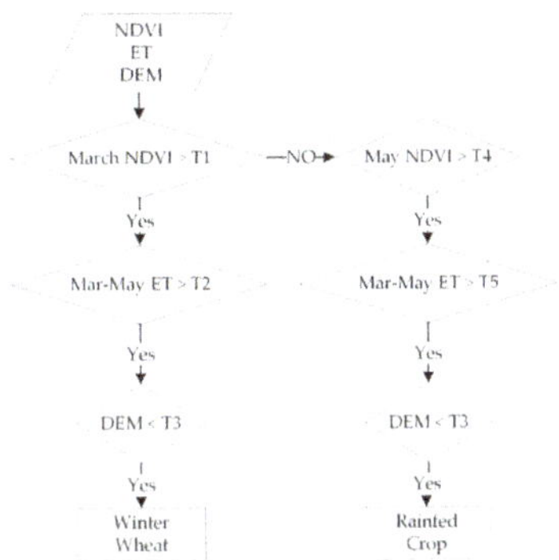

Figure 3. Winter wheat and rainfed crops planting area extraction model. Where March NDVI and Mar-May ET represent the NDVI in March (May NDVI is similar to March NDVI) and cumulative amount of ET from March to May, respectively; DEM is the elevation information; and T is the threshold in different conditions. If the pixel value (such as NDVI and ET) satisfies the threshold, the pixel value is 1, and if it is not satisfied, the pixel value is 0.

The selection of WW samples should be based on SM sites, and more irrigation information can be obtained. Rainfed crop samples should ensure that there are no irrigated crops nearby as much as possible, which can reduce the influence of surrounding crop irrigation on SM. Finally, 11 WW samples and 7 rainfed crop samples were established in the study area, 7 WW samples and 4 rainfed crop samples were used as training samples, and the remaining samples were used as validation samples. These samples are distributed from north to south and can reflect the difference in irrigation time of winter wheat under different latitude conditions.

3.2.2. Irrigation Information Detection and Irrigated Area Downscaling

Extracting the precipitation and SMAP time series data of the meteorological site spatial location can not only be used to evaluate the sensitivity of the SMAP data to the precipitation response but also to support the threshold setting of the irrigation signal detection. The irrigation signal detection is based on the SMAP SM variation. It can be assumed that if the SM of SMAP is increased and the grid has no significant precipitation, the increase in SMAP SM is caused by irrigation. Since the amplitude increase in the SMAP original SM signal is significant, it is difficult to detect irrigation by threshold segmentation and the original signal needs to be processed using the moving average method. In the original SMAP data, due to the existence of signal noise, the SM is may be suddenly reduced (previously without precipitation and irrigation), if this value is calculated with the SM at the latter time, the identified irrigation signal is invalid. SM Value in that time need to be corrected. To reduce the influence of SMAP SM data amplitude on the irrigation signal detection, a 5-point moving average method is used to process the SMAP SM original signal. The 5-point moving average not only ensures the amplitude of the original but also reduces the frequent fluctuations in the original signal. Sun et al. [49] compiled the water requirement for different growth stages of WW in the NCP. In this paper, the daily precipitation of >4 mm was used as the threshold for effective precipitation (referenced by Sun et al.). In this study, the irrigation identification results of the grid were binarized (irrigation is 1, no irrigation is 0).

By accumulating the binarized daily irrigation identification results, the frequency of irrigation in the WW planting region can be obtained. Notably, the irrigation frequency of a grid may be higher than 6 times because the grid (9 km × 9 km) cannot be completed irrigated in one day. After integrating the spatial distribution of the irrigation intensity and the WW planting area, the irrigated area with the irrigation intensity identification was finally obtained. However, the accuracy of the irrigated area recognition results based on a single SMAP data source does not meet the general application requirements. By introducing the previously extracted WW spatial distribution, the downscaled results of irrigation intensity were obtained from the SMAP irrigation intensity results without the influence of non-irrigation pixels (such as rainfed crops and city). The mathematical expression of the method in this section is as follows:

$$IS_{i,j} = SM_{i,j} > T_6 \; and \; Pre_{i,j} < EPre \tag{4}$$

$$II_{i,j} = \frac{\sum_1^t IS_{i,j}}{\max\left(\sum_1^t IS_{m,n}\right)} \tag{5}$$

$$II_{downscale} = \begin{cases} II_{i,j}, \; WW = 1 \\ 0, \; WW = 0 \end{cases} \tag{6}$$

where i and j represent the pixels of the ith row and jth column, respectively; IS is the irrigation signal; SM is the soil moisture derived from SMAP; T_6 is the threshold for soil moisture increase; Pre and $EPre$ represent precipitation and effective precipitation, respectively; II is the irrigation intensity; t is the total number of days in the study period; $\max\left(\sum_1^t IS_{m,n}\right)$ represents the maximum value of the accumulated value of the irrigation signal over the entire event range; and $II_{downscale}$ is the downscaled irrigation intensity. In equation 6, the WW spatial distribution and the irrigation intensity image need to be calculated. If the WW spatial distribution image pixel value is 1, the $II_{downscale}$ pixel value is assigned as the irrigation intensity value. The irrigated area is calculated as the area of the pixel where the irrigation intensity is greater than zero. The algorithm implementation in this section still needs to use the arcpy function based on python 2.7.

3.2.3. Validation and Consistency Analysis

The results of the irrigation signal detection have been validated, and the uncertainty in the irrigated area downscaling has also been analyzed. First, the detection results of the irrigation signal are based on the irrigation record. Since the SMAP SM time series data in this paper used the 5-point moving average method, if the detected WW irrigation signal is different from the irrigation record in three days, the result is correct. Simultaneously, if the non-WW planting area also detects the irrigation signal, it is necessary to reset the irrigation signal detection threshold according to the irrigation signal frequency. The equation for the validation of irrigation timing is as follows:

$$Accuracy = \left(\frac{CDet}{ADet + WRec}\right) * 100\% \tag{7}$$

$$OA = \text{avg}(Accuracy_1 + Accuracy_2 + \ldots + Accuracy_l) \tag{8}$$

where $Accuracy$ is the sample validation accuracy; OA is the overall accuracy and l indicates the total number of validation samples; $CDet$ indicates the number of days that were correctly detected in the irrigation record; $ADet$ represents the number of days for all irrigation detected results; and $WRec$ is the number of days that have not been detected in the irrigation record.

Second, when using the WW data extracted by MODIS to downscale the SMAP irrigation signal detection results, whether the growth of WW covered by one SMAP pixel is consistent must be considered. The selection strategy for the consistent analysis of WW growth is to establish samples in four corners and center points covered by one SMAP pixel as shown in Figure 4. The NDVI daily signal extracted from the samples was subjected to upper envelope processing [23], and the signal was divided according to the growth stage of WW and the change in SM. By counting the number of samples from the consistent growth of WW, the consistency analysis results of WW growth covered by one SMAP pixel were obtained. The consistency analysis results are calculated as follows:

$$P = \left(\frac{RG + J}{10}\right) * 100\% \tag{9}$$

where P is the percentage of growth consistency of WW; RG and J are the number of consistent samples of WW growth in the returning green and jointing stages, respectively; and 10 is the number of samples for all these two stages. Five growth consistency samples can be obtained for each growth stage (corresponding to the red sample point), and 10 consistency analysis samples can be obtained for the two stages of the returning green and jointing stages.

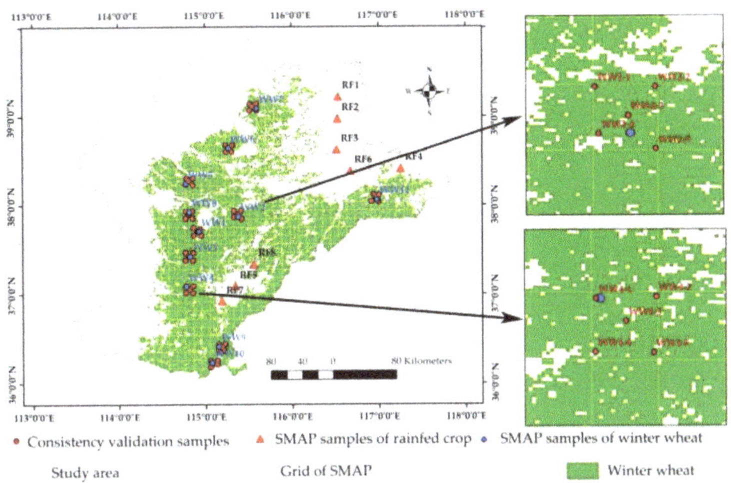

Figure 4. Sample maps. Red triangles and blue points are used to extract the SMAP SM time series signals from different crops; red points are used to extract the winter wheat NDVI time series signal and then compare the consistency of winter wheat growth covered by one SMAP pixel.

4. Results and Validation

4.1. Irrigation Signal Detection

Taking four meteorological stations as examples, the time series of NDVI (8-day maximum synthesis), ET (8-day), precipitation and SM from 2015 to 2017 were plotted in Figure 5. Comparing the time series data of the four meteorological stations, it was found that the vegetation coverage of the Baoding and Nangong stations were rainfed crops and those of the Botou and Raoyang stations were WW. An analysis of the time series changes of NDVI and ET showed that the meteorological stations with WW vegetation cover (Botou and Raoyang) not only had more NDVI peaks than rainfed crop stations (Baoding and Nangong) but also significantly higher ET from March to May. Time series changes of precipitation and SM provide an important basis for irrigation signal detection. During the main growth period of WW (March to May), Botou and Raoyang stations were affected by irrigation and still maintained high SM without precipitation. Simultaneously, the SM observed in the WW growing season was more stable and higher than that of the non-irrigated crops.

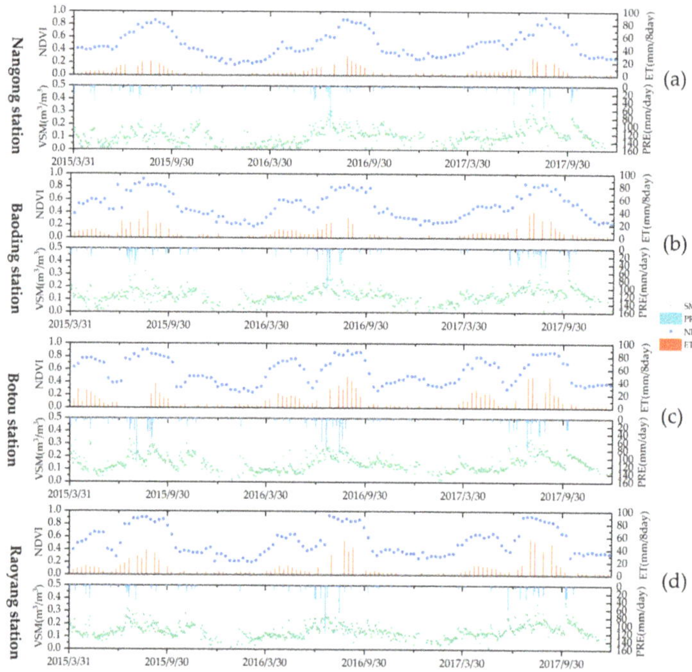

Figure 5. NDVI (8-day maximum synthesis), ET (8-day), precipitation (daily) and SM (daily) time series variations. (**a**) Nangong, (**b**) Baoding, (**c**) Botou and (**d**) Raoyang meteorological stations; and VSM means volume of soil moisture. The land cover at Nangong and Baoding stations was rainfed crops, and the land cover at Botou and Raoyang was winter wheat.

Using the 5-point moving average method for statistical time series SM results, which can reduce the influence of abnormal points on the irrigation signal detection. The smoothed SM results are shown in Figure 6. Figure 6a Changes in SM (blue lines) and effective precipitation events (green lines) in WW samples, and the statistical WW irrigation time is also plotted (Triangle point). Figure 6b Changes in SM and effective precipitation events for rainfed crops. The figure can reflect the response relationship between SM and precipitation, at the same time, by comparing the SM curves of different crops, it can be found that show the WW pixels have a more obvious SM increase than rainfed crop pixels. Comparing WW samples with rainfed crop samples, it was found that both had an increasing trend in SM before the first recorded irrigation. The slowly increasing trend in SM under no precipitation conditions may be caused by seasonal and vegetation water content changes [31]. However, the increasing trend in WW samples with different spatial locations was different before the first irrigation stage. Due to the difference in temperature, the irrigation time was different. The SM of the WW sample in the southern region increased significantly compared to the WW samples in the northern region (top line in Figure 6a is the southern region WW sample, and the bottom is the northern region). Both WW samples and rainfed crop samples have significant SM increase feedbacks under effective rainfall events. The difference is that irrigation events will also significantly increase SM without effective rainfall, which is shown in Figure 6. Setting the threshold for SM increase without an effective rainfall event can be used to detect irrigation signals in the WW region.

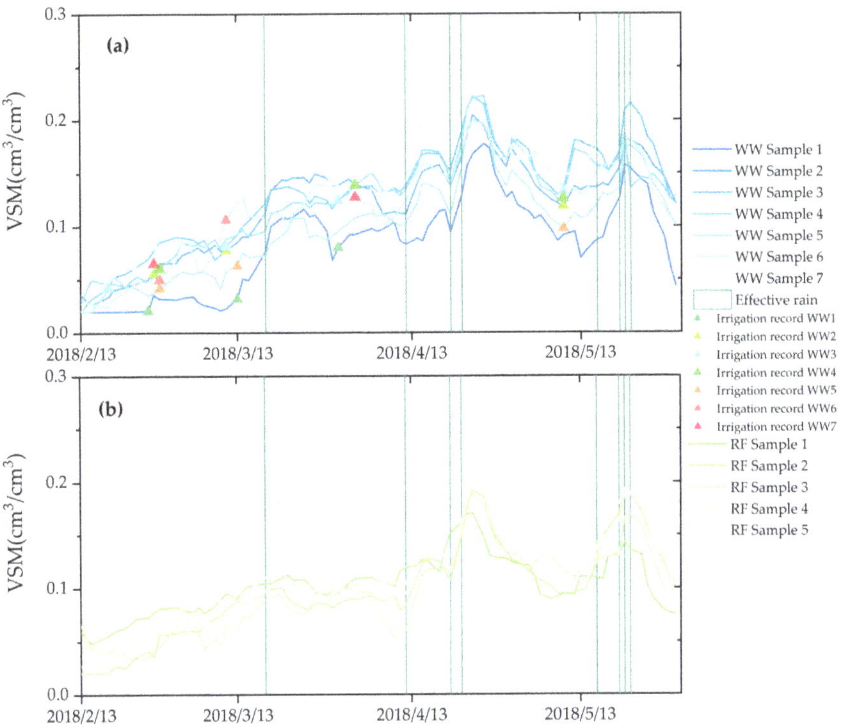

Figure 6. Training samples of irrigation signal detection. (**a**) Winter wheat training samples, and (**b**) rainfed crop training samples. The irrigation record is a summary of the irrigation records of the main irrigation region in the study area and used as a reference for the water supply time for winter wheat.

The irrigation signal detection results of WW and rainfed crops are shown in Figure 7a,b, respectively. By setting the SM change threshold, the time when the SM was significantly increased without effective precipitation is detected as the irrigation time (square point in Figure 7). In the rainfed crop region, only one irrigation signal was detected in this region due to the setting of the SM increase threshold. By comparing the SM trend of WW and rainfed crops, the SM trend in the WW region was more obvious, and there was also a significant increase (it is affected by irrigation) in SM when there was no precipitation. The SM trend in the rainfed crop region is more stable. Under the same precipitation conditions, the SM increase in the rainfed crop region is lower than that in the WW region. According to the results of WW irrigation signal detection, the irrigation frequency was higher from mid-February to mid-March. Due to the high frequency of precipitation in April and May, the irrigation frequency is lower than in February and March. Additionally, in the early WW growth stage (turning green and jointing), the main irrigation water source in the study area is surface water, and the amount of irrigation water will be more than that in the middle and late growth stages of WW. For different study areas, the setting of effective precipitation can be stricter, which may reduce the false detection of irrigation signals. Notably, the results of irrigation signal detection in this paper were large-scale surface water irrigation signals. Due to the small amount of irrigation water and the dispersion of irrigation areas, SMAP pixels do not easily reflect changes in SM amplitude caused by groundwater irrigation.

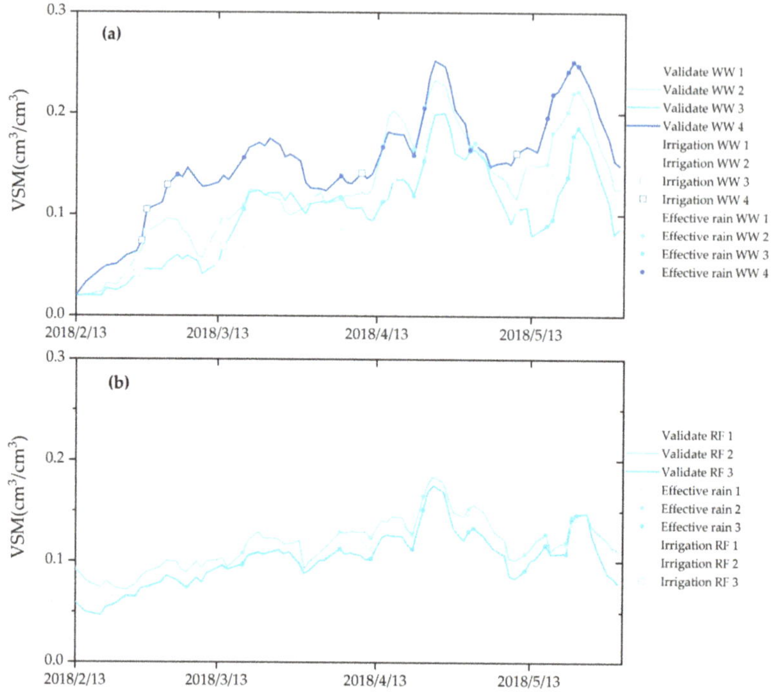

Figure 7. Irrigation signal detection results. (**a**) WW sample detection result and (**b**) RF samples detection result. The time corresponding to the square mark is the irrigation time, and the time corresponding to the circle mark is the effective rain time.

4.2. WW Extraction Results and Irrigated Area

In this paper, irrigation signal detection training samples must refer to both WW and rainfed crops. Figure 8a,b were obtained by daily NDVI using an 8-day maximum synthesis process, and Figure 8c was the cumulative ET from early March to early May. According to the crop growth phenology of the study area, only the WW crop in the study area showed obvious vegetation characteristics in March and early April. Therefore, most of the green areas in Figure 8a characterize the spatial distribution of WW. Since WW is already irrigated, the cumulative ET is significantly higher than that of other crops. Combined with the cumulative ET in Figure 8c, WW pixels with higher precision can be extracted. The vegetation characteristics of rainfed crop pixels appeared later than that of WW, and the cumulative ET was significantly lower than that of WW.

The normalized results of the cumulative irrigation detection signal are downscaled as shown in Figure 9a, wherein all blue areas indicate the spatial distribution of irrigated WW and blue shades indicate the intensity of irrigation. Downscaling normalization results eliminates the effects of non-irrigated pixels and directly expresses the spatial distribution of WW. Figure 9b,c are the results of the irrigated area provided by GIAM and GRIPC, respectively. The largest irrigated area is shown in Figure 9c because the data are classified into only four categories for agricultural areas, and the irrigation area cannot be effectively distinguished, whereas the irrigation area of the two crop rotations is shown in Figure 9b, which is close to the irrigation area identified in this paper. In recent years, due to the problem of overexploitation of groundwater in the NCP, many regions no longer plant high-water-consumption crops, such as WW, which results in Figure 9a irrigated areas being less than that of the GIAM data. Compared with the traditional irrigated area identification results, the proposed method can also reflect the irrigation intensity of the study area.

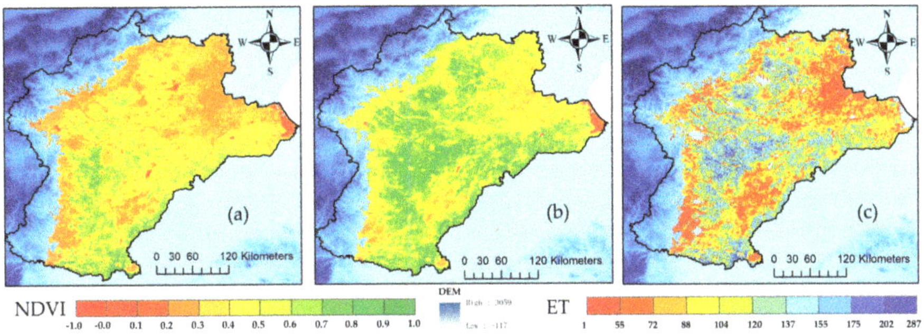

Figure 8. Sample selection based on MODIS NDVI and ET: (**a**) MODIS NDVI of DOY (day of year) 89-97, (**b**) MODIS NDVI of DOY 116-124, (**c**) MODIS ET accumulate from DOY 65-129.

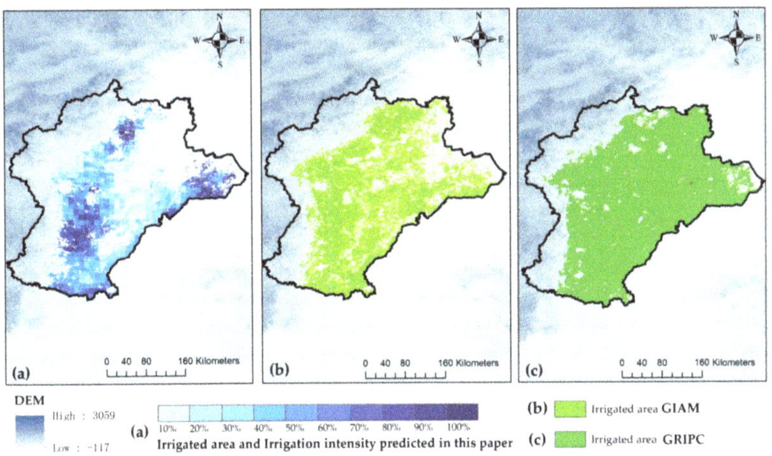

Figure 9. Irrigated area distribution in the study area. (**a**) shows the downscaled irrigated area and irrigation intensity results, (**b**) shows the irrigated area from GIAM, and (**c**) shows the irrigated area from GRIPC.

4.3. Validation and Growth Consistency Analysis

The detection results of irrigation signals in this paper will be validated from two aspects: 1) Validate the time of irrigation according to irrigation record; 2) count the consistent samples of WW growth and validate the effectiveness of the irrigation signal detection result.

In Table 2, the timing of the irrigation signal detection is compared to the timing of the irrigation records. The irrigation detection accuracy of the WW samples WW 1, WW 2, WW 3, and WW 4 used for validation were 50.00%, 100.00%, 75.00%, and 83.33%, respectively. It should be noted this irrigation record corresponds to two detection dates, with the irrigation record recorded for two days to calculate the single sample accuracy validation. Irrigation signals were also detected in the rainfed crop samples, which were added as errors to the calculation of the overall irrigation signal detection accuracy. The overall accuracy of the irrigation timing detection in this paper was 77.08%. The calculation of overall accuracy must consider the detection error of the rainfed crop region.

Since WW presents significant NDVI changes in the returning green and jointing stages and less precipitation during this period, little effect on WW growth is observed. Therefore, the returning green and jointing stages of WW are selected as the key period of growth consistency analysis. WW showed more significant growth consistency in the early stage of returning green and jointing than in other growing stages. The irrigation records show that the irrigation water used in the returning green and jointing stages is surface water, and the irrigation water in other growth stages is irrigated groundwater. Surface water irrigation is a unified supply for water resource management departments, and groundwater irrigation is privately accessible to farmers. Different irrigation times are the main reason for the inconsistency in WW growth. WW is irrigated by surface water during these two growth stages, and surface water irrigation can cover a wide range of WW regions. Irrigation caused an increase in the SMAP pixel value (SM), which was used to identify an irrigation pixel. Due to the low spatial resolution of SMAP data, the consistency of WW growth under the coverage of one SMAP pixel in this study area must be discussed. If most of the WW covered by one SMAP pixel shows a consistent increase in the NDVI, then the spatial distribution of WW is effective for downscaling the irrigation signal. Conversely, if the increase in the NDVI for most WW (covered by one SMAP pixel) is inconsistent, then the irrigation signal identified by the SMAP pixel cannot effectively express WW growth. In Figure 10, SM, NDVI changes (after upper envelop) and irrigation time for different SMAP samples were plotted. Figure 10a–d correspond to Validate WW1, Validate WW2, Validate WW3 and Validate WW4 in Figure 7a, respectively.

Table 2. Overall accuracy of the irrigation timing detection results.

	WW 1		WW 2		WW 3		WW 4		RF 1		RF 2		RF 3	
	Rec	Det	Rec	Det	Rec	Det	Rec	Det	Rec	Det	Rec	Det	Rec	Det
		2/26		2/23		2/26	2/26	2/26	/	3/3	/	/	/	/
	3/13	3/13	2/24	2/25	3/13	3/14		2/27	/	/	/	/	/	/
		3/14		2/27	4/15	4/16		3/3	/	/	/	/	/	/
Dates	3/26	3/27		3/12	5/10	5/10	3/14	3/14	/	/	/	/	/	/
		3/31	3/12	3/13			4/10	4/10	/	/	/	/	/	/
	4/10			3/14			5/10	5/10	/	/	/	/	/	/
			5/11	5/12					/	/	/	/	/	/
Accuracy	50.00%		100.00%		75.00%		83.33%							
Overall accuracy							77.08%							

Det: irrigation detection result. Rec: irrigation records. Units marked in green indicate that the detected irrigation date matches the recorded irrigation date, and units marked in orange indicate the detection irrigation date does not match the recorded irrigation date.

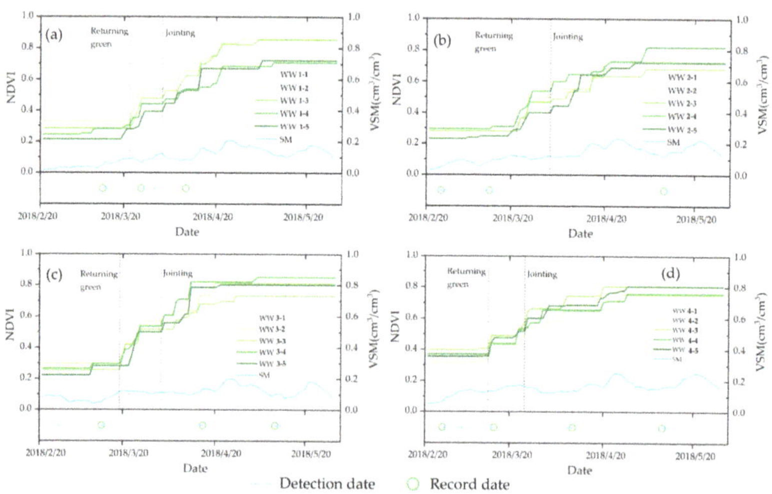

Figure 10. SM, NDVI changes (after upper envelop) and irrigation time for different SMAP samples.

According to the NDVI variation treatment method shown in Figure 10, 55 NDVI samples covered by 11 SMAP WW samples were validated for WW growth consistency. The number of samples with same increase trend of WW NDVI in the returning green and jointing stages was counted separately. For example, at the time of the returning green stage, the simultaneous increase in the NDVI indicates consistency among the WW growth samples, and vice versa. By counting the number of consistent WW samples covered by different SMAP pixels, the percentage of WW growth consistency covered by SMAP pixels can be calculated, and the results are shown in Table 3. In Table 3, the ratio of the consistent growth of WW covered by SMAP pixels is greater 70%, and in some regions, it can reach 100%. The overall consistency result reached 83%, and the results show that the irrigated area after downscaling can effectively express the true WW irrigation situation.

Table 3. Statistical results of the winter wheat sample consistencies.

	WW1	WW2	WW3	WW4	WW5	WW6	WW7	WW8	WW9	WW10	WW11
RG	3	5	4	2	3	5	5	4	3	5	4
J	4	4	4	5	5	5	5	4	4	3	5
P	70.00%	90.00%	80.00%	70.00%	80.00%	100.00%	100.00%	80.00%	70.00%	80.00%	90.00%
OA						82.72%					

RG: returning green stage; J: jointing stage; P: percentage; OA: overall accuracy.

5. Discussion

5.1. Comparison with Other Studies

Lawston et al. [33] proposed a method for detecting irrigation signals based on SMAP data. In this method, dates are first selected during the crop growing season and then the characteristics of SM are compared at irrigated and non-irrigated points. Finally, the method uses time integrated and SM normalized metrics of SM and precipitation to detect irrigation signals. According to the method, the precipitation and SM processing results are obtained, as shown in Figure 11a,b. Since this method does not deduct the effect of precipitation from the SM changes, in the southern part of the study area, sufficient precipitation affects the detection of irrigation signals. Simultaneously, the detection results of the proposed method are normalized, which is more conducive to the comparison of the two methods. In Figure 11, the amplitude change in (c) is more obvious than (d), and some obvious regions in the calculation results are marked. Region 1 contains two large reservoirs adjacent to the Taihang Mountains. Region 2 is the southern part of Beijing. Region 4 is a large wetland named Baiyangdian. The type of underlying surface may affect the monitoring of time series SM changes. Notably, region 3 is the main irrigation area in the southern part of the Hebei Province. However, the irrigation signal for this irrigated area is not significant in (d). Therefore, the method proposed in this paper is more suitable for irrigation signal detection in the study area.

The method proposed in this paper can acquire daily irrigation signal detection result, so the research can describe the irrigated information in the study area in more detail. In order to display the irrigation details more abundantly, the monthly irrigation signals were accumulated to acquire a monthly distribution of WW irrigation (as shown in Figure 12). At the end of February, the southern part of the study area warmed up, and the irrigated area of WW was mainly concentrated in the southern part. In March, a wide range of WW was irrigated, and irrigation in April and May was concentrated in the central and western regions. Compared with existing studies, Chen et al. statistically analyzed the climate distribution characteristics of WW growing season in the NCP for many years, which is consistent with the monthly spatial distribution of irrigation in this paper [50]. Yang et al. collected information on crop planting and irrigated area in the NCP for many years, and acquired crop and irrigation spatial distribution characteristics in this region [51]; the results of Yang's study are similar to the results acquired in this paper, but due to the change of crop pattern in the eastern region, inconsistencies have been caused. Overall, the results of this paper are consistent with existing research findings.

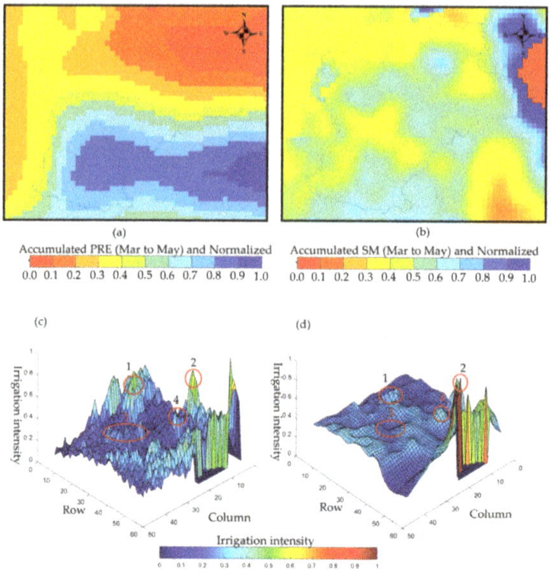

Figure 11. Comparison of the method proposed in this paper with the time-integrated and SM normalized irrigation signal detection methods. (**a**) Accumulated PRE and normalized result, (**b**) accumulated SM and normalized result, (**c**) irrigation intensity calculated by this paper proposed method, and (**d**) time-integrated and SM normalized irrigation signal detection methods. Both normalized results and irrigation intensity are dimensionless variables.

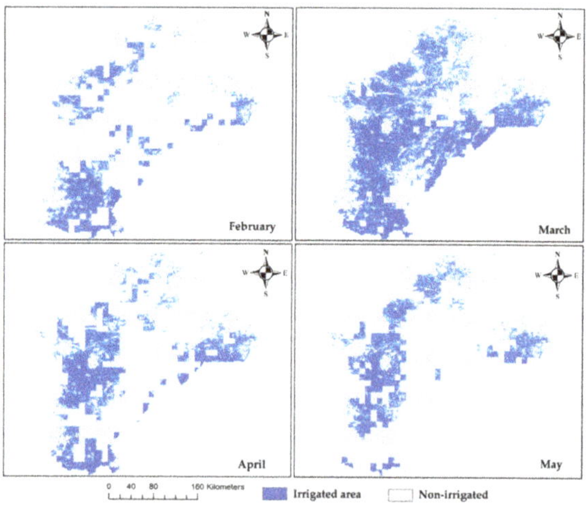

Figure 12. Spatial distribution of winter wheat irrigated area.

In the study of irrigated area extraction without considering SM changes, most of the research extraction methods are based on time series vegetation index changes and supervised classification to identify irrigated areas [13,34,41]. These methods for identifying irrigated areas through optical remote sensing datasets were based on identifying the type of crop to distinguish whether the area is irrigated [52]. The common advantage of these methods is that they can obtain a high resolution crop spatial distribution, and the accuracy can be increased as the spatial resolution of remote sensing images increases, and validated in many areas [16]. Based on the SMAP data extraction irrigation signal, the spatial distribution of crops with high spatial resolution is was introduced as irrigation reference area, which not only maintain irrigation time and frequency information, but also increases precision of SMAP data recognition irrigated area.

5.2. A Rational Discussion of the Irrigation Signal Detection Model

The establishment of the irrigation signal detection model in this paper was based on irrigation records and SMAP SM data. Since the SM change data in the irrigation record is measured every 10 days, the data does not express time-continuous SM variations, so the SMAP SM data is not compared with the in-situ data. However, validation of SMAP SM data and irrigation-induced SM increase researches can demonstrate that variations in SMAP SM data can be used to establish irrigation signal detection models. A validation study of SMAP SM data has been described in the first section, and this section will discuss the relationship between irrigation and SM variation.

Chen et al. [23] analyzed the continuous variation of SM before using the MODIS Greenness Index to detect irrigation signals in Gansu Province. Combined with irrigation and precipitation records, it was found that the sudden increase of SM generally originated from irrigation and effective precipitation. At the same time, the irrigation time was estimated using the continuous SM variation data in year 2016. Under the condition of no in-situ SM data, Lawston et al. [33] obtained the SMAP SM variation of different crop types based on the location information of irrigation and rainfed crops, and according this, they extracted the irrigated area of many regions in the United States. Is the phenomenon of SM sudden increase caused by irrigation also obvious in the NCP region? Some studies based on the effects of different irrigation patterns on WW yield provide a reliable basis. Wang et al. [53] collected SM variation in different irrigation patterns of winter wheat. The data show that although the SM (soil depth 0–80 cm) covered by WW in drip irrigation is slightly lower than level-basin, there is obvious SM increase after WW irrigation. Zia et al. [54] collected more detailed time series SM variation data (soil depth 10 cm and 40 cm); at the soil depth of 10 cm, irrigation will cause significant SM increase, while at 40cm, irrigation will maintain a higher level of SM, and the sudden increase is not significant. In this study, when the in-situ SM data is insufficient, the SMAP SM data can be used to analyze the SM variation characteristics of WW and rainfed crops. Referring to number of studies on the relationship between irrigation and SM response, this paper suggests that irrigation records and SM increase can be used to detect irrigation signals in agricultural areas.

It should be noted that the thresholds in the irrigation signal detection model proposed in this paper are not universal. For example, in the study area of this paper, there are significant differences in SM increase caused by different irrigation patterns. In areas with more complicated irrigation patterns, the irrigation pattern of sample points needs to be considered. In addition, the SMAP SM data of 9 km resolution is acquired by 36 km data downscaling, and the uncertainty of scale conversion may also affect the application of the model. If necessary, consider using multiple filtering methods for data optimization.

6. Conclusions

Based on multisource remote sensing data, including SMAP, MODIS, and an irrigation map, the 5-point moving average method was used to detect irrigation signals in southern Hebei. Then, irrigation record data were used to validate the accuracy of the irrigation signal detection results. The accuracies of the four WW samples used for validation are 50.00%, 100.00%, 75.00%, and 83.33%, and the overall accuracy is 77.08%. The consistency analysis of 55 WW growth samples showed that the growth consistency of WW reached 82.72% in two large-scale surface water irrigation areas. Based on a consistency analysis, the downscaling method can be used to downscale the WW irrigation signal detected by the SMAP data. The proposed irrigation signal detection and downscaling method are more suitable for the detection of large-scale surface water irrigation signals. Limited by the spatial resolution of SMAP data and continuous in situ measured SM data, small-scale groundwater irrigation signal detection is difficult to establish. In future research, small-scale groundwater irrigation signal detection will be further studied.

Author Contributions: Z.H. and H.Z. designed the experiment; C.Z. and Y.J. provided revisions to the paper; H.W. and H.Z. provided financial support; and Z.H. wrote the manuscript.

Funding: National Key R&D Program of China (Grant No. 2018YFC0407705), Fundamental Research Funds for the China Institute of Water Resources and Hydropower Research (WR0145B012017), and Fundamental Research Funds for the China Institute of Water Resources and Hydropower Research (WR0145B272016).

Conflicts of Interest: The authors declare no conflict of interest.

References

1. Li, J.; Inanaga, S.; Li, Z.; Eneji, A.E. Optimizing irrigation scheduling for winter wheat in the North China Plain. *Agric. Water Manag.* **2005**, *76*, 8–23. [CrossRef]
2. Changming, L.; Jingjie, Y.; Kendy, E. Groundwater exploitation and its impact on the environment in the North China Plain. *Water Int.* **2001**, *26*, 265–271. [CrossRef]
3. Min, Z.; Jianbin, D.; Houze, X.U.; Peng, P.; Haoming, Y.A.N.; Yaozhong, Z.H.U. Trend of China land water storage redistribution at medi- and large-spatial scales in recent five years by satellite gravity observations. *Chin. Sci. Bull.* **2009**, *54*, 816–821. [CrossRef]
4. Feng, W.; Shum, C.K.; Zhong, M.; Pan, Y. Groundwater storage changes in China from satellite gravity: An overview. *Remote Sens.* **2018**, *10*, 674. [CrossRef]
5. Xu, C.; Tao, H.; Tian, B.; Gao, Y.; Ren, J.; Wang, P. Field Crops Research Limited-irrigation improves water use efficiency and soil reservoir capacity through regulating root and canopy growth of winter wheat. *Field Crop Res.* **2016**, *196*, 268–275. [CrossRef]
6. Wang, X.; Li, X. Irrigation Water Availability and Winter Wheat Abandonment in the North China Plain (NCP): Findings from a Case Study in Cangxian County of Hebei Province. *Sustainability* **2018**, *10*, 354. [CrossRef]
7. Ali, M.H.; Hoque, M.R.; Hassan, A.A.; Khair, A. Effects of deficit irrigation on yield, water productivity, and economic returns of wheat. *Agric. Water Manag.* **2007**, *92*, 151–161. [CrossRef]
8. Burney, J.; Woltering, L.; Burke, M.; Naylor, R.; Pasternak, D. Solar-powered drip irrigation enhances food security in the Sudano—Sahel. *Proc. Natl. Acad. Sci. USA* **2009**, *10*, 1848–1853. [CrossRef]
9. Phogat, V.; Skewes, M.A.; Cox, J.W.; Sanderson, G.; Alam, J.; Šimůnek, J. Seasonal simulation of water, salinity and nitrate dynamics under drip irrigated mandarin (Citrus reticulata) and assessing management options for drainage and nitrate leaching. *J. Hydrol.* **2014**, *513*, 504–516. [CrossRef]
10. Egea, G.; Diaz-espejo, A.; Fernández, J.E. Soil moisture dynamics in a hedgerow olive orchard under well-watered and deficit irrigation regimes: Assessment, prediction and scenario analysis. *Agric. Water Manag.* **2016**, *164*, 197–211. [CrossRef]
11. Zhang, R.; Kim, S.; Sharma, A. Remote Sensing of Environment A comprehensive validation of the SMAP Enhanced Level-3 Soil Moisture product using ground measurements over varied climates and landscapes. *Remote Sens. Environ.* **2019**, *223*, 82–94. [CrossRef]
12. Yi, Z.; Zhao, H.; Jiang, Y. Continuous Daily Evapotranspiration Estimation at the Field-Scale over Heterogeneous Agricultural Areas by Fusing ASTER and MODIS Data. *Remote Sens.* **2018**, *10*, 1694. [CrossRef]
13. Yang, Y.; Tao, B.; Ren, W.; Zourarakis, D.P.; Masri, B.E.; Sun, Z.; Tian, Q. An Improved Approach Considering Intraclass Variability for Mapping Winter Wheat Using Multitemporal MODIS EVI Images. *Remote Sens.* **2019**, *11*, 1191. [CrossRef]
14. Autovino, D.; Minacapilli, M.; Provenzano, G. Modelling bulk surface resistance by MODIS data and assessment of MOD16A2 evapotranspiration product in an irrigation district of Southern Italy. *Agric. Water Manag.* **2016**, *167*, 86–94. [CrossRef]
15. Hao, Z.; Zhao, H.; Zhang, C.; Zhou, H.; Zhao, H.; Wang, H. Correlation Analysis Between Groundwater Decline Trend and Human-Induced Factors in Bashang Region. *Water* **2019**, *11*, 473. [CrossRef]

16. Ozdogan, M.; Gutman, G. Remote Sensing of Environment A new methodology to map irrigated areas using multi-temporal MODIS and ancillary data: An application example in the continental US. *Remote Sens. Environ.* **2008**, *112*, 3520–3537. [CrossRef]
17. Salmon, J.M.; Friedl, M.A.; Frolking, S.; Wisser, D.; Douglas, E.M. International Journal of Applied Earth Observation and Geoinformation Global rain-fed, irrigated, and paddy croplands: A new high resolution map derived from remote sensing, crop inventories and climate data. *Int. J. Appl. Earth Obs. Geoinf.* **2015**, *38*, 321–334. [CrossRef]
18. Brown, J.F.; Pervez, S. Merging remote sensing data and national agricultural statistics to model change in irrigated agriculture. *Agric. Syst.* **2014**, *127*, 28–40. [CrossRef]
19. Gumma, M.K. Mapping Irrigated Areas of Ghana Using Fusion of 30 m and 250 m Resolution Remote-Sensing Data. *Remote Sens.* **2011**, *3*, 816–835. [CrossRef]
20. Pervez, S.; Budde, M.; Rowland, J. Remote Sensing of Environment Mapping irrigated areas in Afghanistan over the past decade using MODIS NDVI. *Remote Sens. Environ.* **2014**, *149*, 155–165. [CrossRef]
21. Ambika, A.K.; Wardlow, B.; Mishra, V. Data Descriptor: Remotely sensed high resolution irrigated area mapping in India for 2000 to 2015. *Nature* **2016**. [CrossRef]
22. Pervez, S.; Brown, J.F. Data and National Agricultural Statistics. *Remote Sens.* **2010**, *2*, 2388–2412. [CrossRef]
23. Chen, Y.; Lu, D.; Luo, L.; Pokhrel, Y.; Deb, K.; Huang, J.; Ran, Y. Detecting irrigation extent, frequency, and timing in a heterogeneous arid agricultural region using MODIS time series, Landsat imagery, and ancillary data. *Remote Sens. Environ.* **2018**, *204*, 197–211. [CrossRef]
24. Xiao, X.; Boles, S.; Liu, J.; Zhuang, D.; Frolking, S.; Li, C.; Salas, W.; Moore, B. Mapping paddy rice agriculture in southern China using multi-temporal MODIS images. *Remote Sens. Environ.* **2005**, *95*, 480–492. [CrossRef]
25. Peng, D.; Huete, A.R.; Huang, J.; Wang, F.; Sun, H. International Journal of Applied Earth Observation and Geoinformation Detection and estimation of mixed paddy rice cropping patterns with MODIS data. *Int. J. Appl. Earth Obs. Geoinf.* **2011**, *13*, 13–23. [CrossRef]
26. Abuzar, M.; Mcallister, A.; Whitfield, D. Mapping Irrigated Farmlands Using Vegetation and Thermal Thresholds Derived from Landsat and ASTER Data in an Irrigation District of Australia. *Photogramm. Eng. Remote Sens.* **2015**, *81*, 229–238. [CrossRef]
27. Liu, Y.Y.; Dorigo, W.A.; Parinussa, R.M.; de Jeu, R.A.; Wagner, W.; Mccabe, M.F.; Evans, J.P.; Dijk, A.I.J.M. Van Remote Sensing of Environment Trend-preserving blending of passive and active microwave soil moisture retrievals. *Remote Sens. Environ.* **2012**, *123*, 280–297. [CrossRef]
28. Hutchinson, J.M.S. Estimating Near-Surface Soil Moisture using Active Microwave Satellite Imagery and Optical Sensor Inputs. *Trans. ASAE* **2003**, *46*, 225–236. [CrossRef]
29. Pellarin, T.; Laurent, J. Soil moisture mapping over West Africa Soil moisture mapping over West Africa with a 30-min temporal resolution using AMSR-E observations and a satellite-based rainfall product. *Hydrol. Earth Syst. Sci.* **2010**, *13*, 1887–1896. [CrossRef]
30. Colliander, A.; Jackson, T.J.; Bindlish, R.; Chan, S.; Das, N.; Kim, S.B.; Cosh, M.H.; Dunbar, R.S.; Dang, L.; Pashaian, L.; et al. Validation of SMAP surface soil moisture products with core validation sites. *Remote Sens. Environ.* **2017**, *191*, 215–231. [CrossRef]
31. Chan, S.K.; Member, S.; Bindlish, R.; Neill, P.E.O.; Njoku, E.; Jackson, T.; Colliander, A.; Member, S.; Chen, F.; Burgin, M.; et al. Assessment of the SMAP Passive Soil Moisture Product. *IEEE Trans. Geosci. Remote Sens.* **2016**, *54*, 4994–5007. [CrossRef]
32. Kumar, S.V.; Dirmeyer, P.A.; Peters-lidard, C.D.; Bindlish, R.; Bolten, J. Remote Sensing of Environment Information theoretic evaluation of satellite soil moisture retrievals. *Remote Sens. Environ.* **2017**, *204*, 392–400. [CrossRef]
33. Lawston, P.M.; Joseph, A.; Santanello, J.A., Jr.; Sujay, V.K. Irrigation Signals Detected from SMAP Soil Moisture Retrievals. *Geophys. Res. Lett.* **2017**, *44*, 860–867. [CrossRef]
34. Thenkabail, P.S.; Biradar, C.M. International Journal of Remote Global irrigated area map (GIAM), derived from remote sensing, for the end of the last millennium. *Int. J. Remote Sens.* **2009**, *30*, 3679–3733. [CrossRef]
35. Wu, D.; Fang, S.; Li, X.; He, D.; Zhu, Y.; Yang, Z.; Xu, J. Spatial-temporal variation in irrigation water requirement for the winter wheat-summer maize rotation system since the 1980s on the North China Plain. *Agric. Water Manag.* **2019**, *214*, 78–86. [CrossRef]

36. Zhao, Z.; Qin, X.; Wang, Z.; Wang, E. Agricultural and Forest Meteorology Performance of di ff erent cropping systems across precipitation gradient in North China Plain. *Agric. For. Meteorol.* **2018**, *259*, 162–172. [CrossRef]
37. Zou, J.; Zhan, C.; Xie, Z.; Qin, P.; Jiang, S. Climatic impacts of the Middle Route of the South-to-North Water Transfer Project over the Haihe River basin in North China simulated by a regional climate model Jing. *J. Geophys. Res. Atmos.* **2016**, *121*, 8983–8999. [CrossRef]
38. Entekhabi, D.; Yueh, S.; O'Neill, P.E.; Kellogg, K.H.; Allen, A.; Bindlish, R.; Brown, M.; Chan, S.; Colliander, A.; Crow, W.T. *SMAP Handbook–Soil Moisture Active Passive: Mapping Soil Moisture and Freeze/Thaw from Space*; JPL Publication: Pasadena, CA, USA, 2014.
39. Pan, M.; Entekhabi, D.; Lu, H.; Akbar, R.; Peng, B.; McColl, K.A.; Short Gianotti, D.J.; Wang, W. Global characterization of surface soil moisture drydowns. *Geophys. Res. Lett.* **2017**, *44*, 3682–3690. [CrossRef]
40. O'Neill, P.E.; Chan, S.; Njoku, E.G.; Jackson, T.; Bindlish, R. *SMAP Enhanced L3 Radiometer Global Daily 9 Km EASE-Grid Soil Moisture, Version 1*; NASA National Snow and Ice Data Center Distributed Active Archive Center: Boulder, CO, USA, 2016.
41. Thenkabail, P.S.; Biradar, C.M.; Noojipady, P.; Dheeravath, V.; Li, Y.J.; Velpuri, M.; Reddy, G.P.O.; Cai, X.L.; Gumma, M.; Turral, H. *A Global Irrigated Area Map (GIAM) Using Remote Sensing at The End of the Last Millennium*; International Water Management Institute Colombo: Battaramulla, Sri Lanka, 2008; ISBN ISBN 9290906464.
42. Ramankutty, N.; Huang, X.; Sulla-Menashe, D.; Friedl, M.A.; Sibley, A.; Tan, B.; Schneider, A. MODIS Collection 5 global land cover: Algorithm refinements and characterization of new datasets. *Remote Sens. Environ.* **2009**, *114*, 168–182. [CrossRef]
43. Running, S.; Mu, Q.; Zhao, M. *MOD16A2 MODIS/Terra Net Evapotranspiration 8-Day L4 Global 500m SIN Grid V006*; NASA EOSDIS L. Process. DAAC: Sioux Falls, SD, USA, 2017.
44. Vermote, E.; Wolfe, R. *MOD09GA MODIS/Terra Surface Reflectance Daily L2G Global 1 km and 500 m SIN Grid V006*; NASA EOSDIS L. Process. DAAC: Sioux Falls, SD, USA, 2015.
45. Van Leeuwen, W.J.; Huete, A.R.; Laing, T.W. MODIS Vegetation Index Compositing Approach: A Prototype with AVHRR Data. *Remote Sens. Environ.* **1999**, *280*, 264–280. [CrossRef]
46. Dwyer, J.; Schmidt, G. The MODIS Reprojection Tool. In *Earth Science Satellite Remote Sensing*; Springer: Berlin, Germany, 2006.
47. Zhong, Y.; Zhong, M.; Feng, W.; Wu, D. Groundwater Depletion in the West Liaohe River Basin, China and Its Implications Revealed by GRACE and in Situ Measurements. *Remote Sens.* **2018**, *10*, 493. [CrossRef]
48. Zhang, X.; Chen, S.; Sun, H.; Shao, L.; Wang, Y. Changes in evapotranspiration over irrigated winter wheat and maize in North China Plain over three decades. *Agric. Water Manag.* **2011**, *98*, 1097–1104. [CrossRef]
49. Sun, H.; Shen, Y.; Yu, Q.; Flerchinger, G.N.; Zhang, Y.; Liu, C.; Zhang, X. Effect of precipitation change on water balance and WUE of the winter wheat-summer maize rotation in the North China Plain. *Agric. Water Manag.* **2010**, *97*, 1139–1145. [CrossRef]
50. Chen, C.; Baethgen, W.E.; Wang, E.; Yu, Q. Characterizing spatial and temporal variability of crop yield caused by climate and irrigation in the North China Plain. *Theor. Appl. Climatol.* **2011**, *106*, 365–381. [CrossRef]
51. Yang, Y.; Yang, Y.; Moiwo, J.P.; Hu, Y. Estimation of irrigation requirement for sustainable water resources reallocation in North China. *Agric. Water Manag.* **2010**, *97*, 1711–1721. [CrossRef]
52. Velpuri, N.M.; Thenkabail, P.S.; Gumma, M.K.; Biradar, C.; Dheeravath, V.; Noojipady, P.; Yuanjie, L. Influence of resolution in irrigated area mapping and area estimation. *Photogramm. Eng. Remote Sens.* **2009**, *75*, 1383–1395. [CrossRef]
53. Wang, J.; Gong, S.; Xu, D.; Yu, Y.; Zhao, Y. Impact of drip and level-basin irrigation on growth and yield of winter wheat in the North China Plain. *Irrig. Sci.* **2013**, *31*, 1025–1037. [CrossRef]
54. Zia, S.; Wenyong, D.; Spreer, W.; Spohrer, K.; Xiongkui, H.; Müller, J. Assessing crop water stress of winter wheat by thermography under different irrigation regimes in North China Plain. *Int. J. Agric. Biol. Eng.* **2012**, *5*, 24–34. [CrossRef]

remote sensing

MDPI

Article

Quantifying the Evapotranspiration Rate and Its Cooling Effects of Urban Hedges Based on Three-Temperature Model and Infrared Remote Sensing

Zhendong Zou, Yajun Yang and Guo Yu Qiu *

Lab of Environmental and Energy Information Engineering, School of Environment and Energy, Peking University Shenzhen Graduate School, Shenzhen 518055, China; zouzd09@gmail.com (Z.Z.); yangyajun1112@126.com (Y.Y.)

* Correspondence: qiugy@pkusz.edu.cn

Received: 19 December 2018; Accepted: 18 January 2019; Published: 21 January 2019

Abstract: The evapotranspiration (ET) of urban hedges has been assumed to be an important component of the urban water budget and energy balance for years. However, because it is difficult to quantify the ET rate of urban hedges through conventional evapotranspiration methods, the ET rate, characteristics, and the cooling effects of urban hedges remain unclear. This study aims to measure the ET rate and quantify the cooling effects of urban hedges using the 'three-temperature model + infrared remote sensing (3T + IR)', a fetch-free and high-spatiotemporal-resolution method. An herb hedge and a shrub hedge were used as field experimental sites in Shenzhen, a subtropical megacity. After verification, the '3T + IR' technique was proven to be a reasonable method for measuring the ET of urban hedges. The results are as follows. (1) The ET rate of urban hedges was very high. The daily average rates of the herb and shrub hedges were 0.38 mm·h^{-1} and 0.33 mm·h^{-1}, respectively, on the hot summer day. (2) Urban hedges had a strong ability to reduce the air temperature. The two hedges could consume 68.44% and 60.81% of the net radiation through latent heat of ET on the summer day, while their cooling rates on air temperature were 1.29 °C min^{-1} m^{-2} and 1.13 °C min^{-1} m^{-2}, respectively. (3) Hedges could also significantly cool the urban underlying surface. On the summer day, the surface temperatures of the two hedges were 19 °C lower than that of the asphalt pavement. (4) Urban hedges had markedly higher ET rates (0.19 mm·h^{-1} in the summer day) and cooling abilities (0.66 °C min^{-1} m^{-2} for air and 9.14 °C for underlying surface, respectively) than the lawn used for comparison. To the best of our knowledge, this is the first research to quantitatively measure the ET rate of urban hedges, and our findings provide new insight in understanding the process of ET in urban hedges. This work may also aid in understanding the ET of urban vegetation.

Keywords: three-temperature model; infrared remote sensing; urban hedges; evapotranspiration; cooling effects

1. Introduction

Due to rapid urbanization, the urban thermal environment has worsened, and urban heat islands (UHI) have become a common problem in most cities around the world [1,2]. From 1961 to 2000, air temperature has increased 0.16 °C per decade in large cities in northern China [3]. Among 419 large cities around the world, the average annual daytime surface urban heat island is 1.500 ± 1.200 °C [4]. High temperatures in urban areas not only lead to more energy consumption for cooling [5] but also affect human health [6–8]. High temperatures and heat waves could even increase the mortality rate. In 27 European countries, over 28,000 people die every year due to exposure to extreme heat, which accounts for 0.61% of all deaths in these regions [9]. Therefore, studying how to efficiently mitigate urban thermal issues is essential for adaptive strategy under climate change and rapid urbanization.

In recent decades, various methods including changes to underlying surface materials, optimizing urban planning and designing, and the addition of vegetation have been proposed to mitigate UHI [10–12]. Among them, vegetation is considered one of the most effective mitigating methods [13,14]. Many studies have been conducted on the cooling effects of urban vegetation. Urban parks, urban forests, urban lawns, and green roofs can provide different degrees of cooling [15–18]. Research has shown that just a single tree could save 12–24% of cooling energy for a single-story building [19]. Sixty-three large *Eucalyptus camaldulensis* per hectare could reduce air temperature by 1 °C in Mexico City, while 24 large *Liquidambar styraciflua* trees could even reduce the air temperature by 2 °C [20]. A 147-hm^2 park in Nagoya was also found to reduce the air temperature by 1.9 °C on hot days [21]. These studies showed that vegetation area, vegetation shapes and vegetation compositions could affect the microclimate [22–25], and the cooling effects of vegetation could be attributed to its shading, reflection, and evapotranspiration [26,27]. However, most of these studies focused on the cooling effects under different green space ratios and did not quantitatively estimate their ET rates and energy budget. Therefore, quantitative evaluation of the cooling effects is still a challenge.

Although ET is believed to be the most robust cooling mechanism, as it can consume large amounts of latent heat [12], observing the ET characteristics of urban vegetation is especially difficult [28,29]. As the vegetation is segmented by various artificial underlying surfaces in urban settings, it is difficult to meet the fetch requirements of traditional methods such as the Bowen ratio, eddy covariance and large aperture scintillometers [30]. ET could be estimated on a large scale by satellite remote sensing [31], but its resolution is usually too sparse on the street or neighborhoods scale. Moreover, only one image of an area could be obtained over several days. In contrast, sap flow and lysimeter data can only directly measure individual or small groups of plants [32,33]. Therefore, a fetch-free, high-spatiotemporal-resolution ET estimation method is needed to obtain accurate ET characteristics of urban vegetation.

The three-temperature model was proposed and developed to estimate ET via three temperature data points, net radiation and ground heat flux [34,35]. The surface temperature data could be obtained using thermal infrared images, and the meteorological data are easily available. It has been applied in studies on different scales, including the large catchment scale, the field scale and even on a single plant in growth chamber. It has been validated by the Penman–Monteith method, weighing lysimeter, Bowen ratio, eddy covariance, and water budget methods [36–41]. It has also been used to estimate ET of different vegetation types, such as crops, grass, and shrubs [38,42,43]. In urban area, it was used to estimate a small urban lawn's ET and showed great consistency with the Bowen ratio method [30]. These results indicate potential applications for the proposed method to estimate ET of different urban vegetation.

Hedges are narrow bands of woody vegetation and associated organisms that separate fields and are generally composed of low dense vegetation including short woody plants, shrubs, and grasses [44,45]. It is a typical vegetation type in urban areas. During the past several years, attention has been paid to urban hedges because of their ecological functions, such as air quality purification, creation of animal habitat, and more [46,47]. However, their ET characteristics and regulation of the urban microclimate are neglected. Therefore, in this study we aim to (1) investigate the ET characteristics of two common urban hedges using the '3T + IR' method in a subtropical megacity, Shenzhen, and then (2) quantify the cooling effects of the urban hedges and quantify the function of ET. This study could provide a new fetch-free and high-spatiotemporal-resolution method for estimating urban ET. It may contribute to understanding the ET of urban hedges, and then the species selection and landscape design in urban planning for urban heat island mitigation.

2. Material and Methods

2.1. Study Site

The experiment was conducted in Xili University Town, Shenzhen (approximately 22°35′40″N, 113°58′20″E, 17 m above the sea). Shenzhen is a typical costal city in southern China and has a subtropical marine climate affected strongly by the south Asian tropical monsoon. Its mean annual temperature is 22.3 °C. January is the coldest month of the year, while July is the hottest. Its mean annual precipitation is 1924.7 mm, which mainly occurs from May to September. The mean annual sunshine duration is 2060 h, and the solar radiation is as high as 5225 MJ/m^2.

Our study site located in an open square primarily covered by urban hedges (Figure 1). The two hedges consisted of an herb, *Hymenocallis littoralis* (0.4 m high), and a shrub, *Ligustrum quihoui* (0.5 m high); both of these greening species are common in Shenzhen. The area of the *H. lottoralis* hedge and the *L. quihoui* hedge were both about 40 m^2. A *Zoysia matrella* lawn was located nearby and used for comparison with the hedges. The lawn's area was about 2000 m^2.

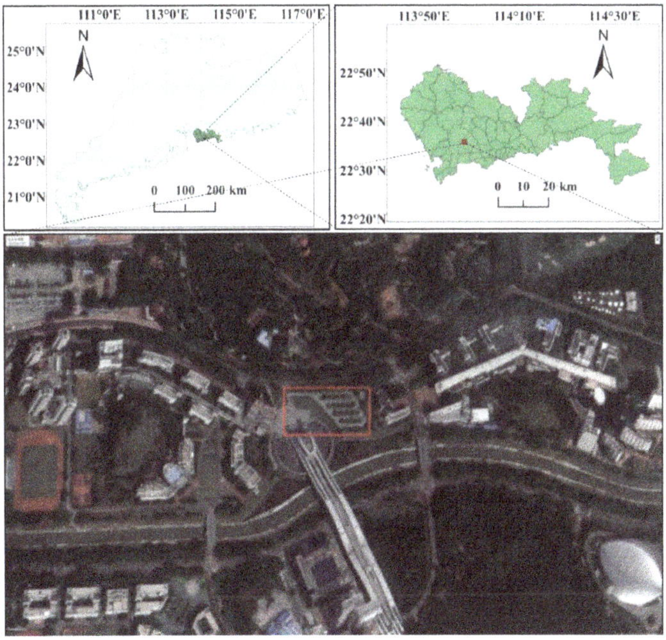

Figure 1. Location of the study area. The upper left figure is the location of Shenzhen City. The upper right figure shows the location of the study area in Shenzhen. The bottom figure is a photo of the studied area by google earth. The studied hedges and lawn are in the red box.

2.2. Three-Temperature Model

The three-temperature model estimates vegetation ET by introducing a reference leaf with no ET [34,48].

$$LE = R_n - R_{np}\frac{T_c - T_a}{T_p - T_a} \quad (1)$$

where LE is the latent heat consumed by vegetation ET. R_n and R_{np} are the net radiation on the vegetation and reference leaf, respectively (W m^{-2}). T_c and T_p are the surface temperature of the vegetation and reference leaf (°C). The surface temperature could be obtained by thermal images, and the maximum T_c in the image is regarded as T_p [37,38]. T_a is air temperature (°C). R_n and R_{np} could be estimated according to [49].

$$R_n = (1 - \alpha_c)R_s + \Delta R_l \quad (2)$$

where R_s is solar radiation (W m^{-2}). α_c is the albedo of the vegetation canopy. To simplify the calculation, the empirical coefficient α_c = 0.22 was used in this study [37]. ΔR_l is the net long-wave radiation (W m^{-2}), which could be estimated by [50,51]

$$\Delta R_l = \left(0.4 + 0.6\frac{R_s}{R_{so}}\right)\left(\varepsilon_a \sigma T_a^4 - \varepsilon_c \sigma T_c^4\right) \quad (3)$$

where R_{so} is the clear day solar radiation (W m^{-2}), which is assumed to equal to R_s in this study as all the experiment were conducted in clear sunny days [41]. ε_c the canopy emissivity, and empirical coefficient ε_c = 0.98 was used here [37]. σ is the Stefan–Boltzman constant (5.67 × 10^{-8} W m^{-2} K^{-4}). ε_a is the atmospheric emissivity and could be estimated according to [52]

$$\varepsilon_a = 0.92 \times 10^{-5} T_a^2 \quad (4)$$

If α_c, ε_c, and T_c are replaced by α_{cp}, ε_{cp} and T_{cp} in Equations (2) and (3), then R_{np} could be estimated. As we use the leaf with the highest surface temperature in the canopy as the reference leaf in this study, α_c, ε_c are assumed to be same to α_{cp}, ε_{cp}.

The analysis procedures were written into a software named "A system to estimate evapotranspiration by infrared remote sensing and the three-temperature model", which can be downloaded and used freely from https://pan.baidu.com/s/19iuz5PIVjZOR96iVObYBqA.

2.3. Field Experiments

The field experiment was carried out over four typical sunny days in four seasons from 2015 to 2016, from 8:00 a.m. to 5:00 p.m. An infrared thermal imager (Fluke Ti55FT, Fluke Corp., Everett, WA, USA) was used to record the surface temperatures vertically down, at a height of 1.5 m. The measuring wavelength of the infrared thermal imager was 8–14 μm, and its resolution was 0.05 °C. The emissivity of the hedges and lawn in our study was set up to be 0.98 according to empirical value [37]. The imager could give out the emissivity-corrected temperature directly. Each thermal infrared image contains 76,800 temperature data points (320 × 240). Three images were taken of each plant at each hour. Before the measurement, the thermal camera was calibrated against a blackbody measurement.

Air temperature and other meteorological factors were recorded by a Bowen ratio system at heights of 2 m and 1.5 m. The system was installed in the middle of the lawn. All the data from the Bowen ratio system were sampled and recorded at intervals of 1 min and 10 min with a Campbell CR1000 data logger. The sensor information is shown in Table 1.

Table 1. Type of sensor, measurement height, and resolutions of the equipment in the Bowen ratio system.

Parameter	Sensor Type	Measuring Height(m)	Sensor Resolution
Humidity and Air Temperature	225-050YA, Novalynx, Grass Valley, CA, USA	2.0; 1.5	±3%, ±0.6 °C
Wind Velocity	200-WS-02, Novalynx, Grass Valley, CA, USA	2.0	±0.2 m s^{-1}
Solar Radiation	PYP-PA, Apogee, Santa Monica, CA, USA	2.0	10–40 μV/W/m^2
Net Radiation	240-100, Novalynx, Grass Valley, CA, USA	2.0	<4%
Soil Heat Flux	HFP01, Hukseflux, Center Moriches, NY, USA	−0.05; −0.02	50 μV/W/m^2

2.4. Verification Experiment

The Bowen ratio energy balance (BREB) method was used as the benchmark to verify the '3T + IR' method. The verification experiment was conducted before the field experiments, on three sunny days (15 July 2014, 16 August 2014, and 13 November 2014). The ET rates of the studied area were simultaneously measured by the '3T + IR' method and the BREB method. The ET rates by the BREB method could be calculated by [53]

$$ET = \frac{R_n - G}{L(1 + \beta)} \tag{5}$$

$$\beta = \frac{C_p \Delta T}{L \Delta q} \tag{6}$$

where L is the latent heat of water vaporization (J kg^{-1}), G is the soil heat flux (W m^{-2}), β is the Bowen ratio, C_p is the specific heat of air at a constant pressure (J kg^{-1} °C^{-1}), and ΔT and Δq are the temperature and humidity difference between the heights of 2.0 m and 1.5 m, respectively. All these parameters were obtained by the Bowen ratio system.

3. Results

3.1. Method Verification

The verification experiments showed quite a coincidence between the ET rates measured by the '3T + IR' method and BREB method. The correlation coefficient of the ET rates of the two methods was 0.958 (significant at the level of 0.01, by SPSS). Moreover, the linear regression demonstrated the consistency of the two methods (Figure 2). The distribution of the data was close to the 1:1 line and the regression line was $ET_B = 1.07ET_3 - 0.06$ ($R^2 = 0.92$), which means the rates measured by '3T + IR' were always close to the rates measured by BREB methods. The RMSE of the two rates was also just 0.03 mm h^{-1}. This finding indicates that the '3T + IR' method could accurately measure the ET rate of urban grass and shrubs. Therefore, we applied this method directly in the field experiments on urban hedges in this study.

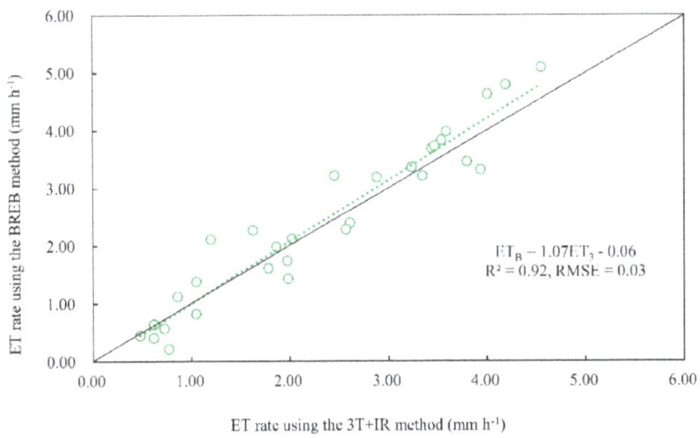

Figure 2. Comparison of the ET rates of urban vegetation estimated by the '3T + IR' method and BREB method. 3T + IR: three-temperature model + infrared remote sensing; BREB: Bowen ratio energy balance; ET_3: ET rate measured by '3T + IR' method; ET_B: ET rate measured by the BREB method.

3.2. Characteristics of Meteorological Conditions

Our field experiments were conducted on a sunny day in each season from 2015 to 2016. The typical days for each season were 22 August 2015 (Summer); 18 December 2015 (Autumn); 4 February 2016 (Winter); and 19 March 2016 (Spring). The season division is according to the Shenzhen Bureau of Meteorology [54]. The daily average temperature of the summer day was as high as 32.32 °C (Figure 3). The temperatures were still high even in the autumn day (14.55 °C) and winter day (17.92 °C). The solar radiation showed an almost single-peak variation in all days. It was the strongest in the summer day, when the daily average reached 555.98 W m^{-2}. The air was the most humid on the spring day (89%) followed by the summer day (72%). The winter day had a low relative humidity (20%). The wind velocity was not high in any of the four days. The highest was during the autumn day, when its daily average was 0.78 m s^{-1}.

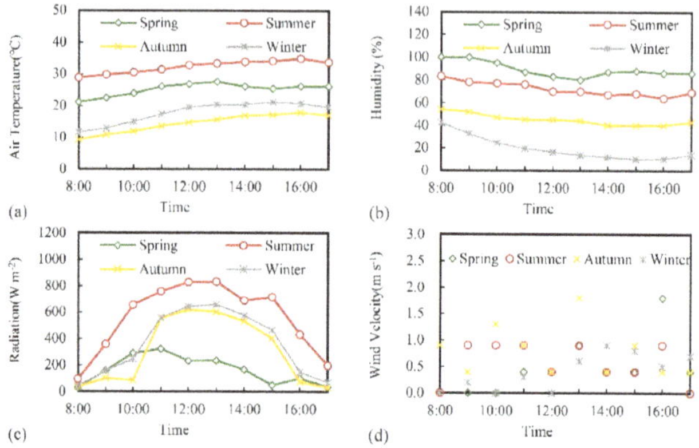

Figure 3. Characteristics of weather in the typical sunny days during each season in the study area. (**a**) the air temperature; (**b**) the relative humidity; (**c**) the solar radiation; (**d**) the wind velocity. Data were measured by the Bowen ratio system in 22 August 2015 (Summer); 18 December 2015 (Autumn); 4 February 2016 (Winter); and 19 March 2016 (Spring). The air temperature and relative humidity are the average of the values measured at 1.5 m and 2.0 m.

3.3. ET Characteristics of Urban Hedges

3.3.1. Surface Temperatures of the Urban Hedges

The infrared images of the hedges and the lawn were taken over four days. Subsequently, the ET rates were calculated by our software (Figure 4).

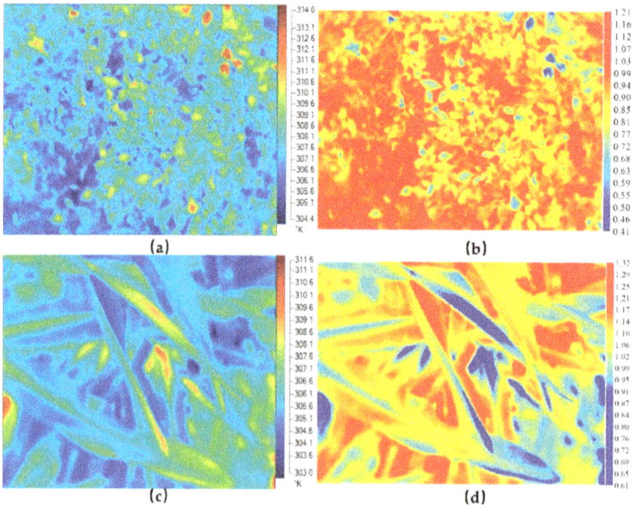

Figure 4. The surface temperatures (K) and ET rates (mm h^{-1}) of the two hedges at 12:00 p.m. in 22 August 2015. (**a**) the surface temperature of the *L. quihoui* hedge; (**b**) the ET rates of the *L. quihoui* hedge; (**c**) the surface temperature of the *H. littoralis* hedge; (**d**) the ET rates of the *H. littoralis* hedge. The ET rates were calculated by our software based on the '3T + IR' method and plotted by ArcGIS.

As depicted in Figure 5, the surface temperatures of the two hedges also showed single-peak variations in all four days, much like the solar radiation. For most of the time during the four days, the surface temperature of the *L. quihoui* hedge was higher than that of the *H. littoralis* hedge. The daily average surface temperature of the *L. quihoui* hedge was 27.43, 34.43, 20.55, and 21.94 °C in each day. At the same time, the surface temperature of the *H. littoralis* hedge was 26.00, 33.66, 19.77, and 20.86 °C, respectively. The surface temperatures of the hedges were slightly higher than the air temperature. The *L. quihoui* hedge was 2.22, 2.11, 6.00, and 4.02 °C higher than the air temperature. The smallest difference between surface and air temperature occurred during the summer day, when the solar radiation and air temperature were the highest. The surface temperature of the lawn used for comparison was much warmer than that of the two hedges. The surface temperature differences between the lawn and the *H. littoralis* hedge were 4.85, 9.14, 5.83, and 3.65 °C over the four days.

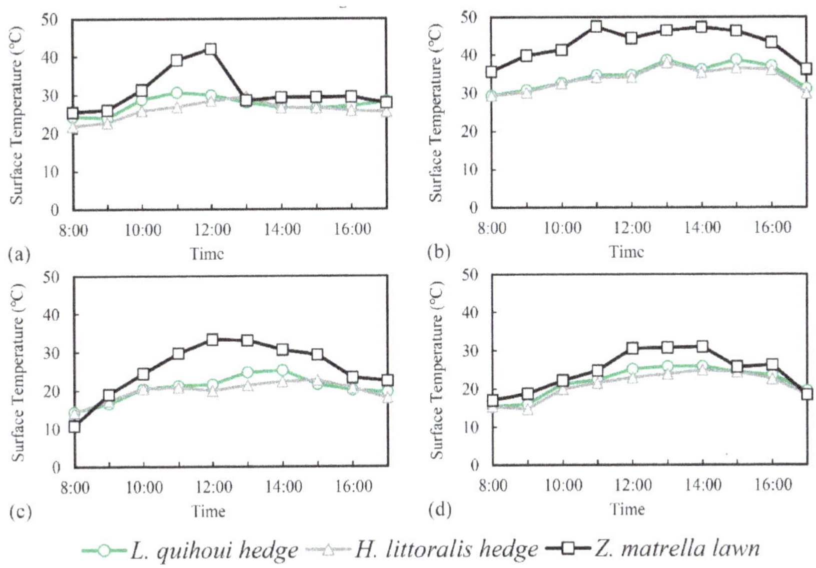

Figure 5. Surface temperature of the hedges and the lawn for comparison over the four days (the temperatures were the average values of three images. (**a**) Spring: 19 March 2016; (**b**) Summer: 22 August 2015; (**c**) Autumn: 18 December 2015; (**d**) Winter: 4 February 2016).

3.3.2. ET Rates of the Urban Hedges

The ET rates of the two hedges showed similar variation trends in the spring, autumn and winter days (Figure 6). They both increased from the morning and began to decrease after reaching peaks in the midday. The sudden drop at 11:00 a.m. during the summer day might be the result of the stomatal closure of the plants due to high surface temperatures. The ET rate was still quite high at 3:00 p.m. on the summer day. The ET of the *H. littoralis* hedge usually reached its maximum when the solar radiation was at its peak (Figure 3). However, the ET of the *L. quihoui* hedge rates reached their peaks at a different time compared to the *H. littoralis* hedge in the spring day. In particular, the ET rate of the *L. quihoui* hedge achieved another peak at 1:00 p.m. during the spring day. We also calculated the vapor pressure deficit (VPD) and found that it increased to its maximum at 1:00 p.m. during that day (data not shown).

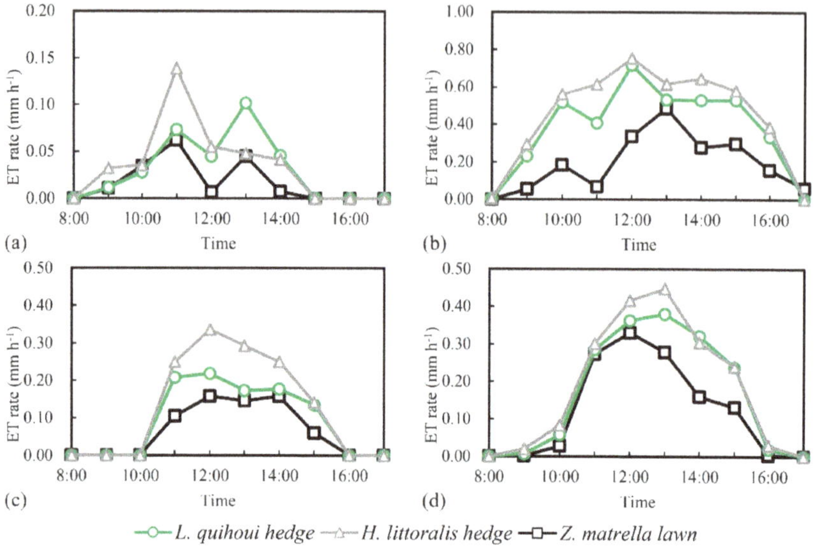

Figure 6. ET rates of the hedges and the lawn for comparison on the four typical sunny days in four seasons. (**a**) Spring: 19 March 2016; (**b**) Summer: 22 August 2015; (**c**) Autumn: 18 December 2015; (**d**) Winter: 4 February 2016.

Figure 6 also showed that the ET rates of the hedges on the summer day were obviously stronger than those of the other three days. The daily average ET rate of the *H. littoralis* hedge was approximately 0.38 mm h^{-1}, while the daily average ET rate of the *L. quihoui* hedge was 0.33 mm h^{-1} (Table 2). Despite a lower level of solar radiation on the winter day, these data showed higher ET rates than on the autumn day, which may be attributed to the lowest relative humidity during this time. The ET rate was the lowest on the spring day with the lowest solar radiation and VPD. Meanwhile, we found that the ET rate of the *H. littoralis* hedge was higher than that of the *L. quihoui* hedge over the four days. The differences were 0.01, 0.05, 0.04, and 0.01 mm h^{-1}. The ET rates of the hedges were always higher than the lawn, especially on the summer day, when the ET rate of the *H. littoralis* hedge was 0.20 mm h^{-1} higher than the lawn. The difference was the smallest on the spring day, when all the three vegetation types had low ET rates.

Table 2. Average ET rates (mm h^{-1}) of the hedges and the lawn for comparison on the four typical sunny days in four seasons. Spring: 19 March 2016; Summer: 22 August 2015; Autumn: 18 December 2015; Winter: 4 February 2016.

Seasons	*H. littoralis* Hedge	*L. quihoui* Hedge	*Z. matrella* Lawn
Spring	0.04	0.03	0.02
Summer	0.38	0.33	0.18
Autumn	0.13	0.09	0.06
Winter	0.18	0.17	0.12

3.3.3. The LE/Rn of the Urban Hedges

It is usually understood that green space could cool the surrounding area through latent heat flux. To reflect the diurnal course of energy exchange, the ratio of latent to net radiation (LE/Rn) was used to illuminate the cooling effect. As shown in Figure 7, the variation of LE/Rn showed different characteristics over the four days. The LE/Rn fluctuated through the day. At 8:00 a.m. on the summer day, the LE/Rn of the two hedges could reach approximately 90%. Their LE/Rn maintained a high value during the summer day, indicating that most of the net radiation was consumed by latent heat. On the autumn and winter day, their LE/Rn had obvious changes in the morning and afternoon because of low latent heat consumption at the beginning and ending of the day.

During the summer day, the hedges could consume over 60% of the net radiation through latent heat. The ET rate was the lowest on the spring day, the LE/Rn of the hedges during that day was also the lowest. Though the LE/Rn of the *L. quihoui* hedge exceeded 50% at 5:00 p.m. during the spring day, its cooling effect was still negligible because the latent heat was only 0.63 W m^{-2}. The LE/Rn of the lawn for comparison had variation trends similar to the LE/Rn of the hedges except for the summer day, which began with a small LE/Rn and was still high at 5:00 p.m.

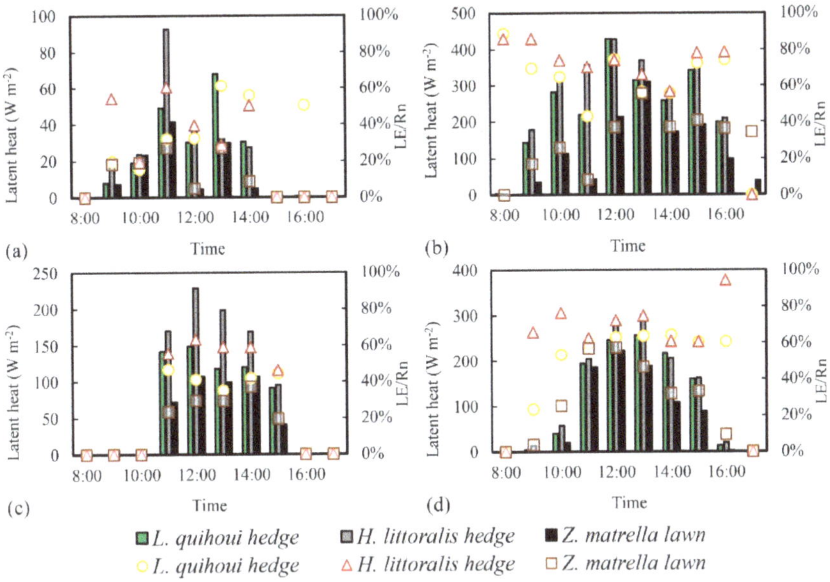

Figure 7. Latent heat flux of the hedges and the lawn for comparison on the four typical sunny days in four seasons. The latent heat flux was figured out directly from the three-temperature model. The LE/Rn is the proportion of the latent heat to the net radiation. (**a**) Spring: 19 March 2016; (**b**) Summer: 22 August 2015; (**c**) Autumn: 18 December 2015; (**d**) Winter: 4 February 2016.

Overall, the daily average LE/Rn of the *H. littoralis* hedge was still higher than that of the *L. quihoui* hedge during all days (Table 3). On the summer day, the *H. littoralis* hedge consumed 68.44% of the net radiation while for the *L. quihoui* hedge it was 60.81%. The LE/Rn of the lawn was lower than that of the two hedges. The largest differences appeared in the summer day and extended to 28.92%, suggesting that the hedges have much better cooling potential than the lawn.

Table 3. Daily average LE/Rn of the hedges and the lawn for comparison on the four typical sunny days in four seasons. Spring: 19 March 2016; Summer: 22 August 2015; Autumn: 18 December 2015; Winter: 4 February 2016.

Seasons	*H. littoralis* Hedge	*L. quihoui* Hedge	*Z. matrella* Lawn
Spring	37.27%	35.72%	28.23%
Summer	68.44%	60.81%	39.52%
Autumn	56.10%	41.45%	34.06%
Winter	65.71%	61.58%	47.28%

3.4. Cooling Effects of Urban Hedges

3.4.1. Cooling Effects on Air Temperature of the Urban Hedges

LE/Rn described the cooling effects of the vegetation in an indirect way. The temperature reduction was also calculated to intuitively evaluate the cooling effect of the hedges henceforth. The cooling effects of plants on air temperature or surface temperature have been widely studied in recent years [55–57]. Most studies on this topic were based on comparing the temperature differences between two sites. However, this method could not divide the cooling effects of the plants and show how much the ET specifically contributes to cooling. Here, we reference a method to calculate the cooling effect of the hedges through ET alone [58]. For the unit volume of air

$$\Delta T_a = 60 * LE/\rho_{air}CV \tag{7}$$

where ΔT_a (°C min^{-1} m^{-2}) is the cooling rate by ET of unit area hedges. LE is the latent heat (W·m^{-2}) and has been analyzed using the '3T + IR' method. C is the specific heat capacity of air, which is 1005 J·kg^{-1}·°C^{-1}. V is the volume of the air and equals 10 m^3 here, following the reference paper [58]. ρ_{air} is the air density (kg·m^{-3}), and it is a function of air temperature (T_a),

$$\rho_{air} = 1.2837 - 0.0039T_a \tag{8}$$

The variation of the cooling rates of the studied hedges always followed the variation of their ET rates (Figure 8). The hedges could cool the air most effectively when the ET rate and radiation reached their maximums. The cooling effects of the hedges were the most robust on the summer day and the weakest on the spring day. Though the cooling effects in the autumn day were stronger than on the spring day, the hedges had a shorter cooling period due to shorter radiation duration. The cooling effect of the *H. littoralis* hedge was slightly stronger than the *L. quihoui* hedge. The daily average cooling rates of the *H. littoralis* hedge were 0.12 °C min^{-1} m^{-2}, 1.29 °C min^{-1} m^{-2}, 0.42 °C min^{-1} m^{-2}, and 0.61 °C min^{-1} m^{-2} over the four days and were 0.10 °C min^{-1} m^{-2}, 1.13 °C min^{-1} m^{-2}, 0.30 °C min^{-1} m^{-2}, and 0.56 °C min^{-1} m^{-2} for the *L. quihoui* hedge. Both hedges had stronger cooling effects than the lawn, especially on the summer day. The cooling rates of the *Z. matrella* lawn were 0.05 °C min^{-1} m^{-2}, 0.63 °C min^{-1} m^{-2}, 0.21 °C min^{-1} m^{-2}, and 0.40 °C min^{-1} m^{-2} over the four days.

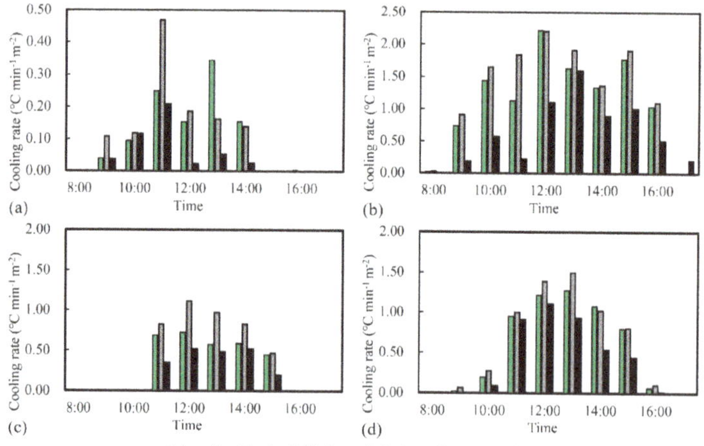

Figure 8. The cooling rates of hedges and the lawn for comparison on air temperature on the four typical sunny days in four seasons. (**a**) Spring: 19 March 2016; (**b**) Summer: 22 August 2015; (**c**) Autumn: 18 December 2015; (**d**) Winter: 4 February 2016.

3.4.2. Cooling Effects of the Urban Hedges on Surface Temperature

Surface temperature can easily be obtained using infrared remote sensing techniques and has therefore become the basis of most studies on the cooling effects of the vegetation. In this study, the cooling effects of the urban hedges on surface temperature at a small scale is discussed. The thermal imager simultaneously photographed the surface temperature of an asphalt road near the study site when the vegetation was photographed. The surface temperature of the asphalt road was always high, especially during the summer day (Figure 9). On that day, it could be as high as 62.73 °C at 3:00 p.m. and the daily average increased to 53.60 °C. The surface temperatures of the asphalt road during the other three days were similar, while the surface temperature of the hedges showed obvious differences (Figure 5). The average daily surface temperatures of the road were 30.22, 28.35, and 27.86 °C in the spring, autumn, and winter day.

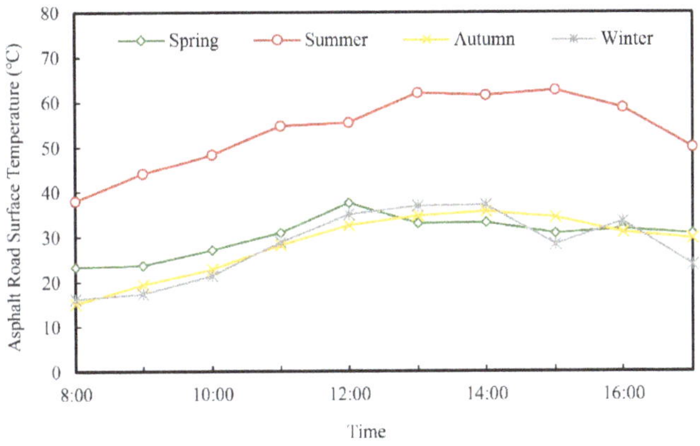

Figure 9. Surface temperature of the asphalt road in the four typical days in four seasons. Spring: 19 March 2016; Summer: 22 August 2015; Autumn: 18 December 2015; Winter: 4 February 2016.

The surface temperature of the asphalt road was higher than the hedges most of the time (Figure 10). The cooling effects of the hedges were more evident in the mid-day, when the underlying surface temperatures were high. During the summer day, the cooling effects on surface temperature could even be over 20 °C between 11:00 a.m. and 4:00 p.m. This means the hedges could significantly reduce the peak surface temperature in a day. The daily average cooling effects of the two hedges during the summer day were 19.17–19.94 °C. They were much weaker on the other three days, especially in the spring day, when the two hedges could only cool the underlying surface by 2.80–4.22 °C. The surface temperature cooling effects were even negative in the morning of the spring day. The low ET rate of the *L. quihoui* hedge restricted its cooling effect at that time. In addition, the asphalt road dissipated heat in the night before becoming cooler in the morning [59]. As a result, the surface temperature of the road could be lower than the hedge.

The *H. littoralis* hedge had better cooling effects on underlying surface temperatures than the *L. quihoui* hedge. The *H. littoralis* hedge cooled the underlying surface temperature by 4.22, 19.94, 8.57, and 7.00 °C on the four days, respectively. Simultaneously, the *L. quihoui* hedge could cool the surface by 2.80, 19.17, 7.80, and 5.92 °C. The hedges always showed better cooling effects than the lawn. The cooling effects of the lawn were −0.62, 10.81, 2.75, and 3.36 °C on the four selected days. The most distinct differences of cooling effects between the hedges and the lawn were during the summer day. The hedges could cool the surface by 9 °C more than the lawn. The minimum differences occurred on the winter day, when the two hedges cooled by more 2.56 °C and 3.64 °C than the lawn.

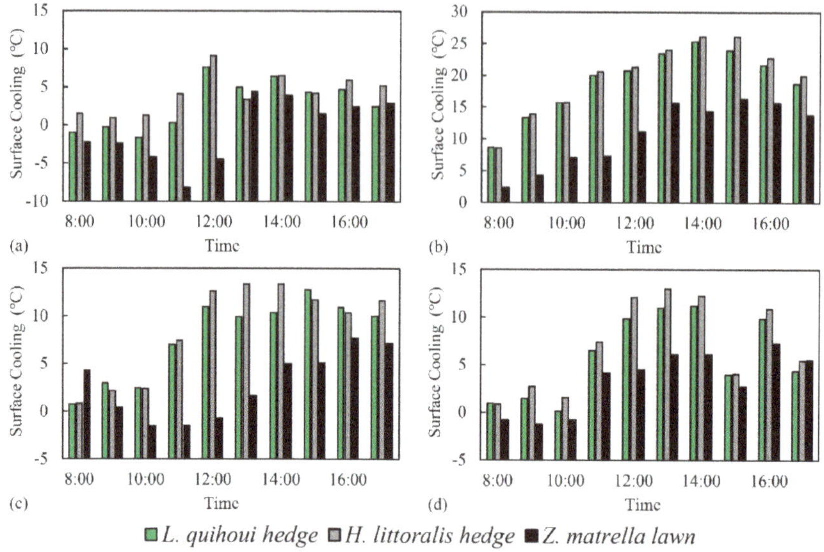

Figure 10. Surface temperature differences between the hedges, lawn and asphalt pavement on the four typical sunny days in four seasons. (**a**) Spring: 19 March 2016; (**b**) Summer: 22 August 2015; (**c**) Autumn: 18 December 2015; (**d**) Winter: 4 February 2016.

4. Discussion

4.1. The ET Characteristics of Urban Hedges

The ET rates of two common urban hedges were estimated in this study, and both were found to be relatively high. The ET rates of the *H. littoralis* hedge were 0.04, 0.38, 0.13, and 0.18 mm h^{-1} during the four typical sunny days from four seasons. The ET rates of the *L. quihoui* hedge were slightly lower: on the four days selected, they were 0.03, 0.33, 0.09, and 0.17 mm h^{-1}, respectively. The ET rates of the two hedges have also been studied in other cities. A study conducted in Hubei, China showed that the ET rate of *H. littoralis* (0.04 mm h^{-1}) was higher than that of six other plants in summer [60]. It also found that the *H. littoralis* had the highest light utilization efficiency and the third highest water use efficiency. Another study conducted in Changsha, China showed that the *L. quihoui* could transport 2576.52 $g \cdot m^{-2} \cdot d^{-1}$ (approximately 0.11 mm h^{-1}) of water into the air through ET in August [61]. It was the third highest out of the 13 studied shrubs. However, the ET rates in the two studies above are much lower than our results, as the two previous cities get less solar radiation compared with our study sites.

The winter in Shenzhen is warm enough to sustain plant growth, so almost all local plants are evergreen [62]. As a result, the ET rate of the *H. littoralis* hedge was still high in the winter day. In addition, with its high light and water utilization efficiency, its ET rate might be slightly higher than that of the *L. quihoui* hedge. The ET rates of the two hedges were both higher than the ET rates of the lawn. LAI might be the predominant reason [63].

4.2. Cooling Effect of the Urban Hedges

Three techniques were used to describe the cooling effects of the urban hedges in this study. Among them, the cooling effects of plants through ET alone was calculated using a reference method. For 10 cubic meters of air, this *H. littoralis* hedge could generate cooling at rates of 0.12 °C min^{-1} m^{-2}, 1.29 °C min^{-1} m^{-2}, 0.42 °C min^{-1} m^{-2}, and 0.61 °C min^{-1} m^{-2} on the four studied days. Meanwhile, the cooling rates of the *L. quihoui* hedge were 0.10 °C min^{-1} m^{-2}, 1.13 °C min^{-1} m^{-2}, 0.30 °C min^{-1} m^{-2}, and 0.56 °C min^{-1} m^{-2}. In our reference research, a 2885-m^2 *S. superba* forest in Guangzhou, another subtropical megacity near Shenzhen, could cool a 10-m^3 air column at rates of 0.15 °C min^{-1} m^{-2} and 0.13 °C min^{-1} m^{-2} in July in 2007 and 2008 [58]. Its cooling rates by per unit area vegetation were lower than our results, as the ET during the day and night in all kinds of weather were included in their study. The daily UHIIs around our study sites over four seasons were approximately 0.76, 1.06, 1.04, and 0.80 °C [64]. For the whole city, the yearly average UHII of Shenzhen was 2.6 °C [65]. Therefore, the urban hedges showed great cooling potential in the mitigation of UHI.

The cooling rate above is the temperature reduction by ET without heat input. The LE/Rn reflects the proportions of the net radiation that ET dissipates. The greater the proportions of net radiation that were consumed by latent heat, the smaller proportion of net radiation could heat the environment through sensible heat. The *H. littoralis* hedge could consume 37.27%, 68.44%, 56.10%, and 65.71% of the net radiation as latent heat over the four days, while for the *L. quihoui* hedge the ratios were 35.72%, 60.81%, 41.45%, and 61.58%. These ratios were significantly higher than artificial underlying surfaces. It was found that only 123 Wh m^{-2} out of the 1949 Wh m^{-2} net radiation reaching to the asphalt roof was consumed by latent heat [66]. Grimmond et al. reported 23% of the LE/Rn in Marseille, where the area fraction of vegetation and water was 10–20%. Meanwhile, areas like Me93 and Vl92 that contained less vegetation had a lower LE/Rn [67]. In Kansas City, the LE/Rn could reach 46–58% in an exurban residential neighborhood, where the vegetation accounted for 58% of the total area [68]. This phenomenon was also demonstrated in a study conducted in Kugahara, Tokyo, where the LE/Rn in the daytime was always larger in hot months and smaller in cooler months [69]. According to LE/Rn, the *H. littoralis* hedge had better cooling effects than the *L. quihoui* hedge. The LE/Rn was larger when the radiation was stronger, which means the cooling effects of ET might be stronger in hotter days.

The albedo differences may result in the surface temperature differences between hedges and asphalt pavements [70,71]. Moreover, the hedges could consume much more heat through ET than artificial underlying surfaces [72]. Compared to the asphalt pavement, the surface temperature of the studied *H. littoralis* hedge were 4.22, 19.94, 8.57, and 7.00 °C lower on the four days. The surface temperature of the *L. quihoui* hedge were 2.80, 19.17, 7.80, and 5.92 °C lower at the same time. The forested land could also cool the surface more than 10 °C in November compared with developed land [73]. Leuzinger et al. found that tree canopies in Basel were 19 °C cooler than roofs in July, and different trees had different canopy temperatures [74]. The land surface temperature differences between land use types of transportation and green spaces in Shenzhen were approximately 4.8 °C in daytime in October [75]. In comparison with the studies above, the cooling effects of the urban hedges in our research are remarkable. The amplitudes of the hedges' surface temperature were also smaller than those of the asphalt road in the daytime, which means the thermal environment was more stable in the urban hedge area (Table 4). Similar results have been found in previous studies [76]. They found that the maximum daily variation of surface temperature was no more than 3 °C, and the maximum surface temperature was only 26.5 °C for *Raphis palm*, while for the hard surface, they were 30 °C and 57 °C, respectively. Latent heat could significantly reduce the maximum surface temperature in a day but showed minimal effects on the minimum temperature [77].

Table 4. Standard deviations of the surface temperatures on the four days (°C).

Underlying Surface	Spring	Summer	Autumn	Winter
Asphalt pavement	4.18	7.90	6.65	7.43
H. littoralis hedge	2.22	2.85	2.43	3.46
L. quihoui hedge	2.06	3.14	3.14	3.69
Z. matrella lawn	5.16	4.19	6.71	5.04

4.3. Applicability of the '3T + IR' Method for the ET Estimation of Urban Hedges

In this study, a new method based on '3T + IR' was applied to accurately estimate urban ET. The applicability of this method on urban vegetation has been verified in this study by comparison with the BREB method. The results showed great reliability in this new method. With this method, the ET rates of the hedges can be calculated by surface temperature, which could be easily obtained by thermal images. Therefore, this method will not be limited by the complexity of urban underlying surfaces, which is the main obstacle for traditional methods. It also has a higher temporal and spatial resolution than traditional satellite remote sensing. Based on accurate ET rates, the specific cooling effects of ET could be obtained.

In this study, to simplify the calculation and measurement, the emissivity of the hedges and the lawn were defined as 0.98 and the net radiation was estimated based on the solar radiation and temperatures. Therefore, there might be some bias of the results. Besides, the leaf with the highest surface temperature was selected to be the reference leaf. The ET of this leaf actually is larger than zero, therefore will leading a little overestimation of the actually ET of the vegetation based on three-temperature model. Though it has been applied and validated in various field experiments out of the city, this is the first application of this new method in the study of urban hedges. Therefore, more research is needed. For example, an idealized reference leaf is still hard to select. The shape, emissivity and albedo of the reference leaf could affect the results. Besides, this method could not be used in continuous measurement at a high frequency, as we could only photo the thermal images and measure the net radiation by hand. Automatic imaging technique of infrared remote sensing and net radiation measurement will be helpful in the future.

5. Conclusions

In this study, the ET characteristics of two urban hedges were measured by a fetch-free method with high spatiotemporal resolution, namely, the 'three-temperature model + infrared remote sensing' method. The method demonstrated high accuracy in the ET estimation for urban vegetation. The results show that: (1) the ET rates of the two studied urban hedges were high. On the summer day, the daily average ET rate of the *H. littoralis* hedge was 0.38 mm h^{-1}, while that of the *L. quihoui* hedge was 0.33 mm h^{-1}. (2) The latent heat of the hedges accounts for a large part of the net radiation. The two hedges consumed 68.44% and 60.81% of the net radiation via ET on the summer day. Therefore, the hedges have great cooling potential in the urban thermal environment. (3) The contribution of ET to the vegetation cooling effects in urban areas could be identified through more accurate ET rates. The daily average cooling rates of the two hedges on air temperature through ET alone could reach 1.13–1.29 °C min^{-1} m^{-2}. (4) The hedges could also significantly cool down the underlying urban surface. The cooling effect was stronger on hotter days. On the hottest day, the cooling effects of the two hedges on the underlying surface were more than 19 °C. (5) The ET rates of the *H. littoralis* hedge were slightly higher than those of the *L. quihoui* hedge and therefore had better cooling effects, while both had much better cooling effects than the lawn used for comparison. These results may contribute to the greening design for urban areas.

This study may be the first research that can quantitatively measure the ET rate of urban hedge and provide a new insight to understand the process of ET in urban hedges, and could also promote the methodology of urban ET studies.

Author Contributions: G.Y.Q. designed the research; Z.Z. and Y.Y. carried out the field experiment; Z.Z. made the data analysis and wrote this paper.

Funding: This research was funded by the Special Fund for National Key Research and Development Plan, grant number 2017FY100206-03.

Acknowledgments: The authors would like to express their great thanks to: Chunhua Yan and Yang Zhang for their assistance in the article writing; Wenli Zhao for coding the software "A system to estimate ET by infrared remote sensing and the three-temperature model"; Peng Mao for help in graphics. Elsevier Language Editing for their efforts to improve the grammar of this paper.

Conflicts of Interest: The authors declare no conflict of interest.

References

1. Tan, J.; Zheng, Y.; Tang, X.; Guo, C.; Li, L.; Song, G.; Zhen, X.; Yuan, D.; Kalkstein, A.J.; Li, F. The urban heat island and its impact on heat waves and human health in Shanghai. *Int. J. Biometeorol.* **2010**, *54*, 75–84. [CrossRef]
2. Gaffin, S.R.; Rosenzweig, C.; Khanbilvardi, R.; Parshall, L.; Mahani, S.; Glickman, H.; Goldberg, R.; Blake, R.; Slosberg, R.B.; Hillel, D. Variations in New York city's urban heat island strength over time and space. *Theor. Appl. Climatol.* **2008**, *94*, 1–11. [CrossRef]
3. Ren, G.; Zhou, Y.; Chu, Z.; Zhou, J.; Zhang, A.; Guo, J.; Liu, X. Urbanization Effects on Observed Surface Air Temperature Trends in North China. *J. Clim.* **2008**, *21*, 1333–1348. [CrossRef]
4. Peng, S.; Piao, S.; Ciais, P.; Friedlingstein, P.; Ottle, C.; Bréon, F.M.; Nan, H.; Zhou, L.; Myneni, R.B. Surface urban heat island across 419 global big cities. *Environ. Sci. Technol.* **2012**, *46*, 696–703. [CrossRef]
5. Lin, W.; Wu, T.; Zhang, C.; Yu, T. Carbon savings resulting from the cooling effect of green areas: A case study in Beijing. *Environ. Pollut.* **2011**, *159*, 2148–2154. [CrossRef]
6. Gobakis, K.; Kolokotsa, D.; Synnefa, A.; Saliari, M.; Giannopoulou, K.; Santamouris, M. Development of a model for urban heat island prediction using neural network techniques. *Sustain. Cities Soc.* **2011**, *1*, 104–115. [CrossRef]
7. Gosling, S.N.; Lowe, J.A.; Mcgregor, G.R.; Pelling, M.; Malamud, B.D. Associations between elevated atmospheric temperature and human mortality: A critical review of the literature. *Clim. Chang.* **2009**, *92*, 299–341. [CrossRef]
8. Kinney, P.L. Climate change, air quality, and human health. *Am. J. Prev. Med.* **2008**, *35*, 459–467. [CrossRef]
9. Merte, S. Estimating heat wave-related mortality in Europe using singular spectrum analysis. *Clim. Chang.* **2017**, *142*, 321–330. [CrossRef]
10. Santamouris, M. Cooling the cities—A review of reflective and green roof mitigation technologies to fight heat island and improve comfort in urban environments. *Sol. Energy* **2014**, *103*, 682–703. [CrossRef]
11. Ketterer, C.; Matzarakis, A. Human-biometeorological assessment of heat stress reduction by replanning measures in Stuttgart, Germany. *Landsc. Urban Plan.* **2014**, *122*, 78–88. [CrossRef]
12. Akbari, H.; Pomerantz, M.; Taha, H. Cool surfaces and shade trees to reduce energy use and improve air quality in urban areas. *Sol. Energy* **2001**, *70*, 295–310. [CrossRef]
13. Xu, C.Y.; Chen, D. Comparison of seven models for estimation of evapotranspiration and groundwater recharge using lysimeter measurement data in Germany. *Hydrol. Process.* **2010**, *19*, 3717–3734. [CrossRef]
14. Takebayashi, H.; Moriyama, M. Study on the urban heat island mitigation effect achieved by converting to grass-covered parking. *Sol. Energy* **2009**, *83*, 1211–1223. [CrossRef]
15. Nastran, M.; Kobal, M.; Eler, K. Urban Heat Islands in Relation to Green Land Use in European Cities. *Urban For. Urban Green.* **2018**, in press. [CrossRef]
16. Imran, H.M.; Kala, J.; Ng, A.W.M.; Muthukumaran, S. Effectiveness of green and cool roofs in mitigating urban heat island effects during a heatwave event in the city of Melbourne in southeast Australia. *J. Clean Prod.* **2018**, *197*, 393–405. [CrossRef]
17. Mariani, L.; Parisi, S.G.; Cola, G.; Lafortezza, R.; Colangelo, G.; Sanesi, G. Climatological analysis of the mitigating effect of vegetation on the urban heat island of Milan, Italy. *Sci. Total Environ.* **2016**, *569–570*, 762–773. [CrossRef]
18. Zheng, S.; Zhao, L.; Li, Q. Numerical simulation of the impact of different vegetation species on the outdoor thermal environment. *Urban For. Urban Green.* **2016**, *18*, 138–150. [CrossRef]
19. Akbari, H.; Dan, M.K.; Bretz, S.E.; Hanford, J.W. Peak power and cooling energy savings of shade trees. *Energy Build.* **1997**, *25*, 139–148. [CrossRef]
20. Ballinas, M.; Barradas, V.L. The Urban Tree as a Tool to Mitigate the Urban Heat Island in Mexico City: A Simple Phenomenological Model. *J. Environ. Qual.* **2016**, *45*, 157–166. [CrossRef]
21. Hamada, S.; Ohta, T. Seasonal variations in the cooling effect of urban green areas on surrounding urban areas. *Urban For. Urban Green.* **2010**, *9*, 15–24. [CrossRef]
22. Feyisa, G.L.; Dons, K.; Meilby, H. Efficiency of parks in mitigating urban heat island effect: An example from Addis Ababa. *Landsc. Urban Plan.* **2014**, *123*, 87–95. [CrossRef]

23. Doick, K.J.; Peace, A.; Hutchings, T.R. The role of one large greenspace in mitigating London's nocturnal urban heat island. *Sci. Total Environ.* **2014**, *493*, 662–671. [CrossRef]
24. Wong, N.H.; Jusuf, S.K.; Win, A.A.L.; Thu, H.K.; Negara, T.S.; Wu, X. Environmental study of the impact of greenery in an institutional campus in the tropics. *Build. Environ.* **2007**, *42*, 2949–2970. [CrossRef]
25. Weng, Q.; Yang, S. Managing the adverse thermal effects of urban development in a densely populated Chinese city. *J. Environ. Manag.* **2004**, *70*, 145–156. [CrossRef]
26. Dimoudi, A.; Nikolopoulou, M. Vegetation in the urban environment: Microclimatic analysis and benefits. *Energy Build.* **2003**, *35*, 69–76. [CrossRef]
27. Scott, K.I.; Simpson, J.R.; Mcpherson, E.G. Effects of tree cover on parking lot microclimate and vehicle emissions. *J. Arboric.* **1999**, *25*, 129–142.
28. Berthier, E.; Dupont, S.; Mestayer, P.G.; Andrieu, H. Comparison of two evapotranspiration schemes on a sub-urban site. *J. Hydrol.* **2006**, *328*, 635–646. [CrossRef]
29. Lazzarin, R.M.; Castellotti, F.; Busato, F. Experimental measurements and numerical modelling of a green roof. *Energy Build.* **2005**, *37*, 1260–1267. [CrossRef]
30. Qiu, G.; Tan, S.; Wang, Y.; Yu, X.; Yan, C. Characteristics of Evapotranspiration of Urban Lawns in a Sub-Tropical Megacity and Its Measurement by the 'Three Temperature Model + Infrared Remote Sensing' Method. *Remote Sens.* **2017**, *9*, 502.
31. Nouri, H.; Glenn, E.; Beecham, S.; Chavoshi Boroujeni, S.; Sutton, P.; Alaghmand, S.; Noori, B.; Nagler, P. Comparing Three Approaches of Evapotranspiration Estimation in Mixed Urban Vegetation: Field-Based, Remote Sensing-Based and Observational-Based Methods. *Remote Sens.* **2016**, *8*, 492. [CrossRef]
32. Pataki, D.E.; Mccarthy, H.R.; Litvak, E.; Pincetl, S. Transpiration of urban forests in the Los Angeles metropolitan area. *Ecol. Appl.* **2011**, *21*, 661–677. [CrossRef]
33. DiGiovanni, K.; Montalto, F.; Gaffin, S.; Rosenzweig, C. Evaluation of Physically and Empirically Based Models for the Estimation of Green Roof Evapotranspiration. *J. Hydrol. Eng.* **2010**, *18*, 99–107. [CrossRef]
34. Qiu, G.Y.; Momii, K.; Yano, T. Estimation of Plant Transpiration by Imitation Leaf Temperature: Theoretical consideration and field verification (I). *Trans. Jpn. Soc. Irrig. Drain. Rural Eng.* **1996**, *64*, 401–410.
35. Qiu, G.Y.; Yano, T.; Momii, K. Estimation of Plant Transpiration by Imitation Leaf Temperature: Application of imitation leaf temperature for detection of crop water stress (II). *Trans. Jpn. Soc. Irrig. Drain. Rural Eng.* **1996**, *64*, 767–773.
36. Wang, Y.Q.; Xiong, Y.J.; Qiu, G.Y.; Zhang, Q.T. Is scale really a challenge in evapotranspiration estimation? A multi-scale study in the Heihe oasis using thermal remote sensing and the three-temperature model. *Agric. For. Meteorol.* **2016**, *230–231*, 128–141. [CrossRef]
37. Tian, F.; Qiu, G.Y.; Lü, Y.H.; Yang, Y.H.; Xiong, Y. Use of high-resolution thermal infrared remote sensing and "three-temperature model" for transpiration monitoring in arid inland river catchment. *J. Hydrol.* **2014**, *515*, 307–315. [CrossRef]
38. Xiong, Y.J.; Qiu, G.Y. Simplifying the revised three-temperature model for remotely estimating regional evapotranspiration and its application to a semi-arid steppe. *Int. J. Remote Sens.* **2014**, *35*, 2003–2027.
39. Xiong, Y.J.; Qiu, G.Y. Estimation of evapotranspiration using remotely sensed land surface temperature and the revised three-temperature model. *Int. J. Remote Sens.* **2011**, *32*, 5853–5874. [CrossRef]
40. Qiu, G.Y.; Benasher, J. Experimental determination of soil evaporation stages with soil surface temperature. *Soil Sci. Soc. Am. J.* **2010**, *74*, 13–22. [CrossRef]
41. Qiu, G.Y.; Zhao, M. Remotely monitoring evaporation rate and soil water status using thermal imaging and "three-temperatures model (3T model)" under field-scale conditions. *J. Environ. Monit.* **2010**, *12*, 716–723. [CrossRef]
42. Qiu, G.Y.; Li, C.; Yan, C. Characteristics of soil evaporation, plant transpiration and water budget of Nitraria dune in the arid Northwest China. *Agric. For. Meteorol.* **2015**, *203*, 107–117. [CrossRef]
43. Qiu, G.Y.; Miyamoto, K.; Sase, S.; Okushima, L. Detection of crop transpiration and water stress by temperature-related approach under field and greenhouse conditions. *JARQ Jpn. Agric. Res. Q.* **2000**, *34*, 29–37.
44. Cao, L.; Zhang, Y.; Lu, H.; Yuan, J.; Zhu, Y.; Liang, Y. Grass hedge effects on controlling soil loss from concentrated flow: A case study in the red soil region of China. *Soil Tillage Res.* **2015**, *148*, 97–105. [CrossRef]
45. Forman, R.T.T.; Baudry, J. Hedgerows and hedgerow networks in landscape ecology. *Environ. Manag.* **1984**, *8*, 495–510. [CrossRef]

46. Gosling, L.; Sparks, T.H.; Araya, Y.; Harvey, M.; Ansine, J. Differences between urban and rural hedges in England revealed by a citizen science project. *BMC Ecol.* **2016**, *16* (Suppl. 1), 45–55. [CrossRef]
47. Gromke, C.; Jamarkattel, N.; Ruck, B. Influence of roadside hedgerows on air quality in urban street canyons. *Atmos. Environ.* **2016**, *139*, 75–86. [CrossRef]
48. Yan, C.; Qiu, G.Y. The three-temperature model to estimate evapotranspiration and its partitioning at multiple scales: A review. *Trans. ASABE* **2016**, *59*, 661–670.
49. Jensen, M.E.; Burman, R.D.; Allen, R.G. Evapotranspiration and irrigation water requirements. In *Manuals and Reports on Engineering Practice No. 70*; American Society of Civil Engineers: New York, NY, USA, 1990; pp. 25–41.
50. Weiss, A. An Experimental Study of Net Radiation, Its Components and Prediction. *Agron. J.* **1982**, *74*, 871–874. [CrossRef]
51. Burman, R.D.; Cuenca, R.H.; Weiss, A. Techniques for estimating irrigation water requirements. In *Advances in Irrigation*; Elsevier: New York, NY, USA, 1983; Volume 2, pp. 335–394.
52. Hatfield, J.L.; Reginato, R.J.; Idso, S.B. Comparison of long-wave radiation calculation methods over the United States. *Water Resour. Res.* **1983**, *19*, 285–288. [CrossRef]
53. Bowen, I.S. The Ratio of Heat Losses by Conduction and by Evaporation from any Water Surface. *Phys Rev.* **1926**, *27*, 779–787. [CrossRef]
54. Meteorological Bureau of Shenzhen Municipality. Available online: http://www.szmb.gov.cn/qixiangfuwu/qihoufuwu/qihouguanceyupinggu/qihougaikuang/201711/t20171109_9584854.htm (accessed on 5 December 2018).
55. Anjos, M.; Lopes, A. Urban Heat Island and Park Cool Island Intensities in the Coastal City of Aracaju, North-Eastern Brazil. *Sustainability* **2017**, *9*, 1379. [CrossRef]
56. Shiflett, S.A.; Liang, L.L.; Crum, S.M.; Feyisa, G.L.; Wang, J.; Jenerette, G.D. Variation in the urban vegetation, surface temperature, air temperature nexus. *Sci. Total Environ.* **2017**, *579*, 495–505. [CrossRef]
57. Park, J.; Kim, J.H.; Dong, K.L.; Park, C.Y.; Jeong, S.G. The influence of small green space type and structure at the street level on urban heat island mitigation. *Urban For. Urban Green.* **2016**, *21*, 203–212. [CrossRef]
58. Zhu, L.W.; Zhao, P. Temporal Variation in Sap-Flux-Scaled Transpiration and Cooling Effect of a Subtropical *Schima superba* Plantation in the Urban Area of Guangzhou. *J. Integr. Agric.* **2013**, *12*, 1350–1356. [CrossRef]
59. Mohan, M.; Kandya, A. Impact of urbanization and land-use/land-cover change on diurnal temperature range: A case study of tropical urban airshed of India using remote sensing data. *Sci. Total Environ.* **2015**, *506–507*, 453–465. [CrossRef]
60. Zhang, X.J.; De-Ming, L.I.; Zhai, K.R. Study on Light and Water Utilization Characteristics of Several Ground Cover Plants. *North Hortic.* **2010**, *12*, 75–78. (In Chinese with English Abstract)
61. Zhu, Y.Q.; Yang, B.S.; Ya-Xin, Y.U.; Jin, X.L. Comparison of temperature decrease and humidity increase effect of familiar shrub species in Changsha City. *Guangdong Agric. Sci.* **2013**, *9*, 42–45. (In Chinese with English Abstract)
62. Liu, Y.; Peng, J.; Wang, Y. Diversification of Land Surface Temperature Change under Urban Landscape Renewal: A Case Study in the Main City of Shenzhen, China. *Remote Sens.* **2017**, *9*, 919. [CrossRef]
63. Saadatian, O.; Sopian, K.; Salleh, E.; Lim, C.H.; Riffat, S.; Saadatian, E.; Toudeshki, A.; Sulaiman, M.Y. A review of energy aspects of green roofs. *Renew. Sustain. Energy Rev.* **2013**, *23*, 155–168. [CrossRef]
64. Qiu, G.Y.; Zou, Z.; Li, X.; Li, H.; Guo, Q.; Yan, C.; Tan, S. Experimental studies on the effects of green space and evapotranspiration on urban heat island in a subtropical megacity in China. *Habitat Int.* **2017**, *68*, 30–42. [CrossRef]
65. Zhang, E.J.; Zhang, J.J.; Zhao, X.Y.; Zhang, X.L. Study on urban heat island effect in Shenzhen. *J. Nat. Disasters* **2008**, *17*, 19–24. (In Chinese with English Abstract)
66. Schmidt, M.; Tu-Berlin, D.I. The interaction between water and energy of greened roofs. In Proceedings of the World Green Roof Congress, Basel, Switzerland, 15–16 September 2005.
67. Grimmond, C.S.B.; Salmond, J.A.; Oke, T.R.; Offerle, B.; Lemonsu, A. Flux and turbulence measurements at a densely built-up site in Marseille: Heat, mass (water and carbon dioxide), and momentum. *J. Geophys. Res.* **2004**, *109*, D24101. [CrossRef]
68. Balogun, A.A.; Adegoke, J.O.; Vezhapparambu, S.; Mauder, M.; Mcfadden, J.P.; Gallo, K. Surface Energy Balance Measurements Above an Exurban Residential Neighbourhood of Kansas City, Missouri. *Bound.-Layer Meteor.* **2009**, *133*, 299–321. [CrossRef]

69. Moriwaki, R.; Kanda, M. Seasonal and diurnal fluxes of radiation, heat, water vapor and carbon dioxide over a suburban area. *J. Appl. Meteorol.* **2004**, *43*, 1700–1710. [CrossRef]
70. Blok, D.; Schaepmanstrub, G.; Bartholomeus, H.; Heijmans, M.M.P.D.; Maximov, T.C.; Berendse, F. The response of Arctic vegetation to the summer climate: Relation between shrub cover, NDVI, surface albedo and temperature. *Environ. Res. Lett.* **2011**, *6*, 035502. [CrossRef]
71. Prado, R.T.A.; Ferreira, F.L. Measurement of albedo and analysis of its influence the surface temperature of building roof materials. *Energy Build.* **2005**, *37*, 295–300. [CrossRef]
72. Crawford, A.J.; Mclachlan, D.H.; Hetherington, A.M.; Franklin, K.A. High temperature exposure increases plant cooling capacity. *Curr. Biol.* **2012**, *22*, R396–R397. [CrossRef]
73. Xie, M.; Wang, Y.; Meichen, F.U.; Zhang, D. Pattern Dynamics of Thermal-environment Effect During Urbanization: A Case Study in Shenzhen City, China. *Chin. Geogr. Sci.* **2013**, *23*, 101–112. (In Chinese with English Abstract) [CrossRef]
74. Leuzinger, S.; Vogt, R.; Körner, C. Tree surface temperature in an urban environment. *Agric. For. Meteorol.* **2010**, *150*, 56–62. [CrossRef]
75. Chen, Z.; Gong, C.; Wu, J.; Yu, S. The influence of socioeconomic and topographic factors on nocturnal urban heat islands: A case study in Shenzhen, China. *Int. J. Remote Sens.* **2012**, *33*, 3834–3849. [CrossRef]
76. Wong, N.H.; Chen, Y.; Ong, C.L.; Sia, A. Investigation of thermal benefits of rooftop garden in the tropical environment. *Build. Environ.* **2003**, *38*, 261–270. [CrossRef]
77. Yang, Y.; Ren, R. On the Contrasting Decadal Changes of Diurnal Surface Temperature Range between the Tibetan Plateau and Southeastern China during the 1980s–2000s. *Adv. Atmos. Sci.* **2017**, *34*, 181–198. [CrossRef]

remote sensing

MDPI

Article

Transition Characteristics of the Dry-Wet Regime and Vegetation Dynamic Responses over the Yarlung Zangbo River Basin, Southeast Qinghai-Tibet Plateau

Liu Liu [1,2,*], Qiankun Niu [1,2], Jingxia Heng [1,2], Hao Li [1,2] and Zongxue Xu [3,4]

1 College of Water Resources and Civil Engineering, China Agricultural University, Beijing 100083, China; nqk@stu.ncwu.edu.cn (Q.N.); hjx125128@cau.edu.cn (J.H.); lihao@cau.edu.cn (H.L.)
2 Center for Agricultural Water Research in China, China Agricultural University, Beijing 100083, China
3 College of Water Sciences, Beijing Normal University, Beijing 100875, China; zxxu@bnu.edu.cn
4 Beijing Key Laboratory of Urban Hydrological Cycle and Sponge City Technology, Beijing 100875, China
* Correspondence: liuliu@cau.edu.cn

Received: 26 April 2019; Accepted: 24 May 2019; Published: 27 May 2019

Abstract: The dry-wet transition is of great importance for vegetation dynamics, however the response mechanism of vegetation variations is still unclear due to the complicated effects of climate change. As a critical ecologically fragile area located in the southeast Qinghai-Tibet Plateau, the Yarlung Zangbo River (YZR) basin, which was selected as the typical area in this study, is significantly sensitive and vulnerable to climate change. The standardized precipitation evapotranspiration index (SPEI) and the normalized difference vegetation index (NDVI) based on the GLDAS-NOAH products and the GIMMS-NDVI remote sensing data from 1982 to 2015 were employed to investigate the spatio-temporal characteristics of the dry-wet regime and the vegetation dynamic responses. The results showed that: (1) The spatio-temporal patterns of the precipitation and temperature simulated by the GLDAS-NOAH fitted well with those of the in-situ data. (2) During the period of 1982–2015, the whole YZR basin exhibited an overall wetting tendency. However, the spatio-temporal characteristics of the dry-wet regime exhibited a reversal phenomenon before and after 2000, which was jointly identified by the SPEI and runoff. That is, the YZR basin showed a wetting trend before 2000 and a drying trend after 2000; the arid areas in the basin showed a tendency of wetting whereas the humid areas exhibited a trend of drying. (3) The region where NDVI was positively correlated with SPEI accounted for approximately 70% of the basin area, demonstrating a similar spatio-temporal reversal phenomenon of the vegetation around 2000, indicating that the dry-wet condition is of great importance for the evolution of vegetation. (4) The SPEI showed a much more significant positive correlation with the soil water content which accounted for more than 95% of the basin area, implying that the soil water content was an important indicator to identify the dry-wet transition in the YZR basin.

Keywords: dry-wet regime; climate change; vegetation dynamics; GLDAS; GIMMS; Yarlung Zangbo River

1. Introduction

The current level of global climate change has been unprecedented in the past decades or even nearly a thousand years. Almost the entire world is experiencing a warming process, which is mainly characterized by rising temperatures, rising sea levels, retreating glaciers, and so on [1,2]. Global warming has exacerbated the global water cycle over the past century, causing a significant increase in the number of extreme weather events, such as storms, heat waves, floods, and droughts [3–9]. Droughts are one of the most threatening natural disasters in the world. They are caused by the below-average level of precipitation over a long period of time and are generally characterized by their high frequency, long duration, and wide range [10,11]. In the context of global warming, the frequency and intensity of droughts have increased significantly [12,13], which has seriously affected agricultural production, water resources, and ecosystems, and have led to economic losses, famines, epidemics, and desertification [14–21]. Because of the complexity of drought variability, it is challenging to objectively quantify the intensity, duration, and spatial extent of droughts [22–24]. Thus, numerous studies have attempted to improve drought detection and monitoring; a few objective indices have been developed on the basis of readily available climate data, such as the Palmer Drought Severity Index (PDSI) [25], Standard Precipitation Index (SPI) [26], and Standard Precipitation Evapotranspiration Index (SPEI) [27]. Among these indices, the PDSI, which is based on the supply and demand in the water balance, is one of the most widely used drought indices in the world. However, the PDSI has several deficiencies, including the strong influence of the calibration period, the limitation in spatial comparability, and the subjectivity in relating drought conditions to the index values [28]. The SPI can effectively represent the multiscale characteristics of droughts; however only the precipitation variability is considered in the SPI calculation, and the role of temperature is ignored. The effect of temperature is evident in initiating droughts, although droughts are primarily caused by a below-average level of precipitation [29]. Therefore, the SPEI was developed by Vicente-Serrano et al. [27], which not only considered the effects of temperature on drought severity but also considered the multiscale characteristics that were incorporated in the SPI. Since it was proposed in 2010, the SPEI has been widely used to monitor and assess the dry-wet conditions around the world [30–33].

Vegetation, linking the atmosphere, hydrosphere, and biosphere [34], is an important component in the terrestrial ecological system and has an obvious relationship with climate change through the physiological responses of plants, such as plant photosynthesis, respiration, and evapotranspiration. The dynamic changes of vegetation play a predictive role in regional climate change [35,36]. Climate change can also affect the spatial-temporal pattern of vegetation. Drought is one of the most frequent natural disasters and the response of vegetation to drought is a considerable scientific problem [37]. In general, an increased frequency of extreme drought was associated with decreased vegetation growth [19,38,39]. For example, Symeonakis et al. (2004) pointed out that drought was the main factor resulting in vegetation and soil degradation in sub-Saharan Africa [39]. Ahmadi et al. (2019) indicated that drought could affect the efficiency of water use in the ecosystem, subsequently disturbing the composition and functionality of terrestrial ecosystems [19]. In Northern China, drought-induced water stress caused a reduction in the terrestrial gross primary production [40]. Studies on the Qinghai-Tibet Plateau (QTP) and the Loess Plateau have revealed that there is a remarkable correlation between vegetation cover reduction and climate change [41,42]. Nevertheless, the magnitude of the response of vegetation to dry-wet conditions remains uncertain due to the complexity of the dry-wet transition and intrinsic drought sensitivity among vegetation types [21,43]. The intensity, duration, and timing of drought partly determine the effect of drought on vegetation productivity, where moderate drought with higher temperatures increases the net primary production (NPP), while severe drought causes a delayed response of NPP to precipitation [22,44]. As a satisfactory indicator of vegetation activities, the Normalized Difference Vegetation Index (NDVI), which was used in this study, has been widely and successfully used to detect the vegetation variations [36–44].

The Yarlung Zangbo River (YZR) basin located in the southeast QTP, is the most important river to understand the water cycle in the QTP because it is not only the largest river system in QTP with the largest mean annual flow (56% flows from the QTP), but also an important moisture transportation channel from the Indian Ocean to the inner region of QTP. Owing to great spatial heterogeneities of climatic conditions and enormous biological diversity, the YZR basin has always been the crucial area of global diversity and ecological protection [45]. A considerable amount of research on the impact of climate change on vegetation variations in the YZR basin has been conducted, revealing that vegetation and precipitation are positively correlated in the whole basin and that the vegetation cover change is restrained by the dry-wet regime, terrain, and other factors simultaneously [46–48].

Song et al. (2011) and Li et al. (2015) have demonstrated that the warming rate of the YZR basin was significantly higher than that of the global average, and the duration and magnitude of the drought have gradually aggravated [49,50]. The water resources problem would become more severe under the impact of precipitation and temperature due to the significant warming of the YZR basin in the future [51]. The rising temperature drives a basin-wide vegetation cover improvement, however it seems that decreasing precipitation does not inverse the overall vegetation greening trend [51]. Similar results occurred in Nepal, bounded by the Tibetan highland and the Himalaya, indicating that the correlation between NDVI and temperature was significantly positive, while NDVI exhibited a negative relationship with precipitation [52]. To sum up, the dry-wet transition is of great importance for the vegetation dynamics, whereas the response mechanism of vegetation to dynamic variations is still unclear. In addition, the special terrain, vulnerable ecological environment, and sensitivity to climate change make the YZR basin one of the hotspot regions for the studies of water-ecology-environment sustainable development under global warming. Therefore, it is crucial to investigate the transition characteristics of the dry-wet regime in the YZR basin and quantify its effect on the vegetation dynamic variations, which could provide a scientific reference for the sustainable development of the environment and ecosystem in the Qinghai-Tibet Plateau. The primary objectives of this study are: (1) to detect the changes of the dry-wet regime in the YZR basin; (2) to quantify the spatial-temporal variations of vegetation from long-term satellite-based NDVI data; and (3) to investigate the dynamic responses of vegetation to possible drivers of the dry-wet transition in the YZR basin.

2. Materials and Methods

2.1. Study Area

The Yarlung Zangbo River originates from the Gyama Langdzom Glacier is one of the highest rivers around the world. It is mainly composed of five tributaries, namely the Duoxiong Zangbo River, the Nianchu River, the Lhasa River, the Nyang River, and the Parlung Tsangpo River. The YZR basin, with a latitude of 28°00 'N–31°16' N and longitude of 82°00 'E–97°07' E, is located in the southeast QTP (Figure 1), with an area of about 24 km^2 and average altitude of more than 4000 m (ranging from 132 m to 7258 m). The climate of the YZR basin is characterized as cold plateau mountain climate with intense solar radiation and low air temperatures. The amount of precipitation gradually increases from northwest to southeast in the basin, which is mainly affected by the warm and humid airflow from the Bay of Bengal and the Indian Ocean. The mean annual precipitation in the basin is 300–500 mm, and the rate increases with elevation by 10–30 mm/100 m [52]. The total rainfall from June to September accounts for 60%–90% of the mean annual precipitation in the whole basin, indicating the precipitation has an uneven distribution within a year. Due to the complex topographical features and high altitude, the vegetation cover within the area exhibits distinct vertical zonality and varies from mountain forest, mountain broad-leaved forest, mountain coniferous forest, and subalpine shrub meadow to alpine meadow along with rising elevation [53].

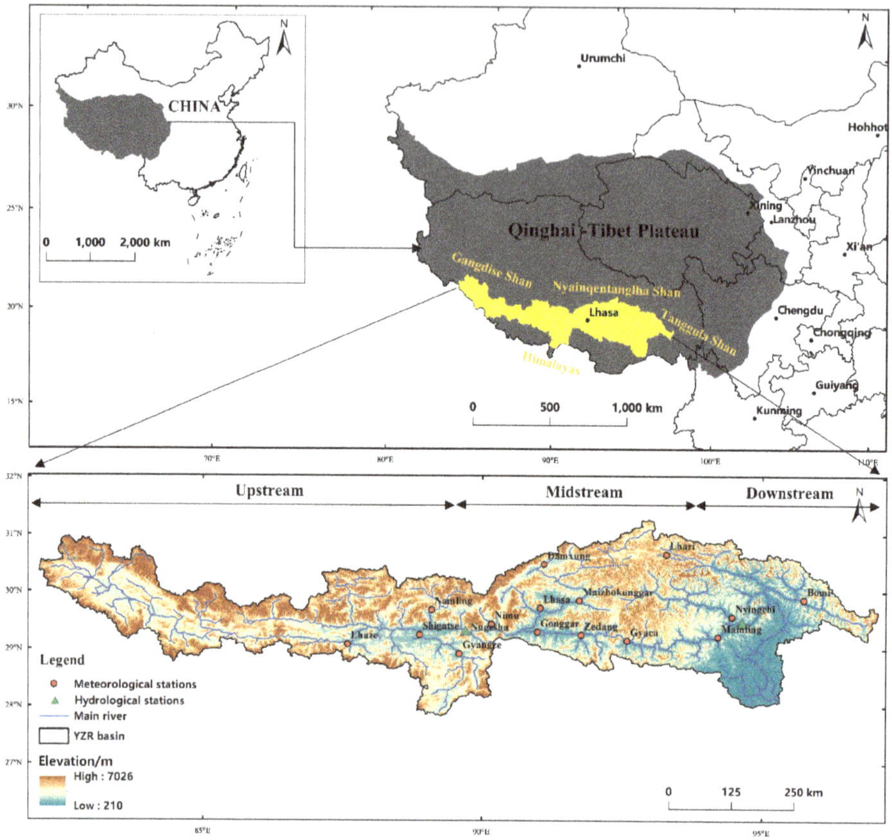

Figure 1. Location of the Yarlung Zangbo River basin and distribution of the hydro-meteorological stations.

2.2. Data

NDVI has been widely applied to detect and quantify the dynamic changes of vegetation in an extensive range [54]. Currently, various remote sensing satellite instruments can provide NDVI data, such as MODIS, SPOT/VEGETATION, and NOAA/AVHRR, etc. Compared with other vegetation index dataset, the GIMMS-NDVI3g, featured by its long time series and wide coverage, has proven to be one of the best datasets in describing vegetation growth dynamic changes [55]. Previous studies have shown that GIMMS NDVI dataset is significantly better than that of MODIS NDVI in reflecting dynamic changes over the Qinghai-Tibet Plateau [56]. In this study, the NOAA/AVHRR GIMMS production with a spatial resolution of 8 km × 8 km was used to calculate the NDVI. The data from January 1982 to December 2015 was derived from the third generation GIMMS NDV3g dataset, developed by the Goddard Aerospace Agency (http://ecocast.arc.nasa.gov/data/pub/gimms/3g/). Meanwhile, to further minimize the impact of clouds, atmosphere, and solar radiation angles on the NDVI values, the GIMMS NDVI3g data was preprocessed by employing S-G Filtering and Maximum Value Composite techniques to ensure the reliability of the research data and the accuracy of the results.

The Global Land Data Assimilation System (GLDAS), consisting of four different land surface models, i.e., CLM, NOAH, MOSAIC, and VIC [57], is a high-resolution land surface data assimilation system that is jointly managed by the American Goddard Space Flight Center and Environmental Forecast Center (http://ldas.gsfc.nasa.gov/gldas/GLDASvegetation.php), with two spatial resolutions

($0.25° \times 0.25°$ and $0.5° \times 0.5°$) and two temporal resolutions (3 hours and 1 month). The dataset with extensive sources is a combination of the surface observed data and the remote sensing satellite data. Compared to other remote sensing datasets, the GLDAS-NOAH data has a higher spatial and temporal resolution, a longer time span (1970 to present), and 28 variables (precipitation, air temperature, and soil moisture content, etc.). In this study, due to the limited number of meteorological gauging stations in the YZR basin, especially in the upper reaches, the monthly GLDAS-NOAH data at the $0.25° \times 0.25°$ spatial resolution from 1982 to 2015 were used to analyze the dry-wet transitions of the YZR basin, and to calculate the SPEI based on the performance evaluation of the GLDAS-NOAH data.

In-situ observations of the precipitation and surface air temperature from twenty meteorological gauging stations (as shown in Figure 1) in the YZR basin were used to evaluate the performance of the GLDAS-NOAH data.

2.3. Methods

The SPEI is a drought index based on the probability model, which was constructed by combining the potential evapotranspiration (PET) with the SPI [58]. Given the input and output of water resources, the calculation results of the SPEI mainly depicted the excess or deficit of water in an ecosystem within a certain period. The computational procedure of the SPEI can be divided into the following—calculation of the potential evapotranspiration (PET) based on the Thornthwaite method; computation of the difference value D between the precipitation and evapotranspiration, and finally, normalization of the value D. The specific calculation methods were as follows:

Firstly, the potential evapotranspiration was calculated and the difference between the potential evapotranspiration and precipitation was defined as:

$$D_i = P_i - PET_i \tag{1}$$

where D_i is the difference between monthly precipitation P_i and potential evapotranspiration PET_i.

Secondly, the D_i data series was normalized—the SPEI, which is similar to SPI, adopts the log-logistic of three parameters to normalize the cumulative values of the sequence of D_i data. The calculation formula was:

$$F(x) = \left[1 + \left(\frac{\alpha}{x-\gamma}\right)^{\beta}\right]^{-1} \tag{2}$$

where $F(x)$ is the probability distribution function, α, β, and γ represent the respective ratio, shape and source parameter, which could all be estimated by the linear distance method.

Finally, the cumulative probability P for a given time scale was derived, and then the normalized value of SPEI was calculated. The equations were as follows:

$$P = 1 - F(x) \tag{3}$$

When P<0.5

$$W = \sqrt{-2ln(p)} \tag{4}$$

$$SPEI = W - \frac{C_0 + C_1 W + C_2 W^2}{1 + d_1 W + d_2 w^2 + d_3 w^3} \tag{5}$$

When P>0.5

$$P = 1 - P \tag{6}$$

$$SPEI = -\left(W - \frac{C_0 + C_1 W + C_2 W^2}{1 + d_1 W + d_2 W^2 + d_3 W^3}\right) \tag{7}$$

The constants included $C_0 = 2.515517$, $C_1 = 0.802853$, $C_2 = 0.010328$, $d_1 = 1.432788$, $d_2 = 0.189269$, and $d_3 = 0.001308$. A negative SPEI value indicates dryness whereas a positive value represents wetness. The Table 1 lists the SPEI-based drought index classification criteria [27].

Table 1. SPEI meteorological drought index classification [27].

SPEI Value	Classification
$[2, +\infty] > 2.00$	Extremely wet
[1.5, 1.99]	Severely wet
[1.0, 1.49]	Moderately wet
[0.5, 0.99]	Slightly wet
[−0.49, 0.49]	Normal
[−0.99, −0.5]	Slightly dry
[−1.49, −1.0]	Moderately dry
[−1.99, −1.5]	Severely dry
$[-\infty, -2]$	Extremely dry

One-dimensional linear regression was employed to analyze SPEI and NDVI in the study area to describe the spatio-temporal trends of SPEI and NDVI between 1982 and 2015 [59]. The calculation formula was:

$$slope = \frac{n\sum_{i=1}^{n}(i \times C_i) - \sum_{i=1}^{n} i \times \sum_{i=1}^{n} C_i}{n \times \sum_{i=1}^{n} i^2 - \left(\sum_{i=1}^{n} i\right)^2} \quad (8)$$

where *Slope* represents the changing trends of NDVI and SPEI, n is the study temporal interval, $n = 34$, and C_i represents SPEI or NDVI for the year i. A significance test was performed on the changing trends of NDVI and SPEI ($P < 0.01$ indicates an extremely significant change, $P < 0.05$ indicates a significant change, and $P > 0.\ 05$ indicated the change is not significant).

The correlation coefficient (R) was used to investigate the linear relationship between NDVI and SPEI at the pixel scale in this study, which was defined as:

$$R = \frac{\sum_{i=1}^{n}[(x - \overline{x})(y - \overline{y})]}{\sqrt{\sum_{i=1}^{n}(x_i - \overline{x})^2 \sum_{i=1}^{n}(y_i - \overline{y})^2}} \quad (9)$$

where x_i and y_i represent the respective annual SPEI and NDVI values for the year i, $\overline{x}$ represents the mean annual SPEI values, and $\overline{y}$ represents the mean annual NDVI values. The significance test was used to illustrate the correlation between SPEI and NDVI ($P < 0.01$ indicates an extremely significant correlation, $P < 0.05$ indicates a significant correlation, and $P > 0.05$ indicates the correlation is not statistically significant).

Four statistical indicators were used in this study to evaluate the performance of the GLDAS-NOAH outputs in the YZR basin, which were the Pearson correlation coefficient (R), mean bias (MB), root-mean-square error ($RMSE$), and Nash-Sutcliffe efficiency coefficient (NSE). The Pearson correlation analysis was used to reflect the strength of the linear relationship between the compared datasets. The MB and $RMSE$ revealed the degree of deviation of the paired data. The MB provided information on the absolute overestimation or underestimation of the two paired datasets, whereas the $RMSE$ was a good reflection of the procedural precision. The NSE ranged from $[-\infty, 1]$, and the credibility of the simulation was much higher when it was approaching 1. These statistical indicators were defined as follows:

$$MB = \frac{1}{n}\sum_{i=1}^{n}(x_i - y_i) \quad (10)$$

$$RMSE = \sqrt{\frac{1}{n}\sum_{i=1}^{n}(x_i - y_i)^2} \quad (11)$$

$$NSE = 1 - \frac{\sum_{i=1}^{n}\left(x_o^i - x_m^i\right)^2}{\sum_{i=1}^{n}\left(x_o^i - \overline{x_o}\right)^2} \quad (12)$$

where n is the number of the data, and x_i nd y_i are the observed data and GLDAS-NOAH data, respectively. In Equation (12), x_o^i and x_m^i separately represent the observed value and model simulated value of the variable; $\overline{x_o}$ is the average value of the observed data. Generally, if the monthly $NSE > 0.5$ and monthly $R > 0.77$ (corresponding to the determination coefficient $R^2 > 0.6$), the model performance was considered to be acceptable [60].

As an effective and practical statistical method recommended by the World Meteorological Organization, the Mann-Kendall nonparametric test was applied to detect the significance of the trend. The detailed information can be obtained in [61–63]. In this study, the 0.1 significance level was used.

3. Results

3.1. Performance Evaluation of the GLDAS-NOAH Data

Due to the strong heterogeneity of the underlying factors over the YZR basin, such as the vegetation, soil type, and elevation, etc., it is essential to evaluate the performance of the GLDAS-NOAH data at both the site scale and the river basin scale. In this study, the observed monthly precipitation and temperature data from 1982–2015 from the twenty meteorological gauging stations covering the upper, middle, and lower reaches of the YZR basin (Figure 1) and extracted model outputs of corresponding grids from the GLDAS-NOAH dataset were utilized to conduct the performance evaluation.

3.1.1. Precipitation Performance at Site Scale

The correlation coefficients between the gridded monthly GLDAS-NOAH precipitation and the corresponding observed monthly precipitation showed high consistency ($R > 0.80$), except the Bomi station, which had a relatively lower R value of 0.57 (Table 2). This can be partly attributed to the dramatic topographic variations around the Bomi station ranging from 3100 m to 5000 m, whereas the elevation of the gauging station was 1300 m, which may have insufficiently represented the regional precipitation within the GLADS-NOAH grid area of approximately 625 km^2 (0.25° × 0.25°). Lv et al. [61] demonstrated a similar conclusion in the study on the performance evaluation of the TRMM satellite precipitation data in the YZR basin. As shown in Table 2, the MB between the GLDAS-NOAH outputs and in-situ measurements ranged from −49.95 mm to 22.99 mm (accounting for 0.07%–17.47% of annual precipitation) and the $RMSE$ ranged from 21.23 mm to 81.45 mm (accounting for 4.20%–29.44% of annual precipitation), which may be attributed to the mismatch of the point and grid scale.

Table 2. Statistical indicators of precipitation between the GLDAS-NOAH and gauging stations.

Gauging Station	*R*	*RMSE*	*MB*	*NSE*
Bomi	0.57	61.17	22.99	0.20
Jiali	0.86	37.72	21.09	0.69
Tsetang	0.89	51.17	−28.76	−0.24
Shigatse	0.91	52.03	−29.66	0.26
Nyingchi	0.9	29.33	−0.5	0.78
Lhasa	0.94	36.51	−18.96	0.54
Jiangzi	0.87	81.27	−48.23	−4.02
Dangxiong	0.93	23.84	−3.65	0.79
Baingoin	0.9	30.26	−9.83	0.31
Dengqen	0.87	31.2	15.76	0.69
Gyaca	0.88	40.46	−17.95	0.48

Table 2. *Cont.*

Gauging Station	*R*	*RMSE*	*MB*	*NSE*
Riwoqe	0.87	30.07	12.97	0.70
Lhunze	0.88	81.45	−49.95	−4.7
Haolong	0.81	30.72	−5.92	0.24
Maltrogangkar	0.94	24.92	−4.20	0.84
Nagqu	0.91	21.23	3.34	0.78
Namling	0.91	46.01	−22.11	0.36
Nyemo	0.90	60.06	−33.43	−0.94
Xainza	0.93	34.65	−14.62	0.25
Sog County	0.81	35.38	15.36	0.58

3.1.2. Temperature Performance at Site Scale

As shown in Table 3, the GLDAS-NOAH data and in-situ measurements showed a high agreement, i.e., $R \geq 0.81$, $-0.63 \leq NSE \leq 0.97$, $-1.01\ ^{\circ}C \leq MB \leq 6.85\ ^{\circ}C$, and $RMSE \leq 7.25\ ^{\circ}C$. In the Bomi station, the GLDAS-NOAH surface air temperature showed a larger discrepancy, i.e., the absolute value of $MB = 6.85\ ^{\circ}C$ and $\leq 4.59\ ^{\circ}C$ for other stations, and $RMSE = 7.25\ ^{\circ}C$ and $\leq 5.37\ ^{\circ}C$ for other stations. Similar to the performance of precipitation, the worse performance of temperature could be partly attributed to the dramatic topographic variations around the Bomi station, which may have caused uncertainties in the temperature estimation. Given that the monthly $NSE > 0.5$ and monthly $R > 0.77$ implied a good simulation, the high R and NSE from all stations indicates the high representation of the GLDAS-NOAH surface air temperature over the YZR basin.

Table 3. Statistical indicators of temperature between the GLDAS-NOAH and gauging stations.

Gauging Station	*R*	*RMSE*	*MB*	*NSE*
Bomi	0.92	7.25	6.85	−0.63
Jiali	0.86	3.71	−1.01	0.88
Tsetang	0.96	4.87	4.59	0.29
Shigatse	0.96	2.43	1.60	0.87
Nyingchi	0.97	4.09	3.83	0.40
Lhasa	0.97	3.34	2.95	0.69
Jiangzi	0.96	2.11	1.09	0.90
Dangxiong	0.93	2.53	0.04	0.97
Baingoin	0.90	3.30	0.65	0.73
Dengqen	0.87	4.51	2.85	0.47
Gyaca	0.81	4.94	2.97	0.39
Riwoqe	0.84	5.37	3.87	0.28
Lhunze	0.87	4.44	3.15	0.47
Haolong	0.83	5.37	3.78	0.28
Maltrogangkar	0.84	3.89	1.07	0.59
Nagqu	0.91	3.32	−0.45	0.78
Namling	0.85	4.56	2.82	0.40
Nyemo	0.83	4.20	1.85	0.57
Xainza	0.90	4.12	2.61	0.60
Sog County	0.88	4.06	1.86	0.63

3.1.3. Spatio-Temporal Patterns at River Basin Scale

In this study, the monthly precipitation and temperature in the entire YZR basin were calculated from 1982–2015 to further investigate the correspondence between the measured and GLDAS-NOAH data. As shown in Figure 2, the R values of precipitation and temperature of the two datasets reached 0.97 and 0.99 respectively, meaning there were good consistencies between the temporal variation patterns of precipitation and temperature, while the GLDAS-NOAH precipitation was larger than the measured precipitation, and the GLDAS-NOAH temperature was smaller than the measured data.

Although the GLDAS-NOAH overestimates and underestimates the precipitation and temperature respectively, such inconsistency is not the key issue in this study because of the following two reasons. Firstly, the observed data were at the point scale, while GLDAS-NOAH data represented the average performance at the pixel scale, approximately an area of 25 km × 25 km. With the high divergence of underlying factors, such as vegetation, soil type, and elevation, etc. over the YZR basin, it is rather hard for the scarce observed data to represent the pixel average features. Th gauge-based precipitation data showed that the difference in annual precipitation between Jiangzi and Shigatse was as much as 40%, while their distance was only 80km [64]. Secondly, this study focused on the spatio-temporal variation trends rather than absolute magnitudes. That is, the systematic overestimation or underestimation could be reasonably eliminated on a tendency or relationship analysis, if the variation patterns of the GLDAS-NOAH data fitted well with those of the in-situ data.

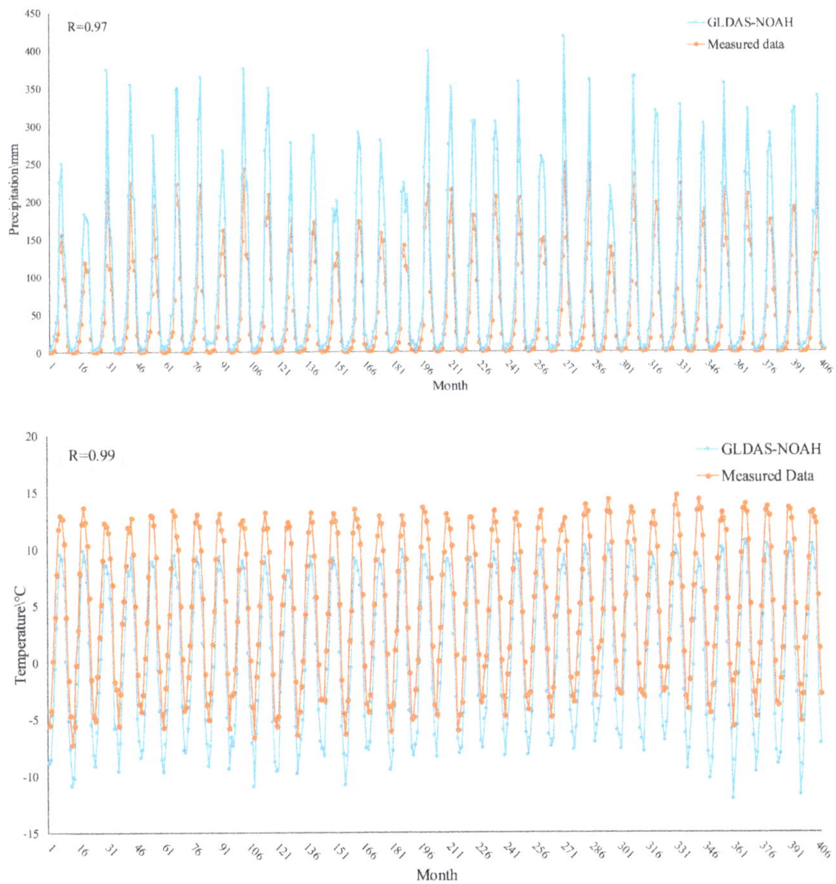

Figure 2. GLDAS-NOAH and measured monthly precipitation and temperature from 1982 to 2015.

In order to identify the spatial performance between the observed and the GLDAS-NOAH data, especially the vertical zonality characteristics, 20 meteorological stations and the corresponding grids were divided into four bands by elevation, i.e., 0–3500 m, 3500–4000 m, 4000–4500 m, and 4500–5000 m. As shown in Table 4 and Figure 3, the precipitation and temperature from the GLDAS-NOAH and in-situ data both decreased with the increase in elevation, implying that the GLDAS-NOAH data could represent the climate characteristics of the vertical zonality in the YZR basin. In terms of precipitation, the mean values of the observed data in 3500–4000 m and 4000–4500 m were 457.6 and 500.4 mm respectively, while those of the GLDAS-NOAH data were 640.4 mm and 619.7 mm respectively. This discrepancy was owing to the large *MB* values of the Lhunze and Nyemo stations as shown in Table 2. However, the variation ranges of the GLDAS-NOAH and in-situ precipitation at the four elevation bands were similar. Compared to the precipitation, the vertical variation characteristic of the GLDAS-NOAH temperature at the four elevation bands was much closer to that of the observed data. Although the observed temperature in the first elevation band seemed much bigger than the GLDAS-NOAH data, which was mainly ascribed to only three gauging stations in this elevation band, the GLDAS-NOAH temperature at the other three elevation bands exhibited a high agreement with the observed data, regardless of whether it was from the perspective of the mean values or from the perspective of the variation ranges. Similar results of the GLDAS performance in the YZR basin and QTP were demonstrated by Zhang et al. [65] and Zhang et al. [66].

Table 4. Elevations of the 20 gauging stations.

Gauging Station	Elevation/m
Bomi	2926
Jiali	4286
Tsetang	4656
Shigatse	3873
Nyingchi	3006
Lhasa	3794
Jiangzi	4025
Dangxiong	4279
Baingoin	4724
Dengqen	3954
Gyaca	3242
Riwoqe	3810
Lhunze	3922
Haolong	3794
Maltrogangkar	3810
Nagqu	4527
Namling	4001
Nyemo	3813
Xainza	4658
Sog County	4078

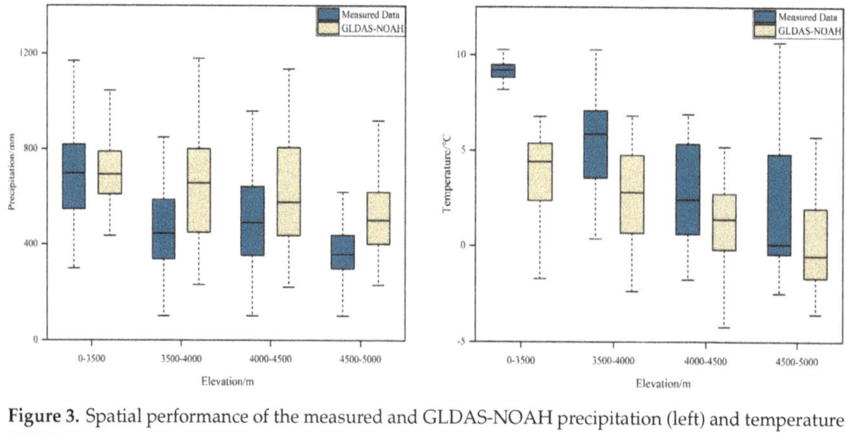

Figure 3. Spatial performance of the measured and GLDAS-NOAH precipitation (left) and temperature (right).

3.2. Transition Characteristics of the Dry-Wet Regime

The SPEI values at four time scales of 1-month, 3-month, 6-month, and 12-month were calculated from 1982 to 2015. SPEI 1 represented that when calculating monthly SPEI, the water deficit condition was taken into account only within the current month. SPEI 3 represented that the water deficit conditions of both the first two months and the current month were considered, and similar representations were used for SPEI 6 and SPEI 12. As shown in Figures 4 and 5, the temporal variations of the SPEI at the four time scales representing the dry-wet conditions in the YZR basin showed that SPEI 12 exhibited the highest agreement with the NDVI at both the annual and growing seasonal (from May to September) scales ($R \geq 0.6$), indicating that the highest dependency of the present dry-wet condition was on that of the preceding 12 months. In this study, SPEI 12 was used as the indicator to analyze the spatio-temporal variations of the dry-wet conditions at the annual and growing season scales in the YZR basin.

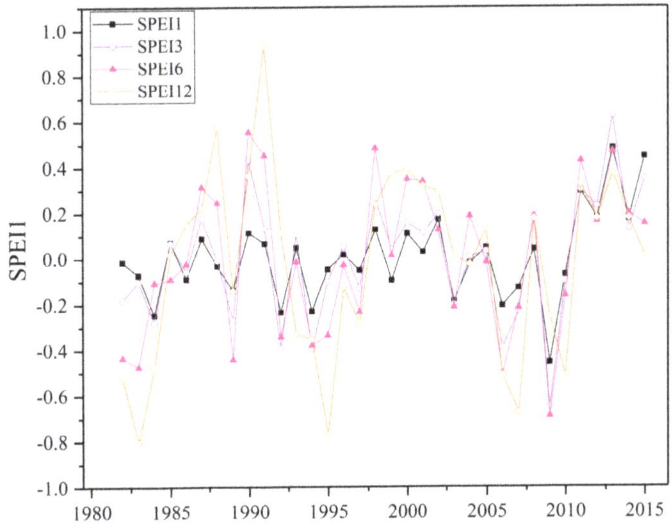

Figure 4. Changes of SPEI at different time scales.

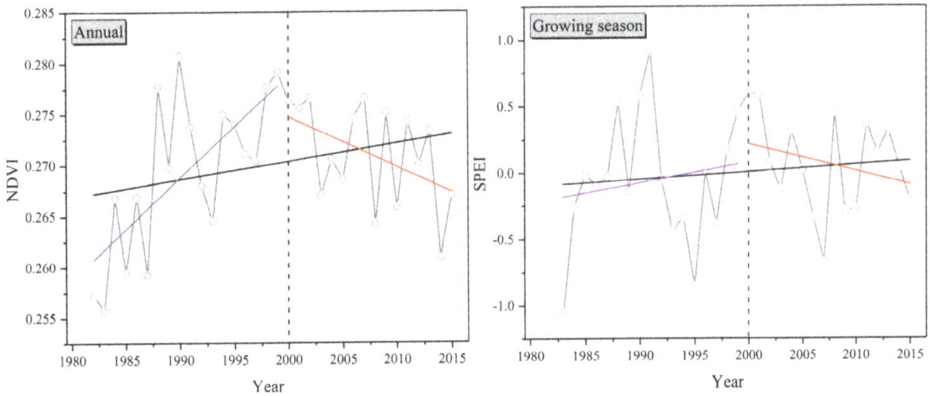

Figure 5. Changes of the annual (**left**) and growing season (**right**) SPEI.

As shown in Figure 5, in terms of the annual time scale, the SPEI showed a significantly increasing trend with a rate of 0.07/decade from 1982 to 2015 ($P < 0.1$, $n = 34$), implying an overall wetting tendency in the YZR basin. However, Wang (2016) and Liu (2015) demonstrated that the precipitation, temperature, and potential evapotranspiration, which were closely related to the dry-wet conditions in the YZR basin, all changed significantly in the late 1990s [67,68]. In order to detect whether there was a transition from wet to dry during the 1990s in the YZR basin, the year of 2000 was taken as the turning point to investigate the variation characteristics of SPEI. Interestingly, the two periods divided by 2000 exhibited the opposite changing trends, i.e., the wetting period was from 1982 to 1999 with a slightly increasing rate of 0.225/decade ($P > 0.1$, $n = 18$), and the drying period was from 2000 to 2015 with a significantly decreasing rate of 0.25/decade ($P < 0.1$, $n = 16$). Compared to the variation trend of the annual SPEI, the growing season SPEI showed a relatively non-significantly increasing trend with a rate of 0.053/decade from 1982 to 2015 ($P > 0.1$, $n = 34$). The opposite changing trend before and after 2000 was also exhibited at the growing season scale, where there was a slightly increasing trend with a rate of 0.16/decade ($P > 0.1$, $n = 18$) from 1982 to 1999 and a slightly decreasing trend with a rate of 0.21/decade ($P > 0.1$, $n = 16$) from 2000 to 2015.

In order to explore the spatial evolution characteristics of the dry-wet regime in the YZR basin, the SPEI was divided into three categories according to Table 1, i.e., SPEI < −0.5 (Dry), −0.5 < SPEI < 0.5 (Normal), and SPEI > 0.5 (Wet). Figure 6 shows the area ratios occupied by the different ranges of SPEI from 1982 to 2015 in the YZR basin at both the annual and growing season scales. The proportions of the annual variation pattern of the dry, wet, and normal areas were consistent with those of the growing season. Unlike the temporal opposite variation trend that occurred before and after 2000, the spatial reversal phenomenon of the dry-wet regime occurred in the three-year wet period from 1999 to 2001. Before this wet period, the proportion of wet areas showed an increasing trend, while the proportion of dry areas declined, causing a wetting condition of the YZR basin. The dry-wet regime reversed after this period, which experienced a drying period as implied by the decrease in wet areas and the increase in dry areas.

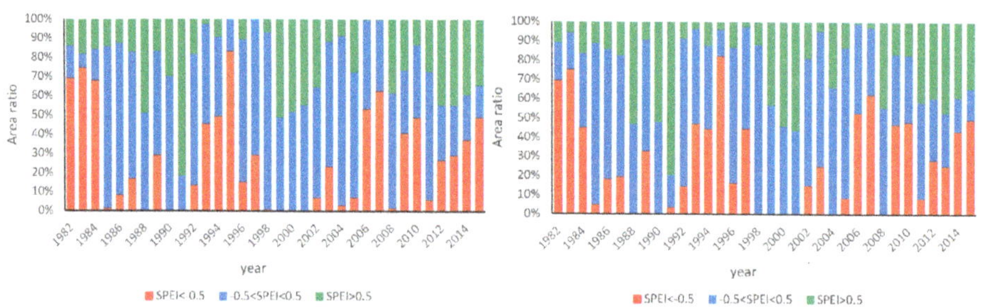

Figure 6. Area proportions of the dry, wet, and normal areas indicated by the annual **(left)** and growing season **(right)** SPEI in different ranges.

The determination of the turning point of the dry-wet regime in the YZR basin was of great importance to conduct further investigation in this study. As the most direct evidence for the dry-wet condition at the river basin scale, runoff has been widely used to represent the dry-wet characteristics [69–71]. The Nuxia hydrological station located downstream of the YZR basin, controls about 80% of the drainage areas of the YZR basin, and its long-term variation of runoff could be used to effectively indicate the dry-wet transition in the YZR basin. In this study, time series of runoff from 1982 to 2015 at the Nuxia hydrological station were adopted to further detect the transition point of the dry-wet regime by using the Mann-Kendall nonparametric test. The results of the Mann-Kendall significance test showed that the runoff showed a significantly increasing trend from 1982 to 1999 ($Z_c = 1.89$, $P < 0.1$), while the runoff exhibited a significantly decreasing trend ($Z_c = -1.76$, $P < 0.1$) from 2000 to 2015. Meanwhile, there was no point detected from 1982 to 2015 that could divide the time series into two parts with a significantly increasing and decreasing trend respectively. Combining with the SPEI spatio-temporal variations, it could be concluded that there was a transition of the dry-wet regime in the YZR basin, which occurred at the year 2000.

To investigate the spatial variation of the dry-wet regime and given the high spatial heterogeneity of the climate conditions in the YZR basin, the spatial distributions of the mean annual and growing season SPEI before and after 2000 were interpolated by using the Kriging method (Figure 7). It could be seen that the annual spatial distribution of the SPEI showed a high consistency with that of the growing season during both periods (1982–1999 and 2000–2015). Before 2000, the dry areas were mainly located in the eastern upstream and midstream regions of the study area, while the western upstream and southeastern downstream regions were relatively wet. Similar to the temporal reversal phenomenon at the year of 2000, the spatial pattern of the SPEI also displayed a reversal phenomenon. The annual and growing season spatial distribution of the SPEI indicated that the spatial distribution of the dry-wet regime in the YZR basin before and after 2000 was opposite. According to Figures 5 and 7, the overall wetting tendency of the YZR basin was mainly attributed to the remarkable wetting trend of the midstream region over the past 34 years, while the prominent trend of drying from 2000 to 2015 in the basin was ascribed to the fact that the western upstream and southeastern downstream turned drier.

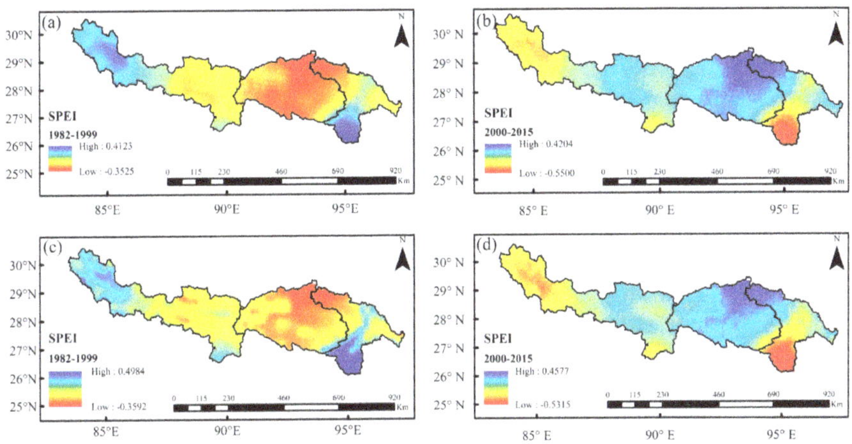

Figure 7. Annual (**a**, **b**) and growing season (**c**, **d**) spatial distributions of SPEI.

To further explore the mechanism of the spatial reversal of the dry-wet regime in the YZR basin around 2000, the annual and growing season *slopes* of the SPEI from 1982 to 2015 with the significance test were analyzed at the pixel scale (Figure 8). It could be found that the western upstream and southeastern downstream regions in the YZR basin presented a gradually drying trend over the past 34 years, while the eastern upstream and midstream regions became wetter, which was consistent with characteristics shown in Figure 7. According to Figure 8a, the mean annual SPEI exhibited an upward trend (*slope* > 0) accounting for 55.07% of the total basin area, indicating a humidification process, whereas an opposite trend (*slope* < 0) occurred in the remainder of the basin. As shown in Table 5, the results of the significance test at the pixel scale indicated that the areas where the annual SPEI showed an extremely significant decrease and a significantly decreasing trend accounted for 10.9% and 2.3% of the total basin respectively, while the areas with an extremely significant increase and a significantly increasing trend of the annual SPEI accounted for 18.05% and 8.25% respectively, inducing an overall wetting process from 1982 to 2015 in the YZR basin. Similar to the annual SPEI, the growing season SPEI indicated a wetting trend (*slope* > 0) across 55.54% of the basin area, while a drying trend (*slope* < 0) occurred in the remaining areas. The results of the significance test for the growing season SPEI at the pixel scale indicated that the area proportions with an extremely significant decrease and a significant decrease in SPEI were 8.7% and 2.03% respectively, while the areas with an extremely significant increase and a significant increase in SPEI occupied 15.21% and 8.87% of the basin respectively.

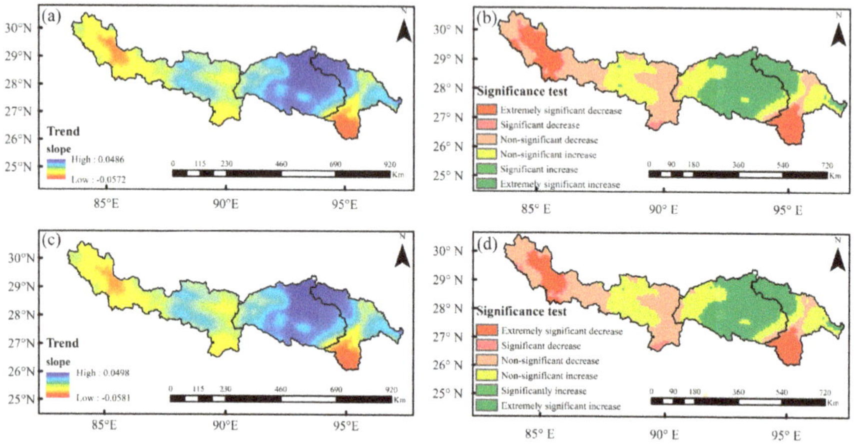

Figure 8. Annual (**a**, **b**) and growing season (**c**, **d**) variation trends of SPEI with the significance test.

Table 5. Area ratios of the SPEI occupied by different trends.

Significance Level	Area Ratio (%)	
	Annual	Growing Season
Extremely significant decrease	10.9	8.7
Significant decrease	2.3	2.03
Non-significant decrease	31.32	33.65
Non-significant increase	29.17	31.54
Significant increase	8.25	8.87
Extremely significant increase	18.05	15.21

Comprehensively taking the results of the temporal variations (Figure 5), spatial distribution characteristics (Figure 7), and trend analysis with the significance test (Figure 8) of the annual and growing season SPEI into account, the annual spatio-temporal variation characteristics of the SPEI were consistent with those of the growth season, which both showed a reversed phenomenon before and after 2000. In terms of the temporal variation, the YZR basin presented a wetting trend before 2000 and a drying trend after 2000, while from the perspective of the spatial pattern, the arid areas became wetter and humid areas became drier.

3.3. Spatio-Temporal Characteristics of Vegetation

Regarding 2000 as a turning point indicated by the transition of the dry-wet regime in the YZR basin, temporal variations of the annual and growing season NDVI were analyzed (Figure 9). The mean annual NDVI was 0.27, which fluctuated between 0.25 and 0.28 over the past 34 years and significantly increased at a rate of 0.002/decade ($P < 0.1$, $n = 34$), implying a gradual improvement of the vegetation cover in the YZR basin. However, the NDVI before and after 2000 showed a completely opposite tendency, that is, the NDVI increased significantly at a rate of 0.01/decade before 2000 ($P < 0.1$, $n = 18$), and decreased significantly at a rate of 0.006/decade after 2000 ($P < 0.1$, $n = 16$). With respect to the variations of the growing season NDVI, the mean NDVI was 0.34, which fluctuated between 0.31 and 0.35 during the past 34 years and increased non-significantly at a rate of 0.002/decade ($P > 0.1$,

$n = 34$). Similar to the changes of the annual NDVI, the growing season NDVI before 2000 showed a significantly increasing trend at a rate of 0.015/decade ($P < 0.1$, $n = 18$), and a significantly decreasing trend at a rate of 0.009/decade after 2000 ($P < 0.1$, $n = 16$). To sum up, the NDVI-indicated vegetation in the YZR basin gradually improved from the perspective of the annual and growing season NDVI variations from 1982 to 2015. However, since the beginning of the 21st century, the vegetation cover has decreased noticeably which corresponds with the simultaneously drying tendency in the whole basin, demonstrating that the improvement of the vegetation cover before 2000 was mainly induced by the gradual wetting of the basin while the degradation of the vegetation cover was attributed to the drying of the basin after 2000.

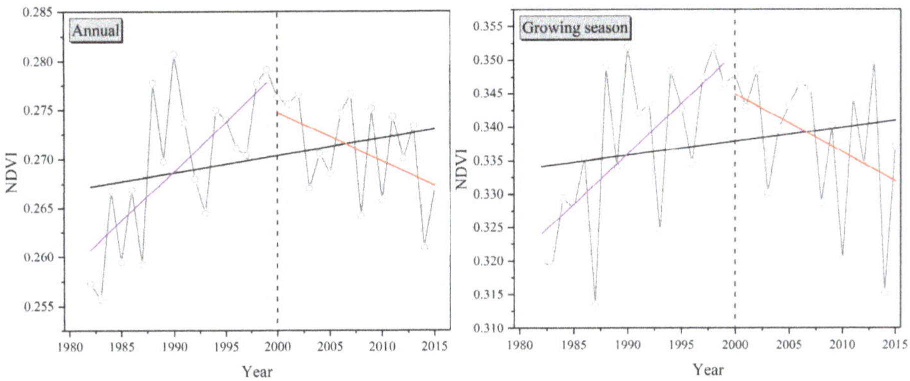

Figure 9. Changes of the annual (**left**) and growing season (**right**) NDVI.

The mean annual and growing season spatial distributions of the NDVI from 1982 to 2015 are portrayed in Figure 10. For the mean annual NDVI, the areas where the NDVI values ranged from 0.1 to 0.3 accounted for approximately 61.08% of the YZR basin, and were mainly located in the upper and middle reaches and high-altitude areas of the downstream regions. Only 5.59% of the area had an NDVI value above 0.7, and were largely concentrated in the midstream and southeastern downstream regions. From the perspective of the whole basin, the mean annual NDVI values gradually increased from northwest to southeast, implying a consistent improvement of the vegetation cover. The growing season vegetation cover indicated by the growing season NDVI exhibited a similar spatial distribution of the mean annual vegetation cover. The spatial variations of the NDVI-indicated vegetation in this study were similar to the vegetation cover dynamic monitoring results in the YZR basin reported by Jiang et al. [45].

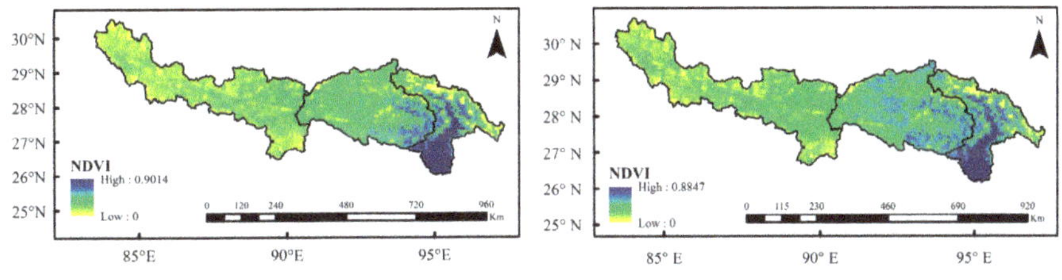

Figure 10. Annual (**left**) and growing season (**right**) spatial distributions of NDVI.

The spatial characteristics of the variation trends for the mean annual and growing season NDVI indicated by the NDVI *slope* in the YZR basin from 1982 to 2015 are depicted in Figure 11a,c respectively. In terms of the mean annual NDVI, approximately 59.4% of the NDVI *slope* in the basin was greater than 0, denoting an increasing trend of the vegetation cover, while the *slope* of the NDVI in the remaining areas was less than 0, implying a degradation of the vegetation cover. As shown in Figure 11b, d and Table 6, the results of the significance test showed that the areas with an extremely significant decrease and significant decrease of NDVI accounted for 7.3% and 4.84% of the whole study area respectively, while the areas with a non-significant increase and non-significant decrease of NDVI took up 31.24% and 27.86% respectively, and the areas with an extremely significant increase and significant increase of NDVI mainly located in the middle reaches occupied 22.76% and 6.1% respectively. The growing season NDVI showed a similar variation characteristic with that of the annual NDVI, i.e., the areas with increased vegetation cover had a proportion of 57.13% while the NDVI *slope* of the other areas was less than 0, indicating the degraded vegetation cover.

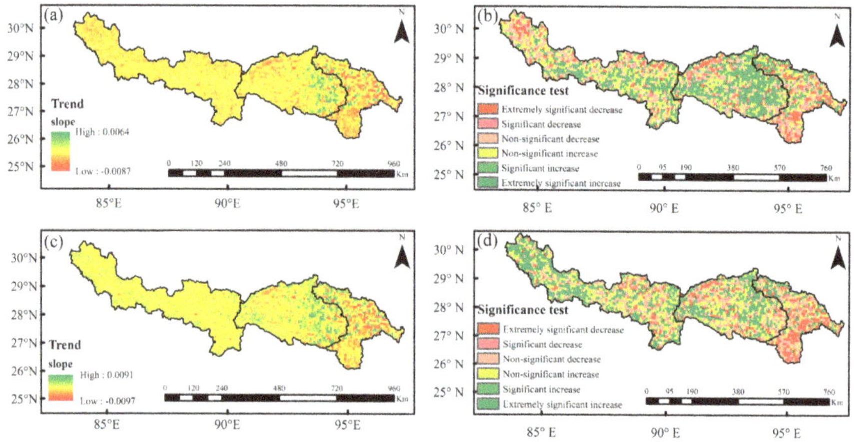

Figure 11. Annual (**a**, **b**) and growing season (**c**, **d**) spatial variations of NDVI with the significance test.

Table 6. Area ratios of NDVI with different change trends.

Significance Level	Area Ratio (%)	
	Annual	Growing Season
Extremely significant decrease	7.3	8.7
Significant decrease	4.84	3.7
Non-significant decrease	27.86	29.26
Non-significant increase	31.24	31.57
Significant increase	6.1	4.5
Extremely significant increase	22.76	22.28

In terms of the spatial variation characteristics for both the mean annual and growing season NDVI, it could be found that the vegetation cover upstream of the YZR basin seldom changed, which could be attributed to the specific land cover types in the upper reaches including the Gobi Desert, glaciers, and plateau meadows, which were less affected by climate change. However, in the midstream region, except for the high-altitude areas at the edge of the basin, the vegetation cover showed a dramatical upward tendency, while in the downstream region, except for the high-altitude areas such as Bomi, the vegetation cover presented a tendency of extremely significant decrease, which were consistent with the results obtained by Lv et al. [72]. Combined with the transition characteristics of the dry-wet regime in the YZR basin, it could be revealed that the vegetation cover increased in the midstream and eastern upstream regions where the climate became wetter, while the vegetation cover in the downstream and western upstream regions decreased where the climate turned drier.

3.4. Response of the Vegetation to the Dry-Wet Transition

Based on the analysis results of the spatio-temporal evolution characteristics of the SPEI and NDVI, it could be seen that the vegetation cover was closely related to the dry-wet regime in the YZR basin. To further explore the influence of the dry-wet conditions on vegetation cover, a correlation analysis between the SPEI and NDVI for both annual and growing season from 1982 to 2015 at the pixel scale was conducted. As shown in Figure 12, about 71.57% of the basin area showed a positive correlation between the mean annual SPEI and NDVI, while the remaining areas exhibited no or negative correlations. The results of the significance test as shown in Table 7 demonstrated that the areas where the NDVI was extremely significantly negatively correlated and significantly negatively correlated with the SPEI accounted for 0.85% and 0.95% respectively, and were mainly located in the eastern midstream and northwestern downstream regions. The areas showing a non-significantly positive correlation and non-significantly negative correlation between the NDVI and SPEI took up 56.28% and 26.37% of the basin area respectively, and were mainly concentrated in the western upstream and part of the midstream regions. In addition, the areas where the NDVI was extremely significantly positive and significantly positively correlated with the SPEI occupied 10.28% and 5.27% of the basin respectively, and were mostly concentrated in the junctions of the middle and upper reaches and the southeastern downstream regions. As for the correlation analysis between the growing season NDVI and SPEI, approximately 65.96% of the total area presented a positive correlation. The results of the significance test indicated that the areas showing an extremely significantly negative correlation and a significantly negative correlation between the NDVI and SPEI took up 1.36% and 1.55% respectively, while the areas where the NDVI showed a non-significantly positive correlation and a non-significantly negative correlation with the SPEI accounted for 49.62% and 30.98% respectively. The extremely significantly positive and significantly positive correlation between the NDVI and SPEI accounted for 10.73% and 5.8% of the total area respectively.

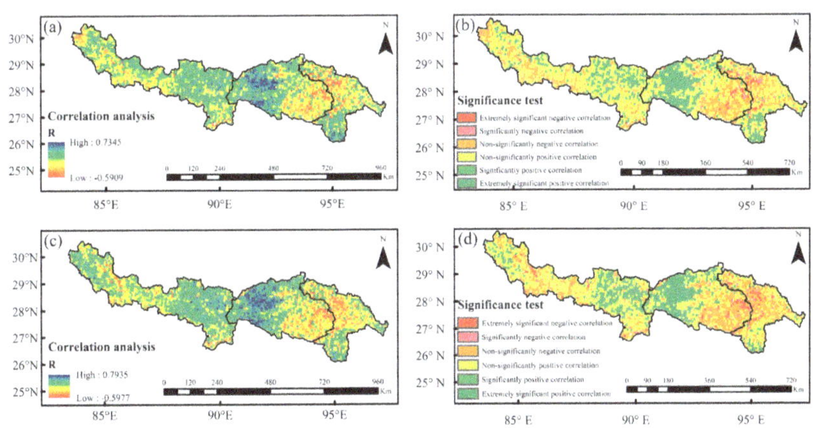

Figure 12. Annual (**a**, **b**) and growing season (**c**, **d**) correlation analysis between the SPEI and NDVI with the significance test.

By combining the temporal variations (Figures 5 and 9) with the spatial distributions of the variation trends (Figures 8 and 11) for the SPEI and NDVI, it was unequivocal that the spatio-temporal variation characteristics of the SPEI were consistent with those of the NDVI in the YZR basin, implying the important role of the dry-wet conditions on the vegetation dynamic variations. In terms of the temporal variation, before 2000, the YZR basin exhibited a wetting tendency and simultaneously the vegetation cover increased, while a tendency of drying was presented after 2000, and the vegetation cover consistently decreased. Such a synchronization phenomenon was also revealed from the perspective of the spatial distribution where the western upstream and southeastern downstream regions showed a drying trend with decreasing vegetation cover, while the midstream and eastern upstream regions displayed a wetting trend with increasing vegetation cover.

Table 7. Area ratios of NDVI and SPEI in different correlation degrees.

Significance Level	Area Ratio (%)	
	Annual	Growing Season
Extremely significant negative correlation	0.85	1.36
Significantly negative correlation	0.95	1.55
Non-significantly negative correlation	26.37	30.98
Non-significantly positive correlation	56.28	49.62
Significantly positive correlation	5.27	5.8
Extremely significant positive correlation	10.28	10.73

4. Discussion

4.1. Implication of the Soil Water Content for the Dry-Wet Regime

The soil water content is the link between the water cycle, carbon cycle, and energy cycle [73]. It is also a key factor in associating vegetation growth with precipitation [74]. Although drought is are caused by the below-average level of precipitation over a long period of time, it is also accompanied by the rising temperature, decreasing atmospheric humidity, increasing evapotranspiration, and declining soil water content [75]. Moreover, the soil water content is an important indicator of drought which could be used to efficiently identify dry-wet conditions, and the acquaintance of the soil water content variation is of great significance to agricultural production, ecological environment, resource allocation, and social-economic development in China [76]. To further identify the spatio-temporal characteristics of the dry-wet transition in the YZR basin, the soil water content variation associated with the SPEI was investigated in this study. The soil water content was characterized by the sum of soil moisture in four soil layers (0–10 cm, 10–40 cm, 40–100 cm, and 100–200 cm) extracted from the GLDAS-NOAH dataset, which was represented by the water depth in millimeters.

4.1.1. Spatio-Temporal Variation of the Soil Water Content

The spatial distributions of the mean annual and growing season soil water content in the YZR basin are shown in Figure 13. It could be seen that the regions with higher soil water content were mainly concentrated in the high-altitude areas such as the western upstream and the boundary of the middle and lower reaches. Except for the above areas, the soil water content decreased gradually from northwest to southeast. To analyze the spatial variation trend of the soil water, the *slope* was calculated at the pixel scale by utilizing the mean annual and growing season soil water content from 1982 to 2015. As shown in Figure 14, the regions with higher soil water content showed a downward tendency, while the regions with lower soil water content exhibited an increasing trend, which was consistent with the spatio-temporal variations of the SPEI. According to the results of the significance test, the areas showing an extremely significant decrease and significant decrease in soil water content were mostly distributed in the upstream and southeastern downstream regions. The areas with a non-significant increase and non-significant decrease in soil water content accounted for a small percentage of the total basin, and were mainly concentrated in Gongga, Lhasa, and Namling, while the areas where the soil water content increased significantly and extremely significantly were mainly situated in the midstream and northern downstream regions.

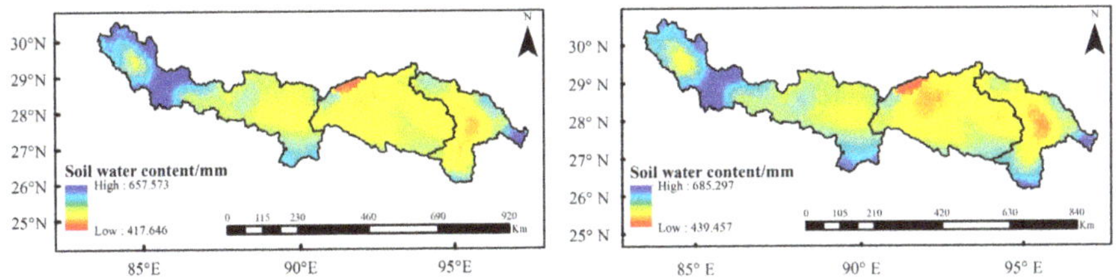

Figure 13. Annual (**left**) and growing season (**right**) spatial distributions of the soil water content.

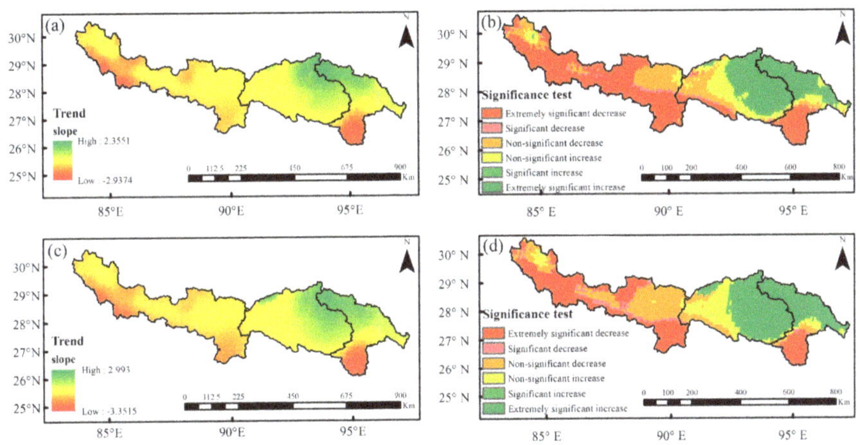

Figure 14. Annual (**a**, **b**) and growing season (**c**, **d**) spatial variation trends of the soil water content with the significance test.

4.1.2. Relationship Between the Soil Water Content and SPEI

In order to further elaborate on the transition of the dry-wet regime in the YZR basin during 1982–2015, the correlation analysis with the significance test between the annual and growing season soil water content and SPEI are illustrated in Figure 15. The correlation coefficient between the mean annual soil water content and the SPEI ranged from −0.204 to 0.923 in the basin, and nearly 99.59% of the total area showed a positive correlation between the soil water content and the SPEI. According to the results of the significance test, the area where the soil water content was extremely significantly positively correlated with the SPEI in the basin accounted for approximately 92.17% of the basin area. The correlation coefficient between the growing season soil water content and SPEI ranged from −0.206 to 0.938, and the soil water content was positively correlated with the SPEI in nearly 99.65% of the basin area, among which the areas with an extremely significant positive correlation reached 97.66%. Such close relationship between the soil water content and SPEI indicates the significantly important role of the soil moisture for identifying the dry-wet condition in the YZR basin [56,67,76].

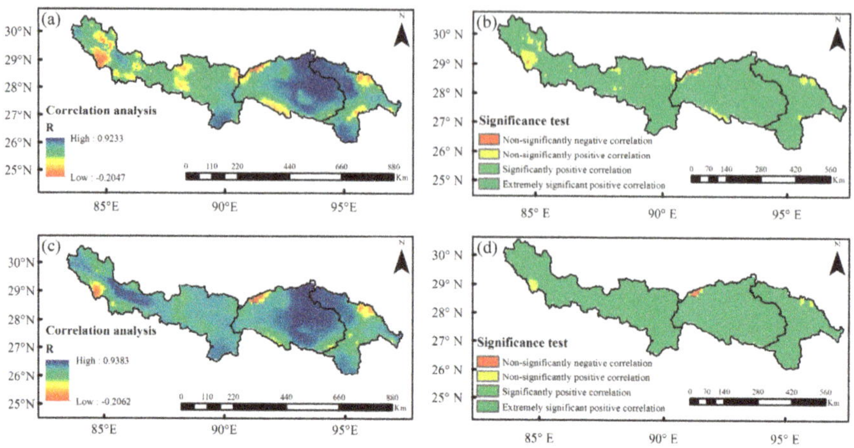

Figure 15. Annual (**a**, **b**) and growing season (**c**, **d**) correlation analysis between the SPEI and soil water content with the significance test.

4.2. Possible Climatic Drivers for the Dry-Wet Regime

The dry-wet transition is a comprehensive effect of climate change. Global warming leads to an increase in the terrestrial temperature, while the rising temperature also generates an increase in evapotranspiration, which results in drought and other meteorological disasters [77]. To further investigate the possible physical mechanism of the reversal phenomenon of the dry-wet regime in the YZR basin before and after 2000, the spatial distributions of variation trends for the mean annual and growing season precipitation, temperature, and potential evapotranspiration (PET) were comparatively analyzed. As shown in Figures 16 and 17, the spatial variation trends of the annual and growing season precipitation, temperature, and PET during 1982–2015 showed identical consistencies in the YZR basin, i.e., the precipitation in the western upstream and southeastern downstream gradually decreased, while the temperature and PET exhibited an increasing trend in the same areas; the increasing trends of the precipitation, temperature, and PET all occurred in the western upstream and eastern midstream; and a tendency for the increase in precipitation in the eastern midstream appeared, while there was a declining tendency for temperature and PET. Compared to the spatial variation characteristics of the SPEI (Figures 7 and 8), in the humid regions within the basin, the precipitation presented an increasing trend while the temperature and PET showed a decreasing trend, whereas, the reversal phenomenon occurred in the arid region within the basin, i.e., there was a decreasing trend in the precipitation and an increasing trend in the temperature and PET. To sum up, the spatial distributions and variation trends with the significance test of the precipitation, temperature, PET, and soil water content showed high consistencies with those of the dry-wet regime indicated by the SPEI in the YZR basin, especially the soil water content.

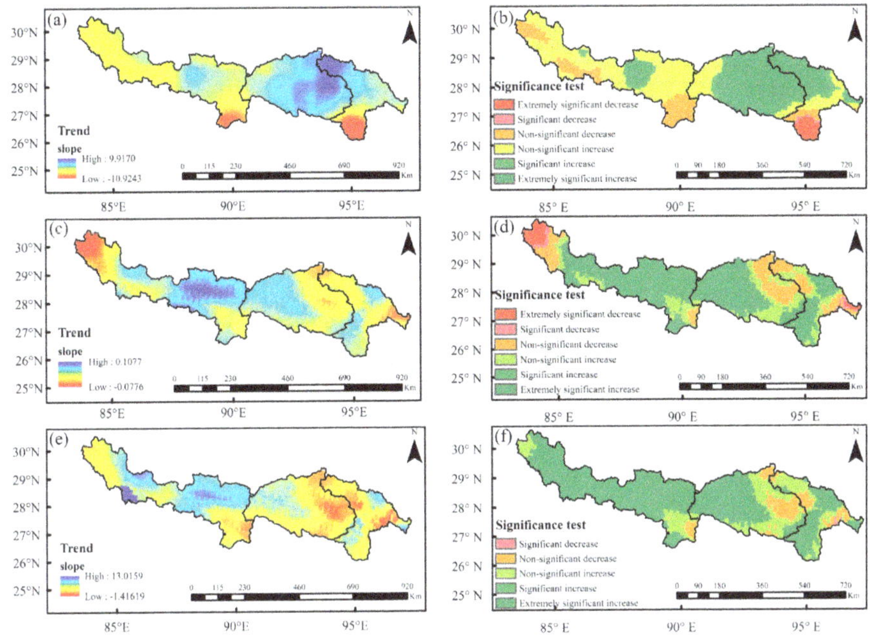

Figure 16. Analysis on the spatial variation **trends** of the annual precipitation (**a**, **b**), temperature (**c**, **d**), and PET (**e**, **f**) with the significance test.

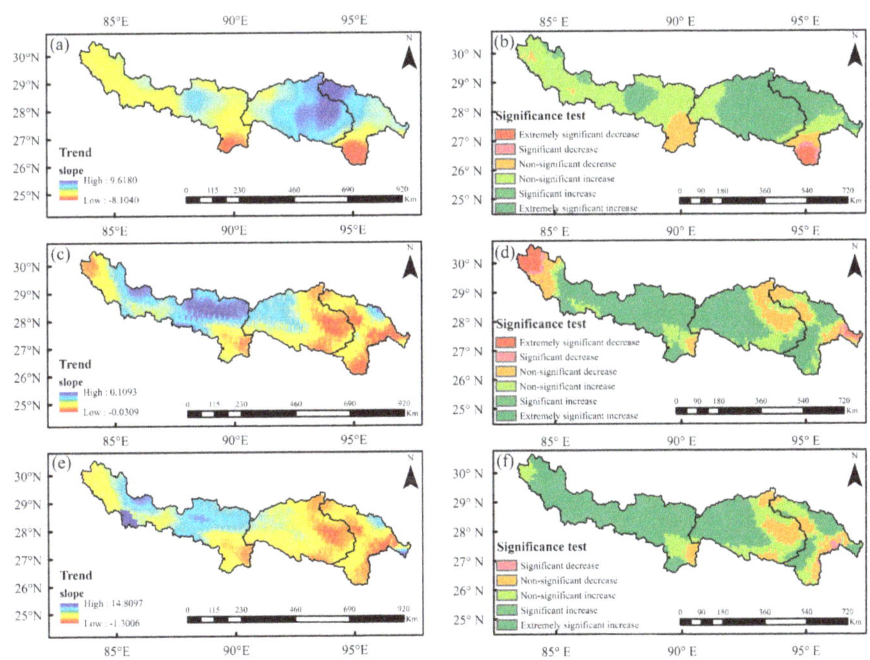

Figure 17. Analysis on the spatial variation trends of the growing season precipitation (**a**, **b**), temperature (**c**, **d**), and PET (**e**, **f**) with the significance test.

5. Conclusions

Based on the performance evaluation of the GLDAS-NOAH data by using the in-situ data from 20 gauging meteorological stations, the transition characteristics of the dry-wet regime in the YZR basin indicated by the SPEI and runoff were identified, which were combined with the satellite-based NDVI to investigate the dynamic responses of the vegetation. The hydro-meteorological factors including the soil water content, precipitation, surface air temperature, and PET were taken into consideration to explore the possible driving mechanisms of the dry-wet transition from 1982 to 2015 and the comprehensive impacts on vegetation dynamic variations. The conclusions are as follows:

(1) During the period of 1982–2015, the precipitation and surface air temperature simulated by the GLDAS-NOAH both showed high consistencies with the in-situ observed data in terms of the spatial and temporal variation patterns, implying a reasonable performance of the GLDAS-NOAH data in the YZR basin.

(2) The transition characteristics of the dry-wet regime indicated by the SPEI from 1982 to 2015 revealed that the YZR basin presented an overall wetting tendency, while the spatial and temporal characteristics of the dry-wet conditions reversed before and after 2000, that is, the basin showed a wetting trend before 2000 and a drying trend after 2000; the arid areas in the basin showed a tendency of wetting, whereas the humid areas exhibited a trend of drying.

(3) The NDVI-based vegetation in the YZR basin exhibited an overall increasing trend during 1982–2015, implying an improvement in the vegetation cover. However, a simultaneously reversal phenomenon of the vegetation variation was also detected before and after 2000, i.e., the vegetation cover in the basin increased before 2000 and decreased after 2000. From the perspective of spatial distribution, the area with higher NDVI values showed a degrading trend of vegetation cover, while the areas with lower NDVI values presented a greening trend with the improvement of vegetation cover.

(4) Approximately 70% of the basin area showed a positive correlation between the SPEI and NDVI, and were mainly located in the middle and lower reaches with high vegetation cover. However, the proportion of areas showing a negative correlation between SPEI and NDVI was small and not statistically significant. The high consistencies between the spatial and temporal variation characteristics of the NDVI and SPEI indicated that the dry-wet conditions played an important role in vegetation variations in the YZR basin.

(5) The investigation of possible driving factors for the dry-wet regime demonstrated that the spatial distributions and variation trends with the significance test of the precipitation, temperature, PET, and soil water content showed high consistencies with those of the dry-wet regime indicated by the SPEI in the YZR basin, emphasizing a great necessity to explore the physical mechanism of the dry-wet transition associated with the vegetation dynamics.

Author Contributions: Conceptualization, L.L.; methodology, L.L. and J.H.; software, J.H.; validation, Q.N., J.H. and L.L.; formal analysis, Q.N. and J.H.; investigation, Q.N. and J.H.; resources, L.L.; data curation, J.H. and H.L.; writing—original draft preparation, L.L and Q.N.; writing—review and editing, L.L. and H.L.; visualization, Q.N., J.H. and H.L.; supervision, L.L.; project administration, L.L. and Z.X.; funding acquisition, L.L. and Z.X.

Funding: This research was financially supported by the National Natural Science Foundation of China (91647202, 41890822, 51509247).

Acknowledgments: Thank the National Meteorological Information Center of China Meteorological Administration for archiving the observed climate data (http://data.cma.cn/). Thank the GIMMS team and the Global Land Data Assimilation System team for making the data freely available. Authors have great thanks to the anonymous reviewers for providing so valuable comments.

Conflicts of Interest: The authors declare no conflict of interest.

References

1. Shen, Y.P.; Wang, G.Y. Key Findings and Assessment Results of IPCC WGI Fifth Assessment Report. *J. Glaciol. Geocryol.* **2013**, *35*, 1068–1076. [CrossRef]

2. Vaughan, D.G.; Comiso, J.C.; Allison, I.; Carrasco, J.; Kaser, G.; Mote, P.; Murray, T.; Paul, F.; Ren, J.; Rignot, E.; Solomina, O.; Steffen, K.; Zhang, T. Observations: Cryosphere 2013. In *Contribution of Working Group I to the Fifth Assessment Report of the Intergovernmental Panel on Climate Change*; Stocker, T.F., Ed.; Cambridge University Press: Cambridge, UK.
3. Ziegler, A.D.; Sheffield, J.; Maurer, E.P.; Nijssen, B.; Lettenmaier, D.P. Detection of Intensification in Global and Continental-Scale Hydrological Cycles: Temporal Scale of Evaluation. *J. Clim.* **2003**, *16*, 535–547. [CrossRef]
4. Allan, R.P.; Soden, B.J. Atmospheric warming and the amplification of precipitation extremes. *Science* **2008**, *321*, 1481–1484. [CrossRef] [PubMed]
5. Yang, H.B.; Yang, D.W. Climatic factors influencing changing pan evaporation across China from 1961 to 2001. *J. Hydrol.* **2012**, *414*, 184–193. [CrossRef]
6. Yang, H.B.; Yang, D.W.; Hu, Q.F.; Lv, H.F. Spatial variability of the trends in climatic variables across China during 1961–2010. *Theor. Appl. Climatol.* **2015**, *120*, 773–783. [CrossRef]
7. Huang, S.Z.; Leng, G.Y.; Huang, Q.; Xie, Y.Y.; Liu, S.Y.; Meng, E.H.; Li, P. The asymmetric impact of global warming on US drought types and distributions in a large ensemble of 97 hydro-climatic simulations. *Sci. Rep. UK* **2017**, *7*, 5891. [CrossRef] [PubMed]
8. Tong, S.; Lai, Q.; Zhang, J.; Bao, Y.; Lusi, A.; Ma, Q.; Li, X.; Zhang, F. Spatiotemporal drought variability on the Mongolian Plateau from 1980-2014 based on the SPEI-PM, intensity analysis and Hurst exponent. *Sci. Total Environ.* **2018**, *615*, 1557–1565. [CrossRef] [PubMed]
9. Gian-Reto, W.; Eric, P.; Peter, C.; Annette, M.; Camille, P.; Beebee, T.J.C.; Jean-Marc, F.; Ove, H.G.; Franz, B. Ecological responses to recent climate change. *Nature* **2018**, *416*, 389–395. [CrossRef]
10. Wang, H.R.; Hong, S.Y.; Qin, D.Q. Discussion on related issues of drought and water shortage. *Water Resour. Prot.* **2017**, *33*, 1–4.
11. Zhang, Q.; Yao, Y.B.; Li, Y.H.; Luo, Z.X.; Zhang, C.J.; Li, D.L.; Wang, R.Y.; Wang, J.S.; Chen, T.Y.; Xiao, G.J.; et al. Research progress and prospect on the monitoring and early warning and mitigation technology of meteorological drouglU disaster in northwest China. *Adv. Earth Sci.* **2015**, *30*, 196–213. [CrossRef]
12. Dai, A. Drought Under Global Warming: A Review. *Wiley Interdiscip. Rev. Clim. Chang.* **2010**, *2*, 45–65. [CrossRef]
13. Dai, A. Increasing drought under global warming in observations and models. *Nat. Clim. Chang.* **2013**, *3*, 52–58. [CrossRef]
14. Beguería, S.; Vicente-Serrano, S.M.; Angulo-Martínez, M. A Multiscalar Global Drought Dataset: The SPEI base: A New Gridded Product for the Analysis of Drought Variability and Impacts. *Bull. Am. Meteorol. Soc.* **2010**, *91*, 1351–1356. [CrossRef]
15. Liu, Z.Y.; Zhou, P.; Zhang, F.Q.; Liu, X.D.; Chen, G. Spatiotemporal characteristics of dryness/wetness conditions across Qinghai Province, Northwest China. *Agric. For. Meteorol.* **2013**, *182–183*, 101–108. [CrossRef]
16. Lobell, D.B.; Field, C.B. Global scale climate crop yield relationships and the impacts of recent warming. *Environ. Res. Lett.* **2007**, *2*, 14002. [CrossRef]
17. Tao, F.L.; Yokozawa, M.; Xu, Y.L.; Hayashi, Y.; Zhang, Z. Climate changes and trends in phenology and yields of field crops in China, 1981–2000. *Agric. For. Meteorol.* **2006**, *138*, 92. [CrossRef]
18. Liu, Y.; Wang, E.L.; Yang, X.G.; Wang, J. Contributions of climatic and crop varietal changes to crop production in the North China Plain, since 1980s. *Glob. Chang. Biol.* **2010**, *16*, 2287–2299. [CrossRef]
19. Ahmadi, B.; Ahmadalipour, A.; Tootle, G.; Moradkhani, H. Remote Sensing of Water Use Efficiency and Terrestrial Drought Recovery across the Contiguous United States. *Remote Sens. Basel* **2019**, *11*, 731. [CrossRef]
20. Huang, L.; He, B.; Han, L.; Liu, J.J.; Wang, H.Y.; Chen, Z.Y. A global examination of the response of ecosystem water-use efficiency to drought based on MODIS data. *Sci. Total Environ.* **2017**, *601–602*, 1097–1107. [CrossRef]
21. Yu, Z.; Wang, J.; Liu, S.; Rentch, J.S.; Sun, P.; Lu, C. Global gross primary productivity and water use efficiency changes under drought stress. *Environ. Res. Lett.* **2017**, *12*, 14016. [CrossRef]
22. Jiang, Y.L.; Wang, R.H.; Peng, Q.; Wu, X.Q.; Ning, H.S.; Cheng, L. The relationship between drought activity and vegetation cover in Northwest China from 1982 to 2013. *Nat. Hazards* **2018**, 1–19. [CrossRef]
23. Duan, W.L.; Hanasaki, N.; Shiogama, H.; Chen, Y.N.; Zou, S.; Nover, D.; Zhou, B.T.; Wang, Y. Evaluation and Future Projection of Chinese Precipitation Extremes Using Large Ensemble High-Resolution Climate Simulations. *J. Clim.* **2019**, *32*, 2169–2183. [CrossRef]

24. Duan, W.L.; He, B.; Takara, K.; Luo, P.P.; Hu, M.C.; Alias, N.E.; Nover, D. Changes of precipitation amounts and extremes over Japan between 1901 and 2012 and their connection to climate indices. *Clim. Dyn.* **2015**, *45*, 2273–2292. [CrossRef]
25. Palmer, W.C. Meteorological Drought. In *Weather Bureau Research Paper*; US Weather Bureau: Washington, DC, USA, 1965; Volume 45, pp. 1–58.
26. McKee, T.B.; Doesken, N.J.; Kleist, J. *The Relationship of Drought Frequency and Duration to Time Scales*; American Meteorological Society: Boston, MA, USA, 1993; Volume 22, pp. 179–183.
27. Vicente-Serrano, S.M.; Beguería, S.; Lópezmoreno, J.I. A multiscalar drought index sensitive to global warming: the standardized precipitation evapotranspiration index. *J. Clim.* **2010**, *23*, 1696–1718. [CrossRef]
28. Chen, H.P.; Sun, J.Q. Changes in Drought Characteristics over China Using the Standardized Precipitation Evapotranspiration Index. *J. Clim.* **2015**, *28*, 75745793. [CrossRef]
29. Yang, Q.; Li, M.X.; Zheng, Z.Y.; Ma, Z.G. Regional applicability of seven meteorological drought indices in China. Science China Earth Sciences. *Sci. China Earth Sci.* **2017**, *60*, 745–760. [CrossRef]
30. Lorenzo-Lacruz, J.; Vicente-Serrano, S.M.; López-Moreno, J.I.; Beguería, S.; García-Ruiz, J.M.; Cuadrat, J.M. The impact of droughts and water management on various hydrological systems in the headwaters of the Tagus River (central Spain). *J. Hydrol.* **2010**, *386*, 13–26. [CrossRef]
31. Kumar, K.N.; Rajeevan, M.; Pai, D.S.; Srivastava, A.K.; Preethi, B. On the observed variability of monsoon droughts over India. *Weather Clim. Extremes* **2013**, *1*, 42–50. [CrossRef]
32. Mcclaran, M.P.; Wei, H. Recent drought phase in a 73-year record at two spatial scales: Implications for livestock production on rangelands in the Southwestern United States. *Agric. For. Meteorol.* **2014**, *197*, 40–51. [CrossRef]
33. Xu, K.; Yang, D.W.; Yang, H.B.; Li, Z.; Qin, Y.; Shen, Y. Spatio-temporal variation of drought in China during 1961–2012: A climatic perspective. *J. Hydrol.* **2015**, *526*, 253–264. [CrossRef]
34. Sun, H.Y.; Wang, C.Y.; Niu, Z.; Bukhosor; Li, B. Analysis of the Vegetation Cover Change and the Relationship between NDVI and Environmental Factors by Using NOAA Time Series Data. *J. Remote Sens.* **1998**, *2*, 204–210. [CrossRef]
35. Zhang, X.L.; Wu, S.; Yan, X.D.; Chen, Z.J. A global classification of vegetation based on NDVI, rainfall and temperature. *Int. J. Clim.* **2016**, 37. [CrossRef]
36. Wang, T.; Peng, S.S.; Lin, X.; Chang, J.F. Declining snow cover may affect spring phenological trend on the Tibetan Plateau. *Proc. Natl. Acad. Sci. USA* **2013**, *110*, E2854–E2855. [CrossRef]
37. Zhang, Q.; Kong, D.D.; Singh, V.P.; Shi, P.J. Response of vegetation to different time-scales drought across China: Spatiotemporal patterns, causes and implications. *Glob. Planet. Chang.* **2017**, *152*, 1–11. [CrossRef]
38. Xu, X.T.; Piao, S.L.; Wang, X.H.; Chen, A.P.; Ciais, P.; Myneni, R.B. Spatio-temporal patterns of the area experiencing negative vegetation growth anomalies in China over the last three decades. *Environ. Res. Lett.* **2012**, *7*, 35701. [CrossRef]
39. Symeonakis, E.; Drake, N. Monitoring desertification and land degradation over sub-Saharan Africa. *Int. J. Remote Sens.* **2004**, *25*, 573–592. [CrossRef]
40. Yuan, W.P.; Liu, D.; Dong, W.J.; Liu, S.G. Multiyear precipitation reduction strongly decrease carbon uptake over North China. *J. Geophys. Res. Biogeosci.* **2014**, *119*, 881–896. [CrossRef]
41. Sun, W.Y.; Song, X.Y.; Mu, X.M.; Gao, P.; Wang, F.; Zhao, G.G. Spatiotemporal vegetation cover variations associated with climate change and ecological restoration in the Loess Plateau. *Agric. For. Meteorol.* **2015**, *209–210*, 87–99. [CrossRef]
42. Chen, Y.N.; Deng, H.J.; Li, B.F.; Zhi, L.; Xu, C.C. Abrupt change of temperature and precipitation extremes in the arid region of Northwest China. *Quat. Int.* **2014**, *336*, 35–43. [CrossRef]
43. Smith, M.D. An ecological perspective on extreme climatic events: A synthetic definition and framework to guide future research. *J. Ecol.* **2011**, *99*, 656–663. [CrossRef]
44. Sun, B.F.; Zhao, H.; Wang, X.K. Effects of drought on net primary productivity: Roles of temperature, drought intensity, and duration. *Chin. Geogr. Sci.* **2016**, *26*, 270–282. [CrossRef]
45. Jiang, L.; Feng, W.L.; Guo, B. Analysis of Dynamic Monitoring of Vegetation Change And the Correlation With Precipitation Factor in Yalu Tsangpo River Basin During the Past 13 Years. *Res. Environ. Yangtze Basin* **2014**, *23*, 1610–1619. [CrossRef]

46. Li, H.D.; Li, Y.K.; Shen, W.S.; Li, Y.N.; Jie, L.; Lu, X.Y.; Xia, X.; Jiang, J. Elevation-Dependent Vegetation Greening of the Yarlung Zangbo River Basin in the Southern Tibetan Plateau, 1999–2013. *Remote Sens. Basel* **2015**, *7*, 16672–16687. [CrossRef]
47. Zhang, W.; Fu, X.F. Analysis and Evaluation of principal climatic factors of NDVI in the Yarlung Zangbo River Basin. *J. Phys. Conf. Ser.* **2015**, *622*, 12048. [CrossRef]
48. Guo, B.; Jiang, L.; Da-Zhuan, G.E.; Shang, M. Driving Mechanism of Vegetation Coverage Change in the Yarlung Zangbo River Basin under the Stress of Global Warming. *J. Trop. Subtrop. Bot.* **2017**, *25*, 209–217. [CrossRef]
49. Li, B.Q.; Zhou, W.; Zhao, Y.Y.; Ju, Q.; Yu, Z.B.; Liang, Z.M.; Acharya, K. Using the SPEI to Assess Recent Climate Change in the Yarlung Zangbo River Basin, South Tibet. *Water* **2015**, *7*, 5474–5486. [CrossRef]
50. Song, M.H.; Ma, Y.M.; Zhang, Y.; Li, M.S.; Ma, W.Q.; Sun, F.L. Analyses of characteristics and trend of temperature variation along the Brahmaputra valley. *Clim. Environ. Res.* **2011**, *16*, 760–766. [CrossRef]
51. Liu, W.F.; Xu, Z.X.; Li, F.P.; Su, L.Q. Climate Change Scenarios in the Yarlung Zangbo River Basin Based on ASD Model. *Plateau Meteorol.* **2014**, *33*, 26–36.
52. Gong, T.L. Mechanism of Hydrological Cycle and Water Resources Management Tragedy in the Yarlung Tsangpo River Basin. Ph.D. Thesis, Beijing Normal University, Beijing, China, 2006.
53. Liu, Z.; Yao, Z.; Huang, H.; Wu, S.; Liu, G. Land use and climate changes and their impacts on runoff in the Yarlung Zangbo River Basin, China. *Land Degrad. Dev.* **2014**, *25*, 203–215. [CrossRef]
54. Zhong, L.; Suhy, M. Assessment of vegetation dynamics and their response to variations in precipitation and temperature in the Tibetan Plateau. *Clim. Chang.* **2010**, *3–4*, 519–535. [CrossRef]
55. Beck, P.S.A.; Goetz, S.J. Satellite observations of high northern latitude vegetation productivity changes between 1982 and 2008: Ecological variability and regional differences. *Environ. Res. Lett.* **2011**, *6*, 45501–45510. [CrossRef]
56. Du, J.Q.; Shu, J.M.; Wang, Y.H.; Li, Y.C.; Zhang, L.B.; Guo, Y. Comparison of GIMMS and MODIS normalized vegetation index composite data for Qing-hai Tibet Plateau. *Chin. J. Appl. Ecol.* **2014**, *25*, 533–544. [CrossRef]
57. Li, X.; Gao, Y.H.; Wang, W.Z.; Lan, Y.C.; Xu, J.W.; Li, K. Climate Change and Applicability of GLDAS in the Headwater of the Yellow River Basin. *Adv. Earth Sci.* **2014**, *29*, 531–540. [CrossRef]
58. Wang, Z.L.; Huang, Z.Q.; Li, J.; Zhong, R.D.; Huang, W.W. Assessing impacts of meteorological drought on vegetation at catchment scale in China based nn SPEI and NDVI. *Trans. Chin. Soc. Agric. Eng.* **2016**, *32*, 177–186. [CrossRef]
59. Lv, Y.; Yang, S.T.; Cai, M.Y.; Zhou, Q.W.; Dong, G.T. The Applicability Analysis of TRMM Precipitation Data in the Yarlung Zangbo River Basin. *J. Nat. Resour.* **2013**, *28*, 1414–1425.
60. Santhi, C.; Arnold, J.G.; Williams, J.R.; Dugas, W.A.; Srinivasan, R.; Hauck, L.M. Validation of the swat model on a large river basin with point and nonpoint sources. *J. Am. Water Resour. Assoc.* **2001**, *5*, 1169–1188. [CrossRef]
61. Yang, Y.H.; Tian, F. Abrupt change of runoff and its major driving factors in Haihe River Catchment, China. *J. Hydrol.* **2009**, *374*, 373–383. [CrossRef]
62. Huang, S.Z.; Chang, J.X.; Qiang, H.; Chen, Y.T. Spatio-temporal Changes and Frequency Analysis of Drought in the Wei River Basin, China. *Water Resour. Manag.* **2014**, *28*, 3095–3110. [CrossRef]
63. Zhang, Q.; Singh, V.P.; Bai, Y.G. SPI-based evaluation of drought events in Xinjiang, China. *Nat. Hazards* **2012**, *64*, 481–492. [CrossRef]
64. Huang, X.; Wang, Z.G.; Sang, Y.F.; Yang, M.Y.; Liu, X.C.; Gong, T.L. Precision of data in three precipitation datasets of the Yarlung Zangbo River Basin. *Prog. Geogr.* **2016**, *35*, 339–348. [CrossRef]
65. Zhang, H.; Zhang, L.L.; Li, J.; An, R.D.; Deng, Y. Climate and Hydrological Change Characteristics and Applicability of GLDAS Data in the Yarlung Zangbo River Basin, China. *Water* **2018**, *10*, 254. [CrossRef]
66. Zhang, H.B.; Zhang, F.; Ye, M.; Che, T.; Zhang, G.Q. Estimating daily air temperatures over the Tibetan Plateau by dynamically integrating MODIS LST data. *J. Geophys. Res. Atmos.* **2016**, *121*, 11, 411–425, 441. [CrossRef]
67. Wang, L. Study on Hydrochemical Characteristics and Its Influencing Factors in Yarlung Tsangpo River Basin. Ph.D. Thesis, Institute of Geographic Sciences and Natural Resources Research, CAS, Beijing, China, 2016.
68. Liu, X.W. Analysis of the Meteorological and Hydrological Charateristics in the Yarlung Zangbo River Basins. Master's Thesis, Tsinghua University, Beijing, China, 2015.

69. Holden, J.; Burt, T.P. Laboratory experiments on drought and runoff in blanket peat. *Eur. J. Soil Sci.* **2010**, *53*, 675–690. [CrossRef]
70. Dai, Z.; Du, J.; Li, J.; Li, W.; Chen, J. Runoff characteristics of the Changjiang River during 2006: Effect of extreme drought and the impounding of the Three Gorges Dam. *Geophys. Res. Lett.* **2008**, *35*, 521–539. [CrossRef]
71. Yuan, B.; Xu, H.; Ling, H. Drought–flood variation and its correlation with runoff in three headstreams of Tarim River, Xinjiang, China. *Environ. Earth Sci.* **2014**, *71*, 1297–1309. [CrossRef]
72. Lv, Y.; Dong, G.T.; Yang, S.T.; Zhou, Q.W.; Cai, M.Y. Spatio-Temporal Variation in NDVI in the Yarlung Zangbo River Basin and Its Relationship with Precipitation and Elevation. *Resour. Sci.* **2014**, *36*, 603–611.
73. Mishra, A.; Vu, T.; Valiyaveettil, A.; Entekhabi, D. Drought Monitoring with Soil Moisture Active Passive (SMAP) Measurements. *J. Hydrol.* **2017**, *552*, 620–632. [CrossRef]
74. Tucker, C.J.; Slayback, D.A.; Pinzon, J.E.; Los, S.O.; Myneni, R.B.; Taylor, M.G. Higher northern latitude normalized difference vegetation index and growing season trends from 1982 to 1999. *Int. J. Biometeorol.* **2001**, *45*, 184–190. [CrossRef]
75. Easterling, D.R.; Wallis, T.W.R.; Lawrimore, J.H.; Heim, R.R. Effects of temperature and precipitation trends on U.S. drought. *Geophys. Res. Lett.* **2007**, *34*, 396. [CrossRef]
76. LIiu, Q.H.; Xin, J.F.; Xin, X.Z.; Tian, G.L.; Yang, G.J. Monitoring Agricultural Drought by Vegetation Index and Remotely Sensed Temperature. *Pap. Remote Sens. Monit.* **2007**, *25*, 12–18. [CrossRef]
77. Callaghan, J.; Power, S.B. Variability and decline in the number of severe tropical cyclones making land-fall over eastern Australia since the late nineteenth century. *Clim. Dyn.* **2011**, *37*, 647–662. [CrossRef]

remote sensing

MDPI

Article

Extended Dependence of the Hydrological Regime on the Land Cover Change in the Three-North Region of China: An Evaluation under Future Climate Conditions

Yi Yao [1,2], Xianhong Xie [1,2,*], Shanshan Meng [1,2], Bowen Zhu [1,2], Kang Zhang [1,2] and Yibing Wang [1,2]

1 State Key Laboratory of Remote Sensing Science, Jointly Sponsored by Beijing Normal University and Institute of Remote Sensing and Digital Earth of Chinese Academy of Sciences, Beijing 100875, China; yaoyibnu@163.com (Y.Y.); mss_bnu@163.com (S.M.); bwzhu@mail.bnu.edu.cn (B.Z.); zhkang@mail.bnu.edu.cn (K.Z.); ybwang@mail.bnu.edu.cn (Y.W.)

2 Beijing Engineering Research Center for Global Land Remote Sensing Products, Institute of Remote Sensing Science and Engineering, Faculty of Geographical Science, Beijing Normal University, Beijing 100875, China

* Correspondence: xianhong@bnu.edu.cn

Received: 27 November 2018; Accepted: 28 December 2018; Published: 4 January 2019

Abstract: The hydrological regime in arid and semi-arid regions is quite sensitive to climate and land cover changes (LCC). The Three-North region (TNR) in China experiences diverse climate conditions, from arid to humid zones. In this region, substantial LCC has occurred over the past decades due to ecological restoration programs and urban expansion. At a regional scale, the hydrological effects of LCC have been demonstrated to be less observable than the effects of climate change, but it is unclear whether or not the effects of LCC may be intensified by future climate conditions. In this study, we employed remote sensing datasets and a macro-scale hydrological modeling to identify the dependence of the future hydrological regime of the TNR on past LCC. The hydrological effects over the period from 2020–2099 were evaluated based on a Representative Concentration Pathway climate scenario. The results indicated that the forest area increased in the northwest (11,691 km^2) and the north (69 km^2) of China but declined in the northeast (30,042 km^2) over the past three decades. Moreover, the urban area has expanded by 1.3% in the TNR. Under the future climate condition, the hydrological regime will be influenced significantly by LCC. Those changes from 1986 to 2015 may alter the future hydrological cycle mainly by promoting runoff (3.24 mm/year) and decreasing evapotranspiration (3.23 mm/year) over the whole region. The spatial distribution of the effects may be extremely uneven: the effects in humid areas would be stronger than those in other areas. Besides, with rising temperatures and precipitation from 2020 to 2099, the LCC may heighten the risk of dryland expansion and flooding more than climate change alone. Despite uncertainties in the datasets and methods, the regional-scale hydrological model provides new insights into the extended impacts of ecological restoration and urbanization on the hydrological regime of the TNR.

Keywords: hydrological cycle; Three-North region; climate change; land cover change; Variable Infiltration Capacity (VIC) model; evapotranspiration; runoff; soil moisture

---

1. Introduction

Global climate has changed dramatically over the past few decades as demonstrated by many studies [1–3]. Mainly due to human activities, land cover also has experienced various and rapid changes, especially in recent decades [4–7]. Both changes in climate and land cover could greatly affect the hydrological cycle [8–12] regarding water balance and energy balance processes at various scales [13]. These changes are more likely to result in serious consequences, such as droughts and floods, in arid and semi-arid regions, where the environment is much more vulnerable than in humid areas [14].

The Three-North region (TNR), which is composed of Northwestern, Northern, and Northeastern China, covers arid, semi-arid, and humid areas. The TNR is also an ecologically fragile area, where land degradation has been very serious due to both human activities and changes in natural conditions since the last century, especially in the Northwestern region [15]. In order to solve this problem, the Chinese government has launched a series of ecological restoration programs, beginning in 1978, including the Three-North Forest Shelterbelt (TNFS) program and the restoration of farmland to forest area [15,16]. Meanwhile, other land cover changes (LCC), such as urban expansion and industrialization, have been accelerating since the population has increased sharply [17]. As a result, land cover in the TNR has radically changed over the past 40 years.

There have been many studies in which hydrological responses to climate change and LCCs have been detected through observation or simulation. Based on observation data, the potential evaporation and actual evaporation in most basins of the TNR, and the resulting annual streamflow, had been decreasing from the 1960s until the start of this century [18]. From the 1960s to the 2010s, the observed streamflow showed a negative trend in the Songhua Basin [19], but a positive one in the Tarim River Basin [20].

Some researchers employed hydrological models, such as the Variable Infiltration Capacity (VIC) model, to simulate the hydrological cycle over this region. Hydrological models have their advantages in considering forces from climate, land cover, soil and topography conditions. In the Yellow River Basin which is in TNR, simulation results of VIC model have indicated that the effects of climate change were stronger than those of LCC [21,22]. Some studies have also focused on future changes in the hydrological cycle and applied the VIC model to detect the hydrological response to future climate change under a Representative Concentration Pathway (RCP8.5) scenario and found that, in Northern China, the evaporation and runoff will increase, while soil moisture will decline [23].

Since ecological restoration programs first began, a few studies have been performed to determine whether or not they have had a positive impact. Results show that from 1970s, with the growing afforestation, windy days and dust storms have declined sharply over a wide range of area. In Northwest, North and Northeast China, the number of windy days had decreased by nearly 50%, and so did the number of dust storms. However, a few studies also show that the effectiveness of these programs may have been overestimated [15]. Although they have had some beneficial impacts on controlling dust storms in arid and semi-arid areas in China, the ecological improvements have been very limited. The desertification rates (fractions of total area that has undergone desertification) did not decline after the construction of the afforestation programs, and even rose in some zones. For example, in northeast China, the rate was more than 40% in early 2000s, which is over four times that in mid-1970s [15]. Furthermore, some simulation results have suggested changes in the hydrological conditions of the TNR are mainly due to climate change, especially the redistribution of precipitation, while the contribution of LCC may be very minimal. From 1989 to 2009, climate change contributed to a loss of over 25 mm in ET and over 14 mm in R, while LCC only resulted in small changes no more than 2 mm in these two elements [24].

The majority of studies on the effects of LCC have focused on historical periods, using different methods, such as observation or stimulation based on historical data. However, in future, the climate may be different, since there has been a warmer trend globally, and the effects of LCC may be also influenced by climate change. So, the question remains: will the effects of past LCC on hydrological cycle be changed under future climate scenario within the TNR? Additionally, the exploration can be seen as an evaluation of ecological programs, since the LCC over TNR were affected significantly by those programs, especially in semi-arid and arid areas. The programs have been in place for over 40 years, yet their influence may not be well-recognized due to the short time series.

In this study, the dependence of the future hydrological regime (2020–2099) on past LCC (from 1986 to 2015) in the TNR was evaluated. A macro-scale hydrological model (i.e., the variable infiltration capacity (VIC) model [25,26]) was employed to simulate the hydrological processes over the entire region. Model simulations were performed based on different vegetation parameters generated by datasets from historical land cover information, and the simulations were forced with climate data sets from global climate models (GCMs) from the Inter-Sectoral Impact Model Inter-comparison Project (ISI-MIP) [27]. As this study was aimed at understanding the implications of past LCC on the future hydrological regime, model simulations were performed using historical land cover data rather than the projected future land cover information. The work can be seen as an evaluation for the effects of the ecological restoration programs constructed in past decades, since these programs played a significant role in altering land cover condition, so it may provide some guidance for the following construction of the programs.

2. Data and Methods

2.1. Study Area

The TNR located in the north of China, covers more than 50% of the land area in China, nearly 5.3 million km^2 (Figure 1). More than 551 counties across 13 provinces are contained in this region, including Beijing-Tianjin-Hebei, Harbin-Changchun, and other metropolitan areas, where the population has risen sharply. The population in TNR has increased from 344.10 million in 2000 to 384.06 million in 2016 (these data are available at the website of National Bureau of Statistics of China: http://www.stats.gov.cn/). As a result, the urban areas in TNR have been expanding rapidly. There are five large river basins in this region: The Inland River Basin (IRB), Yellow River Basin (YRB), Hai River Basin (HRB), Liao River Basin (LRB), and Songhua River Basin (SRB), from west to east. In fact, the study area is slightly larger than the area of the TNFS because we assess hydrological responses at close to the basin level.

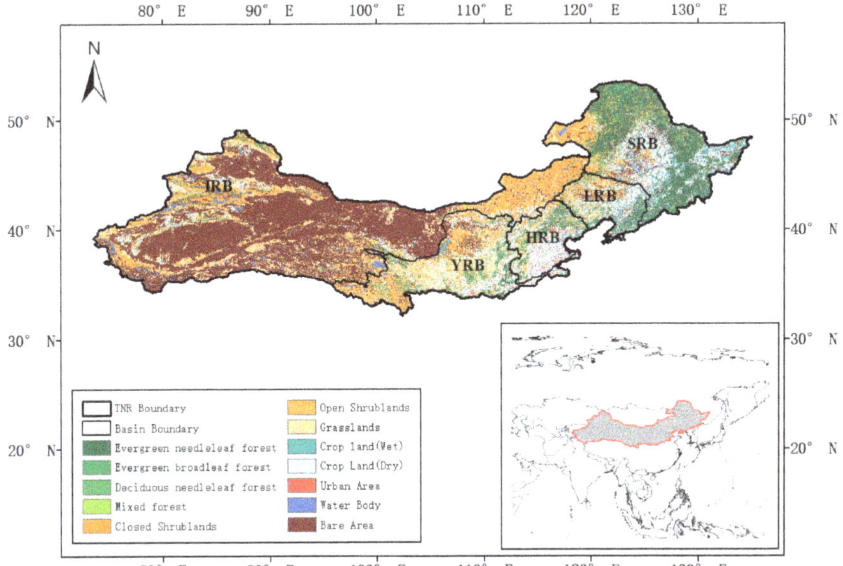

Figure 1. Distribution of land cover types for the year of 2015 in the Three-North region (TNR), where five basins, i.e., the Inland River Basin (IRB), Yellow River Basin (YRB), Hai River Basin (HRB), Liao River Basin (LRB), and Songhua River Basin (SRB), distributed in this region.

The eastern part of this region is mainly influenced by the East Asian Monsoon, and the western part exhibits a typical continental climate [28]. According to the historical datasets from 1959–2009, the climate across the TNR shows great spatial heterogeneity (Figure 2). The annual precipitation reaches more than 1000 mm/year in the southeast, and gradually declines to less than 50 mm/year in the northwest, with an obvious gradient between these endpoints. Mean annual temperature also presents a wide range of spatial variation, from above 12 °C in the south of the HRB and YRB to under 0 °C in the north of the IRB, but without any apparent gradient over the area as a whole.

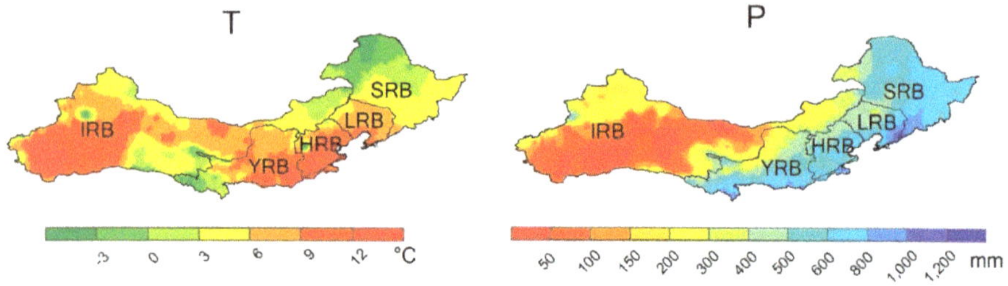

Figure 2. Annual mean temperature and annual precipitation from 1959 to 2009 in the TNR, IRB, YRB, HRB, LRB, and SRB of China.

According to the land cover map for the year 2015 (Figure 1), the land cover over this region also exhibits distinct spatial variability [29]. In the SRB, forestlands account for nearly 50% of the land cover, and in the YRB, grasslands cover nearly half of the area. In the SRB and LRB, which are densely populated, the dominant land cover type is cropland, while in the IRB, over 60% of the area is unutilized. Additionally, the proportion of urban lands is minute in all basins. In the entire region, over 35% of the area is unutilized, including bare lands and bodies of water, and the most abundant vegetative land cover type is grassland, at nearly 30%, with forestland and cropland coverage being approximately equal at 15%, and the least abundant land cover type is the urban area, at no more than 2%.

2.2. Hydrological Model

The VIC model is a semi-distributed physically-based hydrological model. It can simulate water and energy balance. In the model, a study area is divided into grids according to latitudes and longitudes, and all the calculations are done in each grid separately. In each grid, all elements are computed based on different land cover types, before being averaged according to their correspondent fractions [25,26]. Since the model has been proved to perform well over a range of scales [30,31], we employed it in our study. Here, the spatial resolution of modeling is at 0.5° × 0.5° to ensure consistency with the climate forcing data.

2.3. Data Availability

2.3.1. Vegetation and Soil Parameters

To produce the vegetation parameters for VIC, land use maps and the average leaf area index (LAI) over 12 months are necessary. The two land use maps obtained were created by merging Landsat Thematic Mapper (TM) images for the periods ranging from 1983–1986 and 2010–2015. In these studies, land cover conditions were divided into 12 types as in Figure 1. A human–computer interactive interpretation method of remotely sensed land use cover information was used to interpret these maps, and the accuracy was over 90% [29,32]. The LAI datasets were available from the Global Land Surface Satellite (GLASS) products (http://www.bnu-datacenter.com/) that were retrieved from the Moderate Resolution Imaging Spectroradiometer reflectance data (MOD-09A1) using a general regression neural network algorithm [33]. The resulting combinations of land use and LAI datasets for the two collection periods (1983–1986 and 2010–2015) were referred to as "LC1986" and "LC2015," respectively. The other parameters, except for the LAI, in the VIC vegetation library data were set as in the study of a previous study [24]. We assumed that the differences between the two combination datasets could accurately represent the progress accomplished by the ecological programs and other factors from 1986 to 2015. The soil parameters were set according to those of a previous study [24], in which the parameters were derived based on the Food and Agriculture Organization of the United Nations (FAO) digital soil map of the world, and they were previously evaluated at a global scale [34,35].

2.3.2. Bias-Corrected Climate Datasets

The climate data sets used in this study are from the Inter-Sectoral Impact Model Inter-comparison Project (ISI-MIP), and are available at: https://esg.pik-potsdam.de/. This project provides a framework to compare climate impact projections in different sectors and at different scales, which enable quantitative synthesis of climate change impacts at different level of global warming [27]. A new bias-correction method was developed within the first stage of this project, which can reduce the bias between daily or monthly simulated and observed climate data, while maintaining the absolute or relevant long-term trend much better than previous methods [36]. In this study, bias-corrected climate data from five GCMs (HadGEM2-ES, GFDL-ESM2M, IPSL-CM5A-LR, MIROC-ESM-CHEM, and NorESM1-M) with a daily time step were selected in order to lessen the uncertainty of a single model. Considering that the RCP8.5 scenario, in which the anthropogenic radiative forcing will approach 8.5W m^{-2} by 2100 and the concentration of CO_2 will be 3–4 times the present value, estimates extremely severe circumstances in future, evaluation of this scenario was assumed to be more important than for others scenarios [37]. Five meteorological datasets were provided under this scenario: precipitation, maximum temperature, minimum temperature, mean wind speed, and downward shortwave radiation at a $0.5^{\circ} \times 0.5^{\circ}$ spatial resolution, daily time steps from 2006 to 2099 were collected to generate the climate forcing data. These data are accessible at https://www.isimip.org/outputdata/caveats-fast-track/. Among these data, precipitation, temperature, and wind speed are necessary to force the VIC model, and the downward shortwave radiation plays an important role in detecting the effects of land cover on the hydrological cycle, as it will be redistributed differently according to the various vegetation types.

2.4. Experimental Design

In this study, the VIC model was run based on climate forcing datasets from the five GCMs, as described in Section 2.3.2, and the simulation results were averaged to reduce the uncertainties from the forcing data. Moreover, the model parameters described in Section 2.3 have been extensively validated by a previous study [24], with 15 stations of streamflow data and 10 stations of evapotranspiration (ET) data. The model has shown a favorable performance, with an average Nash–Sutcliffe efficiency for streamflow of 0.55 and a Pearson correlation coefficient for the ET of greater than 0.5 [24]. Therefore, we did not further validate the model parameters.

To identify the impact of historical LCC on the future hydrological regime in the TNR, the present land cover state (corresponding to the LC2015 data) was assumed to remain constant throughout the study period. Thus, two simulation experiments were conducted with different land cover and vegetation input data. The first experiment involved running the model with the LC2015 data to explore changes in the hydrological cycle under the influence of climate change alone. We paid the most attention to the ET, runoff (R), and soil moisture (SM), as these three hydrologic features play important roles in the water cycle, and their values can reflect the ecological conditions of the area to some degree. The time series of the climatic (i.e., precipitation and temperature) and the three hydrological variables were divided into four periods: 2020–2039, 2040–2059, 2060–2079, and 2080–2099 to lessen the uncertainties of a single year in the analysis. The spatial distribution for the four periods, as well as the annual mean values over the entire region, were analyzed to determine the hydrological response to climate change across the TNR.

The second simulation experiment ran the model with the LC1986 data for the same period as in the first simulation. Therefore, the two experiments differ only with respect to their land cover and vegetation data, respectively regarded as the land cover conditions after afforestation (LC2015) and before afforestation (LC1986). Comparison of the two experiments can be used to identify the effects of past LCC on the future hydrological regime under RCP8.5 scenario. The comparative analyses between the two experiments were conducted over different basins and periods. The annual differences and the spatial distribution of the annual mean differences between the two experiments were evaluated to measure the effects of afforestation on the hydrology across the TNR. We also analyzed the differences in four seasons, so as to explore the different effects of LCC in different seasons.

3. Results

3.1. Past Changes in Land Cover

We first identified past LCC. According to the data from Liu et al. (2014), we focused on four primary land cover types: forests, grasslands, croplands, and urban areas. As shown in Table 1, urban expansion played an important role over the last 30 years in the TNR, leading to a substantial decrease in the cropland and grassland areas of most basins. In the HRB, the urban area expanded from 2.5% to 9%, resulting in a decrease in cropland areas, and in the LRB, the urban area doubled, which caused a decline in the other land cover types. In the western three basins, the IRB, YRB, and HRB, due to afforestation projects, the forest area has slightly increased. The YRB is the basin where the forestland increased the most, from 12% to 13%, and the increases in the other two basins were under 0.4%. Meanwhile, in the other two basins, the SRB and LRB, the forest area decreased, especially in the SRB, where the reduction was greater than 3%. The reason for the reduction in forestlands in these two basins may be the expansion of croplands and urban areas. For the entire region overall, the proportion of forestlands, grasslands, croplands, and others land cover types all decreased because of the expansion of urban areas.

Table 1. Fractions of primary land cover types of the TNR in 1985 and 2015.

Basin	Year	Urban	Forest	Grass	Crop	Others
TNR	1985	0.5%	15.3%	30.5%	15.9%	37.8%
	2015	1.8%	14.9%	30.0%	15.9%	37.4%
IRB	1985	0.1%	1.8%	33.4%	3.8%	60.9%
	2015	0.5%	2.1%	32.8%	4.4%	60.2%
HRB	1985	2.5%	19.2%	19.8%	56.0%	2.5%
	2015	9.0%	19.4%	19.6%	48.7%	3.3%
YRB	1985	0.7%	12.0%	47.7%	29.2%	10.4%
	2015	3.0%	13.0%	46.9%	26.2%	10.9%
SRB	1985	0.5%	48.8%	17.5%	24.8%	8.4%
	2015	1.8%	45.6%	17.5%	27.0%	8.1%
LRB	1985	1.7%	23.5%	26.7%	39.8%	8.2%
	2015	4.6%	23.2%	24.7%	39.0%	8.4%

Due to afforestation programs and favorable climate conditions, the vegetation in the TNR has grown substantially. We calculated the vegetation index (i.e., the mean LAI from January to December) across the TNR, and the results are shown in Figure 3. Because of the growth of vegetation, the LAI peaked from July to August, and exhibited the lowest values from December to February. The average values between 1983–1986 and 2011–2015 were compared, and the results show that there has been a great improvement in the LAI over the entire region, especially in July and August when it increased by over 0.5.

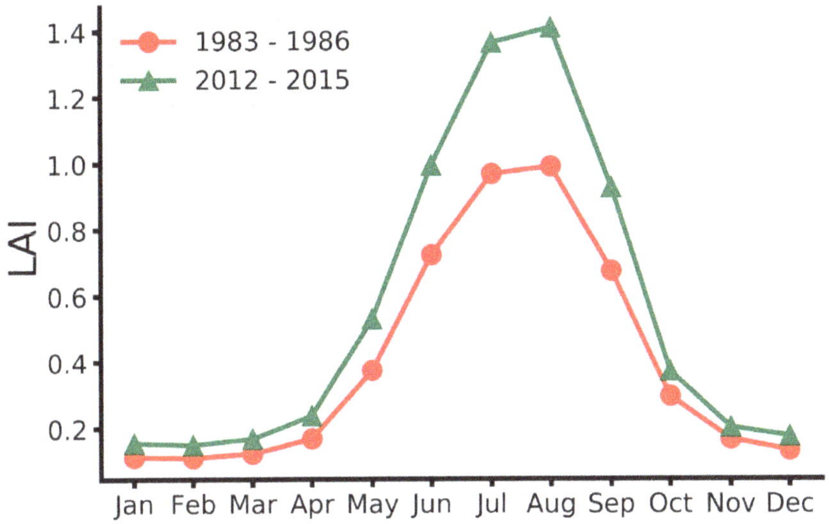

Figure 3. Mean monthly leaf area index (LAI) of the TNR in 1985 and 2015.

3.2. Future Climate Change

To determine the temperature changes projected for the future under the RCP8.5 scenario, we averaged the datasets of five models, and averaged the maximum and minimum temperatures as the mean temperature. In order to correspond to the hydrological changes analyzed, we also selected the period from 2020–2099 as the experimental interval. The mean annual temperatures (MATs) across two different periods, 2020–2039 and 2080–2099, are shown in Figure 4. We found that the spatial pattern across the TNR may not change in future. The HRB will remain the hottest basin, while the SRB will remain the coldest, and the differences between the two periods were very similar in each area. The northeast of the SRB and the north and southwest of the IRB are the areas where temperatures will increase the most, by more than 3.6 °C, and in the south of the YRB, the change will be slightly less at ~3 °C. The MAT of the entire region will rise steadily from nearly 5.5 °C to over 10 °C (Figure 5).

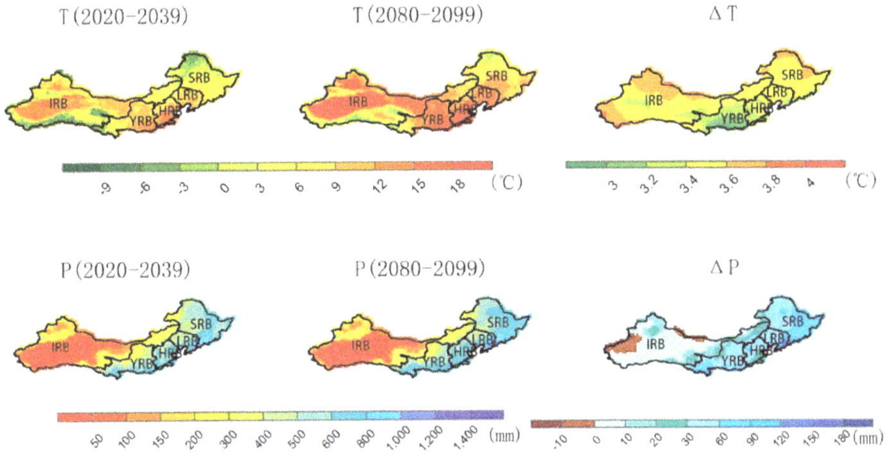

Figure 4. Projected mean annual temperature (MAT) and mean annual precipitation (MAP) over two periods (2020–2039 and 2080–2099) and their differences (Δ).

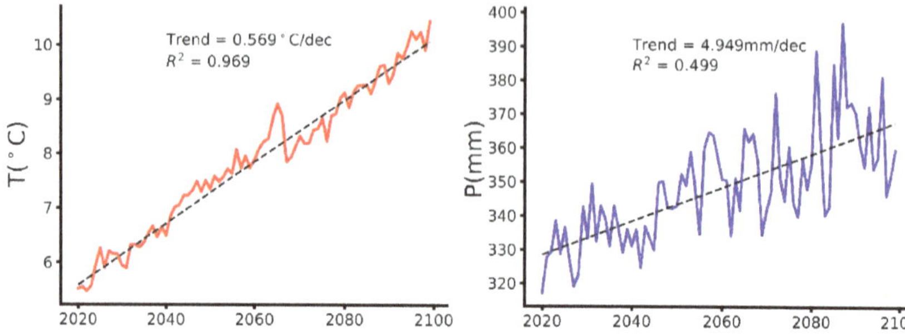

Figure 5. Projected MAT and MAP across the TNR.

The spatial pattern of precipitation also may not exhibit obvious changes obviously; there will remain an obvious declining gradient from the southeast to northeast. As shown in Figure 3, the south of the LRB will remain the wettest area, where the annual precipitation will be over 1400 mm from 2080–2099, and the west of the IRB will be the driest, remaining under 50 mm. Additionally, except in some small zones in the west and north of the IRB, the precipitation will increase from 2020–2039 to 2080–2099 over the entire TNR. The spatial distribution of difference is consistent with that of the precipitation, with the southwest experiencing the most change (more than 180 mm), and the northeast experiencing the least, less than 10 mm or even less than 0 mm. The mean annual precipitation (MAP) of the entire region is shown in Figure 5. There will be a positive, linear trend overall, but the variability will also rise, which means that the discrepancy between the years will also increase. As for wind speed and downward shortwave radiation, no remarkable changes were found (the results were not shown.).

3.3. Hydrological Responses to Future Climate Change

We examined the changes in three hydrological variables, ET, R, and SM, from the VIC model, based on the input of the LC2015 data. As with the analysis of climate change, all values were averaged by the five simulations that were driven by data from the five GCMs, as described in Section 2.3.1. Additionally, we focused on the period from 2020–2099, since the period from 2006–2019 was set as the warm-up period for the VIC model.

The annual mean ET, R, and SM in two corresponding periods (i.e., 2020–2039 and 2080–2099) are shown in Figure 6. The distributions of ET and R are similar to that of precipitation. In the period from 2080–2099, ET decreased from over 1000 mm in the southeast to under 50 mm in the northwest, and R from more than 400 mm to less than 10 mm. Due to the complex spatial variabilities in climate and land surface conditions, the distribution of SM differed from those of ET and R. Across the three eastern basins (i.e., the SR, LR, and HRB), the SM in most areas will be greater than 700 mm, while over the majority of the other two basins, the SM will remain under 400 mm, and even under 200 mm in some areas. In each basin, it was obvious that the SM may rise in those areas where the soil was wetter, but decline in the drier areas.

The three hydrological variables exhibited significant temporal changes. Comparing the results from the two periods (2020–2039 and 2080–2099), we found that ET will increase over the vast majority of the TNR because of the increases in precipitation and temperature. This increase may approach over 150 mm in the southeast, especially in those areas with forest cover. As for SM, the changes appeared closely related to the land cover type. In the entire region, rising temperatures will aggravate the water loss by ET, especially in the areas with forest cover in the SRB, LRB, and HRB, as well as in the areas with grassland cover in the IRB and YRB, leading to a decrease in SM, although precipitation will also greatly increase SM. In other areas, the increase in precipitation will be able to compensate for the loss by ET, so the SM will increase. The center of the SRB and northwest of the IRB are the areas where SM will increase the most; in the southwest of the YRB and as the south of the SRB, the SM will decrease the most, by over 20 mm. The changes in R will be directly influenced by the changes in precipitation, ET, and SM. In most areas, the R will increase, while decrease only in some small zones.

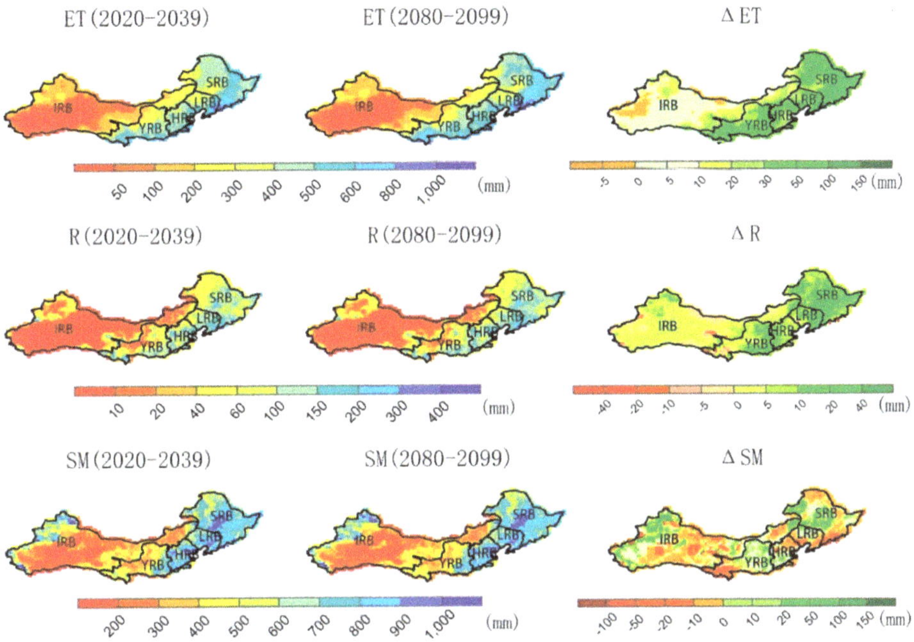

Figure 6. Mean annual evapotranspiration (ET), runoff (R), and soil moisture (SM) during two periods, 2020–2039 (**left**) and 2080–2099 (**middle**), and their differences (**Δ**).

The annual mean ET, R, and SM for the entire TNR are shown in Figure 7. As with the changes in precipitation, the ET and R will both experience a positive and fluctuating trend. Because of rising temperatures, the trend of ET is more pronounced than that of R, and the fluctuations in R are greater than those of ET. The SM will increase in the first 20 years and then remain relatively stable at 510 mm from 2030 onward.

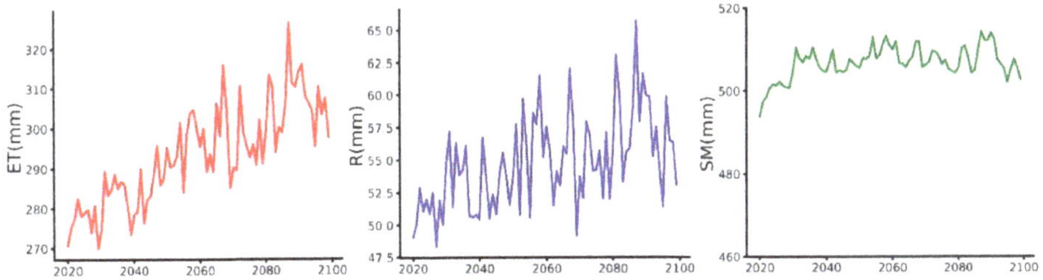

Figure 7. Projected mean annual ET, R, and SM across the TNR.

3.4. *Hydrological Effects of Past Land Cover Changes (LCC)*

Based on the two simulation experiments performed in this study, we can quantify the contribution of past LCC on the future hydrological regime. The differences in ET, R, and SM between the two experiments for the period from 2080–2099 are shown in Figure 8. It is obvious that across nearly all the TNR, LCC will result in a reduction of ET, as the areas with vegetation cover decreased, as illustrated in Section 3.1, though the LAI increased to some degree. In particular, the center and the southeast of the HRB is the area where the ET decreased most distinctly, by over 40 mm. Only in a few grids, such as in the center and the east of the SRB and the center of the YRB, where the ET increase within a range of 40–60 mm. As for the changes in R, the distribution is somewhat complex. Contrary to that of ET, in a majority of the eastern part of the TNR, R will increase, which is mainly because of the reduction of ET. The center and southeast of the HRB will be the place where the difference is the greatest, by over 40 mm. Some grids in the east of the SRB are the areas where R will decrease the most, by more than 60 mm in 2080–2099. As for the west part of TNR, R will increase in nearly half of the grids and decrease in the others grids, while the range may not be greater than 20 mm. As SM is very sensitive to the ET process, its distribution is similar to that of R across the majority of TNR, but the maximum increase will be in the center and southeast of the HRB, where urban expansion is the most intense, and will reach over 80 mm. SM will decrease in some areas in the SRB by over 80 mm. In contrast, in some areas across the northeast of the IRB, SM will decrease by more than 120 mm. Overall, the effects of LCC on R and SM may be positive in the SRB, LRB, HRB, and YRB, but negative in the IRB.

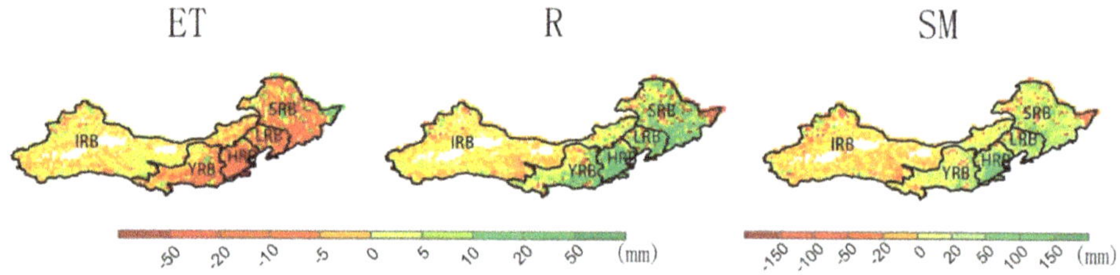

Figure 8. Shifts in annual ET, R, and SM for 2080–2099 driven by past land cover changes (LCC).

The annual mean changes in ET, R, and SM are shown in Figure 9. Obviously, the changes in land cover will cause a decrease in ET, as well as increases in R and SM. Similar to the change in annual precipitation, there is a positive trend in these effects over time, which means that under the RCP8.5 scenario, LCC will more strongly influence the hydrological cycle in the future than at present.

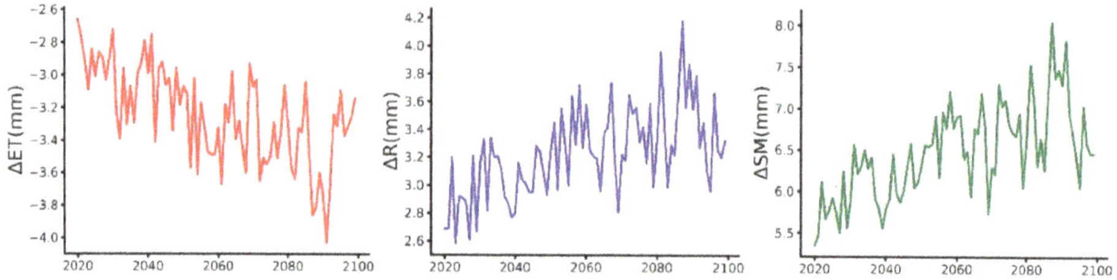

Figure 9. Projected average annual changes in ET (**left**), R (**middle**), and SM (**right**) between the two results, based on different land cover datasets across the TNR.

The shifts in the hydrological regime due to past LCC also exhibit obvious seasonal variability; Figure 10 shows the shifts in ET among the four seasons across every basin. Across most of the basins, the effects in spring, summer, and autumn will be negative, while they will be positive during winter. For example, across the YRB from 2080–2099, the shift will be nearly 0.1 mm in winter, and −1.9 mm, −3.1 mm, and −0.4 mm in spring, summer, and autumn, respectively. Moreover, the HRB will be the most strongly affected, with a change of over 10 mm in the summer; the magnitude of change observed in the HRB is followed by that for the LRB (~9 mm in the summer), and in the IRB, where effects are the weakest, there will be a change of no more than 0.5 mm. The effects of LCC on R and SM across the TNR are shown in Figure 11. In every season, the shifts in R and SM due to LCC will be positive, and stronger from 2080–2099 than in other periods. Especially in the summer, R and SM may significantly by ~2 mm. A similar positive pattern will likely appear in each of the five basins despite different amounts of change.

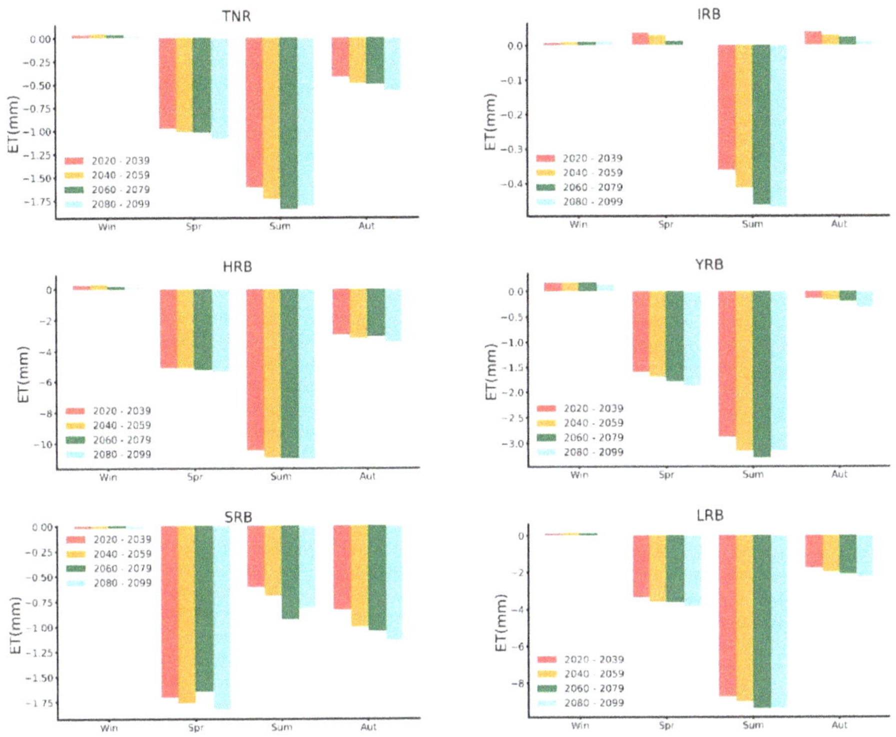

Figure 10. Projected changes in ET between the two results (based on different land cover datasets) of each season and different periods, over the five basins in the TNR.

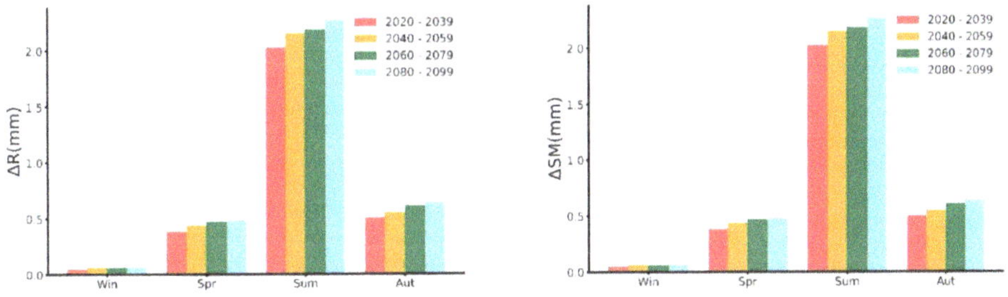

Figure 11. Projected changes in R and SM between the two results, based on the different vegetation parameters of each season.

4. Discussion

4.1. Extended Effects on Hydrological Regimes of Climate Change and LCC

In this study, we detected the future hydrological regime based on current land cover condition (LC2015) at first. In previous studies about the predicted hydrological regime in TNR, especially under the most extreme scenario, some similar results were obtained. Over the entire region, annual ET may increase steadily under RCP8.5 scenario, possibly due to the increasing temperature and precipitation [23]. Other researchers found that annual R would increase in most areas of TNR, especially in eastern basins [38], which is consistent with our study. Additionally, other study also found the interannual changes in ET and R may be expanded under RCP8.5 scenario, which means they may be more fluctuating in future [23], resulting from the fluctuating P. Those studies all predicted an uneven distribution in changes of R. In those humid areas, the increase in R may be much more than that in arid and semi-arid areas, and in some arid areas over Northwest China, the R may even experience a decrease [23,38]. The reason may be that in humid areas, the increase in P could compensate for the water loss caused by ET, while in those arid or semi-arid areas it could not. Many studies based on various scales and time series have summoned a similar result: "Wet get wetter, dry get drier" [39–41]. In our study, the SM in those areas densely covered by vegetation may experience a decrease in future, while in other areas the SM would rise in contrary. It can be attributed to the effects in promoting ET by vegetations, especially driven by increasing temperature. In other study, researchers predicted that over each basin in TNR, the SM would decline in future [23], which is inconsistent with our study, and those differences may be caused by the choose in soil depth.

Then we explored the dependence of the future hydrological regime on past LCC. With the progress of ecological restoration from 1986 to 2015, including afforestation, the forest area increased in North China (i.e., in the HRB) and in Northwest China (i.e., in the IRB), but decreased in Northeast China (i.e., in the HRB and LRB). The TNR also showed a significant increase in urban area of ~1.3% (52,701 km^2) due to rapid urbanization. Results show that compared to the effects of climate change, those of LCC would be far less observable. Based on past climate condition, other study has found the similar conclusion that the LCC did not obviously alter the hydrological regime at a regional scale [24]. There was also study that found the afforestation did not help solve, and even aggravate the desertification in arid and semi-arid areas [15]. However, although the effects may be less observable than those of climate change, our study found that under the projected climate scenario they may also be intensified, and could be able to alter the hydrological regime in some areas. Growing temperature and precipitation may be responsible for the amplification in those effects. We also explored the effects in different seasons. Most improved R and SM may occur in summer, because P in this season may be much more than in other seasons and the abundance of water resources could intensify the effects of LCC.

4.2. Implications

Our results predicted the future hydrological regime under RCP8.5 scenario, and it may bring some consequences. The increasing ET will likely lead to drier conditions in the TNR, especially in those arid regions (i.e., the IRB), where the SM may decrease. This finding imply that climate change may cause dryland expansion and exacerbate the risk of land degradation and desertification. Moreover, R would also rise in future, particularly in areas with dense population (i.e., the HRB), which may enhance the risk of flooding. Besides, the fluctuation in R may lead to the amplification of the inter-annual gap in water resources, so more solutions may need to be done to balance the temporal distribution.

Land cover change due to urbanization may slowly inhibit the increase of ET to some degree, but accelerate R in most areas. These effects are much stronger in the eastern basins than in the western ones. The majority of arid and semi-arid areas are located in the IRB, where effects are limited, while in the humid areas, such as the SRB, LRB, and HRB, increasing R caused by LCC may increase the risk of floods in the summer, especially in those places where the cities are densely structured. Considering that climate change may also have the similar effects, floods in these areas would be severe problems. Additionally, in arid and semi-arid regions, such as IRB, the ecological restoration programs have expanded the areas of forest or grass, and such change may increase the ET and decrease the R and SM, which can further intensify desertification and other forms of environmental degradation, since the rising temperature may have exacerbated these risks.

Overall, the progress of ecological programs may be unable to help balance neither spatial nor temporal distribution. In humid areas, the risk of floods may be heightened, especially in summer, while in arid and semi-arid areas, the desertification and drought could not be solved. Such results could aggravate the results "dry get drier, wet get wetter", which are caused by climate change [39–41].

4.3. Potential Limitations

Potential limitations still exist in this study with respect to the data, the methods, and modeling. First, there are biases within the projected climate datasets. Our study was focused on detecting the regularity of hydrological changes in the future, so such biases could be neglected to some degree. Second, the RCP8.5 scenario that was considered in this study is much more extreme than others scenarios, and we only explored the hydrological response under this scenario. However, when only one scenario is considered, it may result in uncertainties, so additional scenarios should also be considered in future studies. Third, differences in the spatial resolution between the model and land cover data may induce uncertainties in the simulations. The spatial resolution in this study was $0.5° \times 0.5°$, which was limited by the projected climate forcing datasets. Although the vegetation parameters were generated with 1 km $\times$ 1 km land cover maps and LAI datasets, there were also changes in smaller zones, especially some extreme events that could not be captured. Perhaps downscaling the forcing data to a finer resolution can be used to ease this problem to some degree. Finally, in this study we considered climate change and LCC as two independent factors. However, a great number of studies have suggested that there are links between them [42,43]. We kept the land cover condition unchanged when simulating, but the climate would change dramatically in future under RCP8.5 scenario, and the growth of vegetation in some areas may be threatened by the slightly increasing temperature. In our study, we intend to evaluate the hydrological effects of past LCC (or the progress that has been achieved by series of ecological restoration programs), rather than predicting how the hydrological regime would change in future, so this could be acceptable. If we want to do a more accurate projection in future researches, an improved VIC model which can couple dynamic vegetation parameters, reliable predicted land cover condition and LAI should be necessary.

5. Conclusions

This study employed remote sensing data and macro-scale hydrological modeling, and determined the role of past LCC in reshaping the future hydrological regime across the TNR of China. Land cover change may have imposed minor impacts on the hydrological regime over the past three decades. In contrast, it was found in this study that LCC may play a more significant role in altering ET and R in the future. The conclusions are as follows:

(1) There has been a significant change in land cover in the TNR over the past three decades, primarily due to ecological restoration projects, urban expansion, and industrialization. In most basins across the TNR, urban areas expanded, leading to the reduction of other land cover types, and the replaced land cover types vary among basins. Forest areas increased in the south and the west of the TNR (i.e., in the HRB, YRB, and IRB), but decreased in Northeast China (i.e., in the HRB and LRB). With afforestation and favorable climate conditions, LAI exhibited a positive change in all basins, and the most significant changes occurred between July and August.

(2) Climate will experience obvious changes in the TNR under the RCP8.5 scenario. Temperatures will steadily rise in all basins at the rate of ~0.57° per decade from 2020–2099. The spatial pattern of precipitation will remain unchanged, but the mean annual value will increase, except for in some small zones in the IRB. The area where the precipitation will increase the most (by over 120 mm) will be the southeast.

(3) Forced by future climate conditions, the hydrological regime will experience various changes. Similar to precipitation, the ET and R will increase over most of the TNR. However, the changes in SM will vary. Specifically, over those areas with forest cover in the SRB, LRB, and HRB, and areas with grassland cover in the IRB and YRB, the SM will decrease due to the excessive increase in ET. However, over the areas with other land cover types, the SM may increase, mainly due to increased precipitation. Additionally, in all basins, the SM may increase where the soil is wetter, while decreasing in drier areas.

(4) Land cover changes in the TNR will play different roles in influencing ET and R. Specifically, LCC will likely slow the rate of increasing ET, while promoting increases in R, although the strength of these effects will vary across different basins. In the SRB, LRB, and HRB, the effects are much stronger than in other basins, and in the eastern four basins, R and SM will increase due to LCC, while decrease in the IRB. Additionally, the effects of LCC on ET, R, and SM will all increase over time, which means that the effects of LCC will increasingly strengthen in the future.

Compared to the hydrological effects over the past decades, the strengthening role of LCC on the future hydrological regime can be attributed to its long-term, cumulative effects and strengthening climate change (i.e., the rising temperature and increasing precipitation). However, the future hydrological regime in the TNR will be primarily driven by climate change. Based on the SM, the arid region (i.e., the IRB) may become drier and the humid regions (i.e., the east of the YRB and the south of the HRB) may become wetter, and these effects are consistent with the standard catchphrase, "dry gets drier, wet gets wetter" [39–41]. LCC may likely intensify these effects, implying dryland expansion in arid and semi-arid areas and a potentially increased flood risk in humid areas, although it may diminish ET so as to preserve water resources. Uncertainties may exist in our study, including the interactions among climate, water, and vegetation. Coupling land use and vegetation dynamics in the hydrological modeling can improve future hydrological projections.

Author Contributions: Conceptualization, Y.Y. and X.X.; methodology, Y.Y.; validation, X.X.; formal analysis, Y.Y.; investigation, Y.Y.; resources, Y.Y.; data curation, Y.Y., X.X., S.M., B.Z., K.Z. and Y.W.; writing—original draft preparation, Y.Y.; writing—review and editing, X.X.; visualization, Y.Y.; supervision, X.X.; project administration, X.X.; funding acquisition, X.X.

Funding: This research was funded by the National Natural Science Foundation of China (No. 41471019, 61661136006) and the National Key Research and Development Program of China (No. 2016YFC0401404). And The APC was funded by the National Natural Science Foundation of China (No. 41471019).

Acknowledgments: We would like to thank the three anonymous reviewers for their constructive comments and language editing.

Conflicts of Interest: The authors declare no conflict of interest.

References

1. Qin, D.; Stocker, T. Highlights of the IPCC Working Group I Fifth Assessment Report. *Progressus Inquisitiones de Mutatione Climatis* **2014**, *10*, 1–6. [CrossRef]
2. Webster, M.A.; Rigor, I.G.; Nghiem, S.V.; Kurtz, N.T.; Farrell, S.L.; Perovich, D.K.; Sturm, M. Interdecadal changes in snow depth on Arctic sea ice. *J. Geophys. Res. Oceans* **2014**, *119*, 5395–5406. [CrossRef]
3. Notz, D.; Stroeve, J. Observed Arctic sea-ice loss directly follows anthropogenic CO_2 emission. *Science* **2016**, *354*, 747–750. [CrossRef]
4. Lambin, E.F.; Geist, H.J.; Lepers, E. Dynamics Ofland-Use Andland-Coverchange Intropicalregions. *Annu. Rev. Environ. Resour.* **2003**, *28*, 205–241. [CrossRef]
5. Liu, M.; Tian, H. China's land cover and land use change from 1700 to 2005: Estimations from high-resolution satellite data and historical archives. *Glob. Biogeochem. Cycles* **2010**, *24*. [CrossRef]
6. Jepsen, M.R.; Kuemmerle, T.; Müller, D.; Erb, K.; Verburg, P.H.; Haberl, H.; Vesterager, J.P.; Andrič, M.; Antrop, M.; Austrheim, G.; et al. Transitions in European land-management regimes between 1800 and 2010. *Land Use Policy* **2015**, *49*, 53–64. [CrossRef]
7. Olofsson, P.; Holden, C.E.; Bullock, E.L.; Woodcock, C.E. Time series analysis of satellite data reveals continuous deforestation of New England since the 1980s. *Environ. Res. Lett.* **2016**, *11*. [CrossRef]
8. Scanlon, B.R.; Reedy, R.C.; Stonestrom, D.A.; Prudic, D.E.; Dennehy, K.F. Impact of land use and land cover change on groundwater recharge and quality in the southwestern US. *Glob. Chang. Biol.* **2005**, *11*, 1577–1593. [CrossRef]
9. Oki, T.; Kanae, S. Global Hydrological Cycles and World Water Resources. *Science* **2006**, *313*, 1068–1072. [CrossRef]

10. Schilling, K.E.; Jha, M.K.; Zhang, Y.-K.; Gassman, P.W.; Wolter, C.F. Impact of land use and land cover change on the water balance of a large agricultural watershed: Historical effects and future directions. *Water Resour. Res.* **2008**, *44*. [CrossRef]
11. Fu, S.S.Q. A drier future. *Science* **2014**, *343*, 737–739. [CrossRef]
12. Haddeland, I.; Heinke, J.; Biemans, H.; Eisner, S.; Florke, M.; Hanasaki, N.; Konzmann, M.; Ludwig, F.; Masaki, Y.; Schewe, J.; et al. Global water resources affected by human interventions and climate change. *Proc. Natl. Acad. Sci. USA* **2014**, *111*, 3251–3256. [CrossRef] [PubMed]
13. Frans, C.; Istanbulluoglu, E.; Mishra, V.; Munoz-Arriola, F.; Lettenmaier, D.P. Are climatic or land cover changes the dominant cause of runoff trends in the Upper Mississippi River Basin? *Geophys. Res. Lett.* **2013**, *40*, 1104–1110. [CrossRef]
14. Molnar, P. Climate change, flooding in arid environments, and erosion rates. *Geology* **2001**, *29*, 1071–1074. [CrossRef]
15. Wang, X.M.; Zhang, C.X.; Hasi, E.; Dong, Z.B. Has the Three Norths Forest Shelterbelt Program solved the desertification and dust storm problems in arid and semiarid China? *J. Arid. Environ.* **2010**, *74*, 13–22. [CrossRef]
16. Li, W. Degradation and restoration of forest ecosystems in China. *For. Ecol. Manag.* **2004**, *201*, 33–41. [CrossRef]
17. Li, Z.; Zheng, F.-L.; Liu, W.-Z. Spatiotemporal characteristics of reference evapotranspiration during 1961–2009 and its projected changes during 2011–2099 on the Loess Plateau of China. *Agric. For. Meteorol.* **2012**, *154–155*, 147–155. [CrossRef]
18. Zha, Y.; Gao, J. Characteristics of desertification and its rehabilitation in China. *J. Arid Environ.* **1997**, *37*, 419–432. [CrossRef]
19. Li, F.; Zhang, G.; Xu, Y.J. Spatiotemporal variability of climate and streamflow in the Songhua River Basin, northeast China. *J. Hydrol.* **2014**, *514*, 53–64. [CrossRef]
20. Tao, H.; Gemmer, M.; Bai, Y.; Su, B.; Mao, W. Trends of streamflow in the Tarim River Basin during the past 50 years: Human impact or climate change? *J. Hydrol.* **2011**, *400*, 1–9. [CrossRef]
21. Cuo, L.; Zhang, Y.; Gao, Y.; Hao, Z.; Cairang, L. The impacts of climate change and land cover/use transition on the hydrology in the upper Yellow River Basin, China. *J. Hydrol.* **2013**, *502*, 37–52. [CrossRef]
22. Wang, J.; Hong, Y.; Gourley, J.; Adhikari, P.; Li, L.; Su, F. Quantitative assessment of climate change and human impacts on long-term hydrologic response: A case study in a sub-basin of the Yellow River, China. *Int. J. Clim.* **2010**, *30*, 2130–2137. [CrossRef]
23. Leng, G.; Tang, Q.; Huang, M.; Hong, Y.; Ruby, L.L. Projected changes in mean and interannual variability of surface water over continental China. *Sci. China Earth Sci.* **2014**, *58*, 739–754. [CrossRef]
24. Xie, X.; Liang, S.; Yao, Y.; Jia, K.; Meng, S.; Li, J. Detection and attribution of changes in hydrological cycle over the Three-North region of China: Climate change versus afforestation effect. *Agric. For. Meteorol.* **2015**, *203*, 74–87. [CrossRef]
25. Liang, X.; Lettenmaie, D.P.; Wood, E.F.; Burges, S.J. A simple hydrologically based model of land surface water and energy fluxes for general circulation models. *J. Geogr. Res.* **1994**, *99*, 14415–14428. [CrossRef]
26. Liang, X.; Wood, E.F.; Lettenmaier, D.P. Surface soil moisture parameterization of the VIC-2L model: Evaluation and modification. *Glob. Planet. Chang.* **1996**, *13*, 195–206. [CrossRef]
27. Warszawski, L.; Frieler, K.; Huber, V.; Piontek, F.; Serdeczny, O.; Schewe, J. The Inter-Sectoral Impact Model Intercomparison Project (ISI-MIP): Project framework. *Proc. Natl. Acad. Sci. USA* **2014**, *111*, 3228–3232. [CrossRef] [PubMed]
28. Zheng, X.; Zhu, J. A new climatic classification of afforestation in Three-North regions of China with multi-source remote sensing data. *Theor. Appl. Clim.* **2015**, *127*, 465–480. [CrossRef]
29. Liu, J.; Kuang, W.; Zhang, Z.; Xu, X.; Qin, Y.; Ning, J.; Zhou, W.; Zhang, S.; Li, R.; Yan, C.; et al. Spatiotemporal characteristics, patterns, and causes of land-use changes in China since the late 1980s. *J. Geogr. Sci.* **2014**, *24*, 195–210. [CrossRef]
30. Chang, J.; Wang, Y.; Istanbulluoglu, E.; Bai, T.; Huang, Q.; Yang, D.; Huang, S. Impact of climate change and human activities on runoff in the Weihe River Basin, China. *Quat. Int.* **2015**, *380–381*, 169–179. [CrossRef]
31. Treesa, A.; Das, J.; Umamahesh, N.V. Assessment of impact of climate change on streamflows using VIC model. *Eur. Water* **2017**, *59*, 61–68. [CrossRef]

32. Ning, J.; Liu, J.; Kuang, W.; Xu, X.; Zhang, S.; Yan, C.; Li, R.; Wu, S.; Hu, Y.; Du, G.; et al. Spatiotemporal patterns and characteristics of land-use change in China during 2010–2015. *J. Geogr. Sci.* **2018**, *28*, 547–562. [CrossRef]
33. Xiao, Z.; Liang, S.; Wang, J.; Xiang, Y.; Zhao, X.; Song, J. Long-Time-Series Global Land Surface Satellite Leaf Area Index Product Derived from MODIS and AVHRR Surface Reflectance. *IEEE Trans. Geosci. Remote Sens.* **2016**, *54*, 5301–5318. [CrossRef]
34. Nijssen, B.; O'Donnell, G.M.; Lettenmair, D.P. Predicting the Discharge of Global Rivers. *J. Clim.* **2001**, *14*, 3307–3323. [CrossRef]
35. Nijssen, B.; O'Donnell, G.M.; Hamlet, A.F.; Lettenmaier, D.P. Hydrologic Sensitivity of Global Rivers to Climate Change. *Clim. Chang.* **2001**, *50*, 143–175. [CrossRef]
36. Hempel, S.; Frieler, K.; Warszawski, L.; Schewe, J.; Piontek, F. A trend-preserving bias correction-the ISI-MIP approach. *Earth Syst. Dyn.* **2013**, *4*, 219–236. [CrossRef]
37. Canadell, J.G.; Quéré, C.L.; Raupacha, M.R.; Fielde, C.B.; Buitenhuisc, E.T.; Ciaisf, P.; Conwayg, T.J.; Gillettc, N.P.; Houghtonh, R.A.; Marlandi, G. Contributions to accelerating atmospheric CO_2 growth from economic activity, carbon intensity, and efficiency of natural sinks. *Proc. Natl. Acad. Sci. USA* **2007**, *104*, 18866–18870. [CrossRef]
38. Wang, G.Q.; Zhang, J.Y.; Jin, J.L.; Pagano, T.C.; Calow, R.; Bao, Z.X.; Liu, C.S.; Liu, Y.L.; Yan, X.L. Assessing water resources in China using PRECIS projections and a VIC model. *Hydrol. Earth Syst. Sci.* **2012**, *16*, 231–240. [CrossRef]
39. Huang, J.; Yu, H.; Guan, X.; Wang, G.; Guo, R. Accelerated dryland expansion under climate change. *Nat. Clim. Chang.* **2015**, *6*, 166–171. [CrossRef]
40. Huang, J.; Li, Y.; Fu, C.; Chen, F.; Fu, Q.; Dai, A.; Shinoda, M.; Ma, Z.; Guo, W.; Li, Z.; et al. Dryland climate change: Recent progress and challenges. *Rev. Geophys.* **2017**, *55*, 719–778. [CrossRef]
41. Greve, P.; Orlowsky, B.; Mueller, B.; Sheffield, J.; Reichstein, M.; Seneviratne, S.I. Global assessment of trends in wetting and drying over land. *Nat. Geosci.* **2014**, *7*, 716–721. [CrossRef]
42. Luyssaert, S.; Jammet, M.; Stoy, P.C.; Estel, S.; Pongratz, J.; Ceschia, E.; Churkina, G.; Don, A.; Erb, K.; Ferlicoq, M.; et al. Land management and land-cover change have impacts of similar magnitude on surface temperature. *Nat. Clim. Chang.* **2014**, *4*, 389–393. [CrossRef]
43. Pielke, R.A. Land Use and Climate Change. *Science* **2005**, *310*, 1625–1626. [CrossRef] [PubMed]

remote sensing

MDPI

Article

External Groundwater Alleviates the Degradation of Closed Lakes in Semi-Arid Regions of China

Jiaqi Chen [1], Jiming Lv [1], Ning Li [2,*], Qingwei Wang [1] and Jian Wang [1]

1 School of Computer and Information, Hohai University, Nanjing 210098, China; jiaqichen@hhu.edu.cn (J.C.); ljmhhu@hhu.edu.cn (J.L.); qingwei@hhu.edu.cn (Q.W.); jianwang@hhu.edu.cn (J.W.)

2 College of Computer and Information Engineering, Henan University, Kaifeng 475004, China

* Correspondence: hedalining@henu.edu.cn; Tel.: +86-13283780125

Received: 18 November 2019; Accepted: 18 December 2019; Published: 20 December 2019

Abstract: There are a large number of lakes with beaded distribution in the semi-arid areas of the Inner Mongolian Plateau, and some of them have degraded or even disappeared during the past three decades. We studied the reasons of the disappearance of these lakes by determining the way of replenishment of these lakes and the impact of the natural-social environment of the basin, with the aim of saving these gradually disappearing lakes. Based on remote sensing image and hydrological analysis, this paper studied the recharge of Daihai Lake and Huangqihai Lake. The deep learning method was used to establish the time-series of lake evolution. The same method was combined with the innovative woodland and farmland extraction method to set up the time-series of ground classification composition in the basins. Using relevant survey data, combined with soil water infiltration test, water chemical, and isotopic signature analysis of various water bodies, we found that the Daihai Lake area is the largest in dry season and the smallest in rainy season and the other lake is not satisfied with this phenomenon. In addition, we calculated the specific recharge and consumption of the study basin. These experiments indicated that the exogenous groundwater is recharged directly through the faults at the bottom of Daihai Lake, while the exogenous groundwater is recharged in Huangqihai Lake through rivers indirectly. Large-scale exploitation of groundwater for agricultural irrigation and industrial production is the main cause of lake degradation. Reducing the extraction of groundwater for agricultural irrigation is an important measure to restore lake ecology.

Keywords: remote sensing; deep learning; Daihai Lake; Huangqihai Lake; lake degradation

1. Introduction

Lakes are generally facing rapid decline in the arid and semi-arid regions of the world [1–4]. The Inner Mongolia Plateau is a semi-arid area with much more evaporation than precipitation [5], but there are hundreds of lakes with an area of more than 1 km^2 distributed on the plateau [6], such as the Daihai Lake, Huangqihai Lake, Dali Lake, and Wuliangsuhai Lake. These lakes are mainly distributed in a bead chain shape in the Altun boreal margin of Yinshan mountain fracture, which is one of the two giant faults in the China zone [7,8]. Although the area of these lakes of the plateau has decreased by 30.3% between 1987 and 2010 [9], the value was even below 35.3% in Central Asia, another typical arid and semi-arid area, from 1990 to 2007 [6]. In order to figure out this phenomenon, determining the recharge and discharge relationship of lakes is a key step.

Faced the situation of shrinking lakes, there are two different perspectives on the relationship between water recharge and emissions. Firstly, local precipitation is the only source of recharge for these lakes. The supply of lakes receives rivers and groundwater runoff, all of which come from region rainfall in the basins [10]. Secondly, local precipitation cannot recharge groundwater through soil infiltration basically and the main source of recharge of lakes comes from exogenous groundwater. External groundwater recharges lakes through fault zones and other transmission channels of water [11,12]. For the first view, researchers calculated the amount of water that recharges into groundwater by precipitation infiltration into soil based on a hydraulic water balance model [13]. Nevertheless, the second view negated this opinion. Researchers verified their standpoint depending on isotope and hydrochemistry analysis. They assumed that groundwater recharge in the basin comes from precipitation, so the weighted average of hydrogen and oxygen isotopes of precipitation should be the same as that of groundwater. However, through experimental analysis, the isotopes of these lakes are obviously different from those of local precipitation and overly high soil salinity in the basin. It has been concluded that precipitation cannot recharge groundwater and lakes to receive a recharging of external water [14].

The current research methods have some shortcomings. The first method assumed that groundwater transportation can only be restricted to the basin. This statement is difficult to satisfy for the plateau region where the basal fault zone develops. In the second method, due to the limited sampling, it is difficult to collect continuous data on changes of lake water, river water, groundwater, and precipitation. The error of data analysis results is large and it is hard to quantitatively determine the groundwater recharge.

To overcome the above problems in previous studies, this paper will study the continuous change of lake area by remote sensing combined with on-site observation data and isotope analysis results to determine whether these lakes accept an external groundwater supplement. Daihai Lake (the third largest inland lake [15]) and Huangqihai lake (formerly the fourth largest inland lake in the Inner Mongolia Plateau [16]) are selected as typical research objects. Moreover, we explored the causes of lake shrinkage and the impact of lake shrinkage on the surrounding ecological environment through the surface changes of the basin.

2. Study Region

The Daihai basin and Huangqihai basin are located in Inner Mongolia [17], northwest of China, which possess arid and semi-arid climate environment Figure 1. Daihai and Huangqihai are ancient inland tectonic lakes [18], which were born in the early Quaternary and the distance between the two lakes is about 64 km [19,20]. The main characteristics of this region are a dry climate, sparse precipitation, low surface wetness, and poor ecological stability. The average precipitation and evaporation in this area are 350 to 450 mm and 1800 to 2100 mm, respectively [21]. Lakes in this difficult natural environment often play an extremely important role in the survival of animals and plants and in human activities.

The area of Daihai basin is about 2312 km^2. There are Liangcheng County in the Figure 1 and the total population of the area was about 249000 as of 2013. Daihai is the third largest inland lake in Inner closed Mongolia with a maximum depth of 19.1 m [22]. Due to the large number of dams in the Daihai basin, a small number of surface paths flow into the lake [23]. Therefore, ground surface precipitation and groundwater are the main water supply for Daihai Lake. The Daihai Lake is almost shrinking every year, which has become the focus of concern for the local people.

The area of Huangqihai basin is about 4480 km^2. There are Ulanchab City, Chaharyouyiqianqi County and Chaharyouyizhongqi County Figure 1 [23]. The total population of the region was about 720,000 as of 2012. The lake is a closed lake with an average water depth of three meters in 1986 [24]. The main water supply is surface precipitation, seven seasonal rivers, and groundwater. It was dried out completely in 2006 [16]. The dry Huangqihai Lake has had significant impacts on local biological, ecological, and human activities.

The surface runoffs flowing into Huangqihai Lake come from spring water. The hydrogen and oxygen isotope of groundwater of the basin are significantly different from that of precipitation and spring water. Researchers have stated that the multi-year average value of surface precipitation isotopes in Daihai basin and Huangqihai basin is $\delta^{18}O = -5.4‰$, $\delta D = -35‰$; in addition, they took 44 groundwater samples (including deep well water and spring water) in the study region, with the average value $\delta^{18}O = -9.4‰$, $\delta D = -74.1‰$ [14]. It demonstrates that the main source of supply of the lake is groundwater. The massive exploitation of groundwater for agricultural irrigation is the main reason for the shrinking, or even disappearing, of the lake.

In the third section, this paper would explain that we processed the remote sensing images through deep learning method. In the fourth section, the use of remote sensing data would be combined with other related data for analysis. Finally, we discussed and summarized the analysis result.

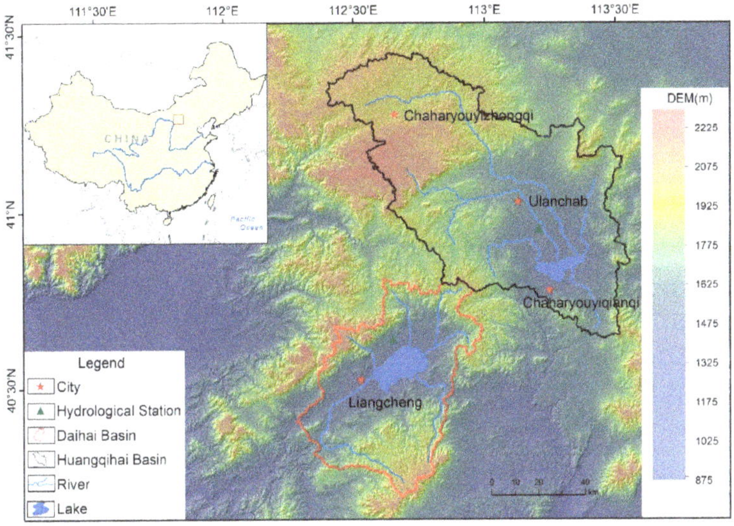

Figure 1. Location and ground observation station distribution map of Daihai basin and Huangqihai basin, China.

3. Materials and Methods

3.1. Flowchart and Datasets

This study had two fundamental aspects. The processing flowchart is shown in Figure 2. There are four parts in the article (different background colors are used to distinguish). First, the light orange part is the data and method of the article. Second, the light blue part is the main evidence of the analysis. Third, the light yellow part is the main angle of the analysis. Fourth, the light purple part is the expansion analysis of the conclusion of the article. The first was to explore on time-series of two different lakes' surface area. From former study [23], we realized the lakes area had changed dramatically. In this study, three satellites are applied, namely Landsat-5 TM satellite, Landsat-7 ETM+ satellite, and Landsat-8 OLI_TRIS satellite respectively. All Landsat data used in this study are obtained from the United States Geological Survey (USGS) website (http://glovis.usgs.gov/) and the Geospatial Data Cloud, Computer Network Information Center, Chinese Academy of Sciences website (http://www.gscloud.cn/). The necessary image preprocessing steps include radiation calibration and atmospheric correction (top-of-atmosphere, TOA) [25], which are carried out through ENVI 5.3 software. In order to observe the variation of lake area in more detail, we selected 22 remote sensing images about each lake during the past three decades from 1984 to 2018. We selected the images of production time as far as possible from April to June, because the water storage was relatively stable during this period. The lake surface data sources in the study are listed in Table 1.

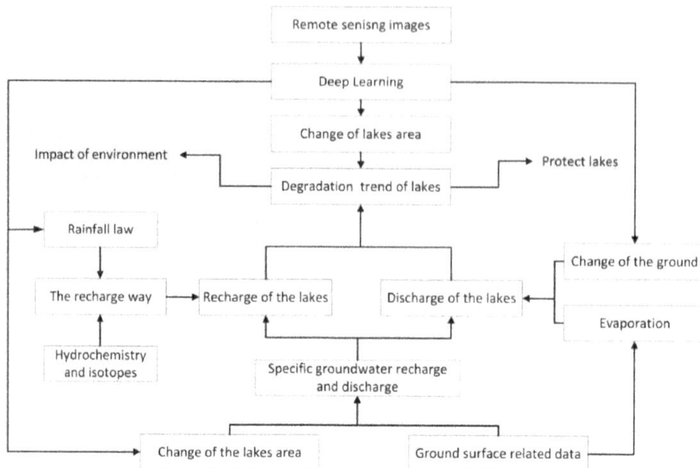

Figure 2. Processing flowchart of this study.

Table 1. Remote sensing data source of study region

Daihai Lake			Huangqihai Lake		
Index	**Time**	**Data Source**	**Index**	**Time**	**Data Source**
1	1984/5/26	Landsat 5(TM)	1	1984/5/26	Landsat 5(TM)
2	1986/5/16	Landsat 5(TM)	2	1986/5/16	Landsat 5(TM)
3	1987/9/15	Landsat 5(TM)	3	1987/9/15	Landsat 5(TM)
4	1989/9/29	Landsat 5(TM)	4	1989/9/29	Landsat 5(TM)
5	1990/8/22	Landsat 5(TM)	5	1990/7/14	Landsat 5(TM)
6	1992/6/17	Landsat 5(TM)	6	1992/6/17	Landsat 5(TM)
7	1994/4/4	Landsat 5(TM)	7	1994/4/4	Landsat 5(TM)
8	1996/6/3	Landsat 5(TM)	8	1996/6/3	Landsat 5(TM)
9	1998/5/24	Landsat 5(TM)	9	1998/5/24	Landsat 5(TM)
10	1999/7/14	Landsat7(ETM) SLC-on	10	1999/7/31	Landsat7(ETM) SLC-on
11	2000/5/22	Landsat7(ETM) SLC-on	11	2000/5/22	Landsat 7(ETM) SLC-on
12	2002/4/10	Landsat7(ETM) SLC-on	12	2002/4/10	Landsat 7(ETM) SLC-on
13	2004/6/9	Landsat 5(TM)	13	2004/6/9	Landsat 5(TM)
14	2006/6/15	Landsat 5(TM)	14	2006/6/15	Landsat 5(TM)
15	2008/9/1	Landsat 5(TM)	15	2008/9/1	Landsat 5(TM)
16	2010/5/2	Landsat 5(TM)	16	2010/5/2	Landsat 5(TM)
17	2013/4/15	Landsat 8(OLI)	17	2013/4/15	Landsat 8(OLI)
18	2014/5/20	Landsat 8(OLI)	18	2014/5/20	Landsat 8(OLI)
19	2015/5/16	Landsat 8(OLI)	19	2015/5/16	Landsat 8(OLI)
20	2016/4/23	Landsat 8(OLI)	20	2016/4/23	Landsat 8(OLI)
21	2017/4/3	Landsat 8(OLI)	21	2017/4/3	Landsat 8(OLI)
22	2018/3/5	Landsat 8(OLI)	22	2018/3/5	Landsat 8(OLI)

The other primary field was to go into the relationship between surface material composition and lake surface area in basin. Due to the limitations of clouds and available time, we selected eight images about each basin during three decades from 1986 to 2018. These images are about four to five years apart. The months of the selected samples were concentrated from May to September because this part of the time was in a relatively stable state of summer vegetation and the amount of water was relatively abundant and easy to observe. The surface material composition data sources in the study are listed in Table 2.

Table 2. Remote sensing data source of study basin

Dahai Basin			Huangqihai Basin		
Index	**Time**	**Data Source**	**Index**	**Time**	**Data Source**
1	1986.06	Landsat 5(TM)	1	1986.06	Landsat 5(TM)
2	1993.09	Landsat 5(TM)	2	1993.09	Landsat 5(TM)
3	1998.05	Landsat 5(TM)	3	1998.05	Landsat 5(TM)
4	2001.08	Landsat 7(TM) SLC-on	4	2001.08	Landsat 7(TM) SLC-on
5	2006.06	Landsat 5(TM)	5	2006.06	Landsat 5(TM)
6	2010.06	Landsat 5(TM)	6	2010.06	Landsat 5(TM)
7	2014.08	Landsat 8(OLI)	7	2014.08	Landsat 8(OLI)
8	2018.05	Landsat 8(OLI)	8	2018.05	Landsat 8(OLI)

3.2. Calculation of the Lakes Area

At present, there are many mature methods for extracting waters from optical images [26]. In this study, we chose a deep learning approach to process images. Deep learning is a domain that has been prevalent in recent years, especially in image classification. The continuous development of many high-quality models has brought higher accuracy to image classification. Considering the multispectral properties of Landsat images and the accuracy of deep learning models, we decided to use the Pyramid Scene Parsing Network (PSPNet) [27]. All experiments are conducted on Python with tensorflow-gpu 1.14.0 and the desktop computer we used is equipped with Windows 10, Intel(R) Core (TM) i7-6800K CPU and NVIDIA GeForce GTX 1080 8G GPU.

The traditional semantic analysis is only to obtain each pixel label of the known object, while ResNet is based on the semantic segmentation of scene analysis, which is to obtain the category label of all pixels in the image. Its integrated global features are more conducive to the accurate acquisition of target pixel tags, and its algorithm effect is better than traditional methods [27]. For this work, it is necessary to parse all the pixels in the whole image, so this method was adopted. However, this method has so far been used less in processing optical remote sensing images, so this experiment is a combination of optical image processing and computer vision methods.

The Pyramid Pooling Module combines features of four different scales Figure 3 [27]. The coarsest level highlighted in red square frame in the Figure 3 that is a single bin output generated by global pooling. The remaining three levels divide the input feature map into several different sub-areas, pool each sub-area, and finally combine the pooled single bins containing the location information. In the pyramid pooling module, different levels output different levels of feature maps. In order to maintain the weight of the global features, we employed a 1×1 convolution kernel after each pyramid level. If a level dimension is N, this model can reduce the dimension of context feature to 1 / N of original feature. Then, the low dimensional feature map is directly upsampled to be the same as the original feature map by bilinear interpolation. Finally, the feature maps of different levels are stitched into the final pyramid pooled global features.

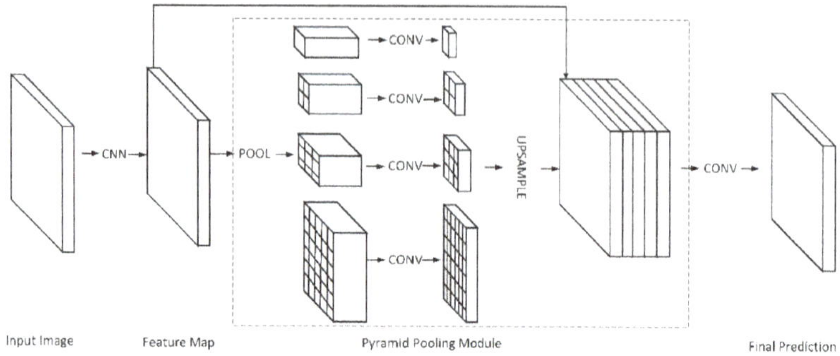

Figure 3. Neural network framework of PSPNet.

Though neural networks can provide good performance through deep pre-training, the increase of network depth may bring additional optimization difficulties for image classification. As a module in PSPNet, the ResNet is used to extract the feature map of the input image, and ResNet solves this problem by using a skip connection in each block. In the deep ResNet, the latter layer mainly learns the residuals thrown by the previous layer. Based on the original ResNet, PSPNet adds an auxiliary loss in the fourth stage in addition to the main branch of the final classifier using softmax loss. Finally, it adds weights to balance the auxiliary loss function. The two losses are then combined, using different weights to optimize the parameters together.

Therefore, for the input image in Figure 3, we used a pre-trained ResNet model with an extended network strategy to extract the feature map. The final feature size is 1/8 of the input image, as shown above Figure 3. This work utilized the pyramid pooling module Figure 3 to obtain context information for the above feature map. The pyramid pooling module was divided into four levels, and the pooled kernel size is the whole, half, and small parts of the image. Eventually, they can be merged into global features. Then, in the final part of the Figure 3 module, this study connects the global features and the original feature map. Finally, the final prediction map was generated by convolutional layers in Figure 3.

Created data set is basic step, and we need to make a data set for the pre-processed Landsat images for training and testing [28]. The pseudo color with the combinations of Band Red, Band SWIR1, and Band SWIR2 are selected for the training process [29]. We selected eight images for each lake as training samples, and selected two images for the test sample as water data sets and manually labeled them. The annotated data set was then cut into 600 sample blocks of size 256 * 256 * 3. There is no doubt that the number of these samples is too small to train. Therefore, we used the operations of pan, rotate, zoom, add noise, etc. to expand the number of image samples to 30,000.

Finally, in the experiment, adopt overall accuracy (OA) to assess the accuracy of water extraction area. The overall accuracy (OA) can be obtained by the equation

$$OA = \frac{TP + TN}{TP + FN + FP + TN} \tag{1}$$

where TP and TN represent the pixel points whose labels are positive or negative and corresponding result is predicted to be the same while FN and FP represent the opposite. The OA is about 98.9%.

3.3. Land Surface Classification

Similarly, we employed PSPNet method in land surface classification. OA1 and OA2 represent the accuracy of land surface classification of Daihai basin and Huangqihai basin respectively. Classification accuracy of each category is listed in Table 3.

Table 3. Overall accuracy for ground classification

	Construction	Farmland	Woodland	Saline–alkali	Nudation	Water
OA1	96.33%	67.54%	64.59%	95.99%	96.87%	98.65%
OA2	95.88%	72.31%	69.82%	96.91%	97.02%	97.83%

Obviously, classification accuracy of farmland and woodland is low. We summarized the following reasons, including low training samples because of the general quality original images and farmland pixels and woodland pixels with too many similar features. In addition, the most important reason is over reclamation in forest areas produce complex forest-staggered areas with many error classification pixels. Thus, we had to establish an innovative means to distinguish between two similar cells.

We found those farmlands are relatively gentle area, while woodlands are relatively rugged through an investigation into the basin. In other words, the classification of farmland and woodland can convert to the recognition of land type. Based on this feature, we used ASTER GDEM 30 m data to look for the relationship between DEM and land type. In this part, the random unclassified pixel is selected as the central point and then extended around to form a large cell measuring 5 × 5 (Figure 4).

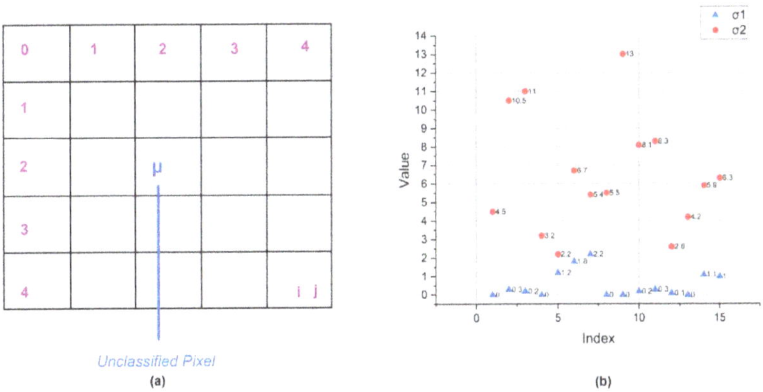

Figure 4. (**a**): Schematic diagram of unclassified pixel processing; (**b**) Standard deviation of farmland (σ1) and woodland (σ2) with DEM.

Taking the central point as the mean value and the peripheral point as the sample value, calculated the fluctuation of the target pixel and the surrounding pixel by the standard deviation of variation (1).

$$\sigma = \frac{1}{25} \sum_{i=0,j=0}^{i=4,j=4} (X_{ij} - \mu), \quad (2)$$

As showed in Figure 4, calculated σ for farmland pixel and woodland pixel with 15 sample points for each in this work. In this picture, σ1 and σ1 represent farmland and woodland respectively. We draw the conclusion that the farmland pixel standard deviation value is generally less than two, and the pixel value of the woodland is the opposite. According to the above method, the OA of farmland and woodland classification is 93.7%.

4. Results

4.1. Variation for Lake Area

After the above stage, construction of a yearly time-series of the water surface areas on Daihai Lake and Huangqihai Lake are displayed in Figure 5.

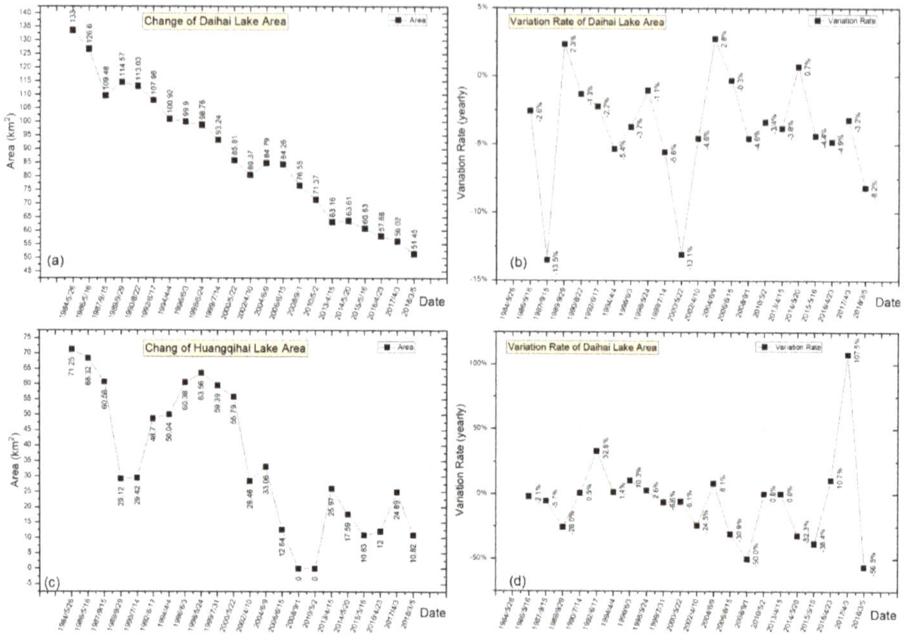

Figure 5. Time-series for variation of two lakes.

In this experiment, 22 pictures were used and a set of time-series data was made for each lake during from 1984 to 2018. Overall, the trend of surface area in Daihai Lake is declining. Daihai Lake area reached a maximum level of 133.5km^2 in 1984 and then declined at a rate of about three per cent. The entire lake area shrank to a staggering 82.05 km^2, which accounted for 61.5% of the water surface in 1984. This shrinkage data is more than 30.3% of the total shrinking area of lakes in Inner Mongolia for nearly 30 years [6].

If Daihai Lake is a typical example of serious shrinkage of the water surface as considered, then the complete drying up of the Huangqihai Lake is regrettable. Overall, the water area of the Huangqihai Lake showed a downward trend until it ran dry completely. The area reached a peak of 71.25 km^2 in 1984, but declined sharply in 1989, then rose to a rising trend from 1994 to 1998. After 1999, the lake fell sharply again until it dried up completely from 2006 to 2013. In the following years, water area has recovered slightly, but it is limited to the region that was located in the northwest and northeast of the lake. The recovery of the water area is mainly owing to the inflow of the river. There is a contrast between the two adjacent lakes. Daihai Lake still maintains a certain water quantity, but Huangqihai Lake has experienced drying up. This is important evidence that there are other sources of water supply in Daihai Lake and the decline of Daihai has been alleviated.

In the above paragraphs, we described the changes in the two lakes in terms of time distribution. Then, we continued to observe the characteristics of water surface changes for each lake from a spatial perspective. We selected some year images with significant changes and drew diagrammatic sketch of the water surface change on Daihai Lake and Huangqihai Lake respectively. Figures 6 and 7 show the spatial changes of each lake.

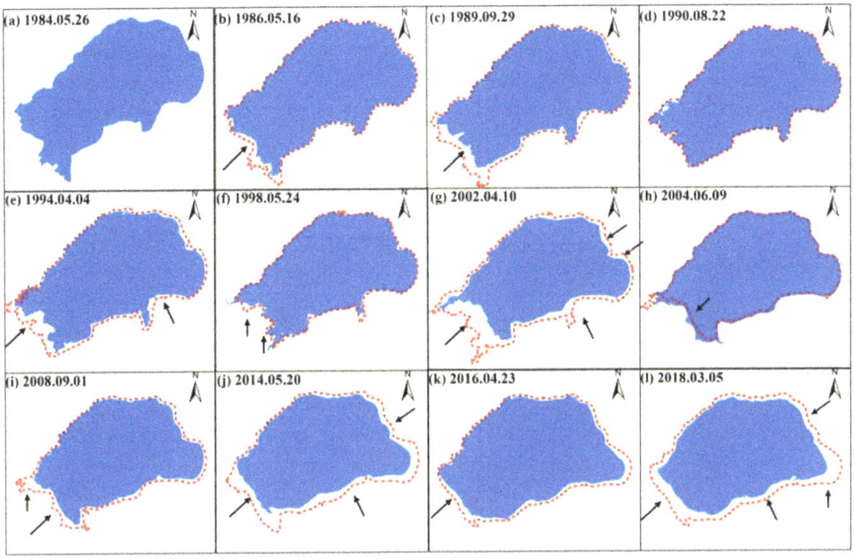

Figure 6. Spatial variation of Daihai Lake surface.

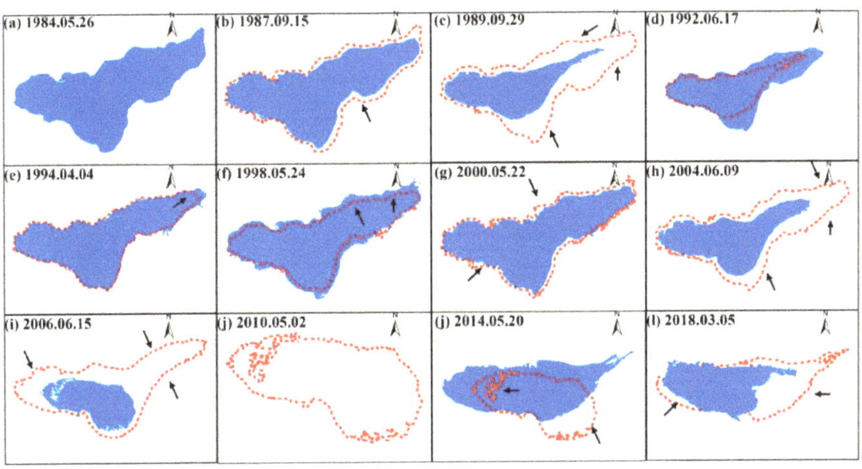

Figure 7. Spatial variation of Huangqihai Lake surface.

Figure 6 shows that the whole Daihai Lake mainly shrank from south to north, while the north and northeast also showed a small shrinkage. Before 1989, the lake shrank mainly in the south, and the water surface in the north and northeast was stable. During the later 1990s, the south and northeast of Daihai Lake began to shrink sharply. After 2004, the main shrinking parts of water surface have gathered around in the south and northeast of the lake.

Compared with the steady shrinkage of Daihai Lake, the water surface of Huangqihai Lake fluctuates more violently. Before 1994, the surface of the lake shrank only slowly inward along the shoreline. However, between 1994 and 2002, the southern and northeastern parts of Huangqihai Lake showed drastic contraction. From 2002 to 2008, the lake dried up for the first time. This result is basically consistent with the previous results [30]. Although the lake recovered in the northwest region between 2014 and 2018, as shown in Figure 1, it was just the result of the river flowing into the lake.

4.2. Ground Composition

Using the above methods, this experiment carries on the ground pixel classification to Daihai basin and Huangqihai basin. In order to be more suitable for the study of the causes of lake water changes, the surface categories are divided into six categories. Specific classified data are shown in Tables 4 and 5.

Table 4. Statistics of ground classification in Daihai basin.

Daihai basin Ground classification (km^2)						
Date	**Construction**	**Farmland**	**Woodland**	**Nudation**	**Water**	**Saline–alkali**
1986.06	1.39	182.72	72.73	1927.00	138.94	0.00
1993.09	2.66	123.21	95.14	1975.89	116.98	0.00
1998.05	3.06	349.93	222.41	1635.72	103.68	0.00
2001.08	5.90	230.25	19.61	1972.26	84.74	0.00
2006.06	6.54	134.93	102.21	1961.41	99.97	9.48
2010.06	7.19	145.89	115.46	1971.03	75.56	0.00
2014.08	50.89	291.80	64.41	1837.62	70.48	0.00
2018.05	86.99	377.24	145.40	1642.66	59.98	4.65

Table 5. Statistics of ground classification in Huangqihai basin.

Huangqihai Basin Ground classification (km^2)						
Date	**Construction**	**Farmland**	**Woodland**	**Nudation**	**Water**	**Saline–alkali**
1986.05	40.30	107.45	23.35	4221.00	78.30	0.00
1993.09	53.86	68.04	10.28	4226.29	60.60	47.63
1998.05	67.35	317.75	108.33	3904.10	82.97	0.00
2001.08	76.23	694.73	81.77	3572.57	55.20	0.00
2006.09	79.67	71.15	17.84	4265.09	28.24	2.29
2010.07	89.86	197.69	19.65	4137.20	25.23	11.07
2014.05	123.78	269.59	32.54	3961.07	58.91	0.00
2018.05	234.38	632.50	184.37	3375.68	31.08	50.49

The area of the construction increased in Daihai basin year by year. Before 2010, the area was stable relatively and expanded 10-fold from 2010 to 2018 (Figure 8). Distribution of the construction are mainly located at Liangcheng county in southwest of the lake, which began to scatter around the lake and the county after 2010. The farmland area was stable about 150 km^2 until 2010, but the area of 1998 showed a three times increase compared to 1993. Because the local government adopted the project of the expansion of planting area in order to increase economic income. The woodland area has been stable relatively between 50 km^2 and 150 km^2 except in 1998. The local government's excessive afforestation plan gave rise to the burst increase of the water consumption in the basin. Regrettably, the growth resulted in excessive water consumption in the region, which cannot sustain the growth of these trees. This is the reason why trees withered in a large area in 2001. The nudation area, accounting for more than 85% of the whole basin, includes bare mountains and arable land without growing crops, which is enough to show that the soil and water conservation capacity of the land is low. In addition to the main Daihai Lake, the water area also includes reservoirs, dams, and some water in the river, which shows a downward trend. Saline–alkali area are more volatile, mainly around the lake and part of the dry beach, which is related to direct human production and vegetation cover at that time.

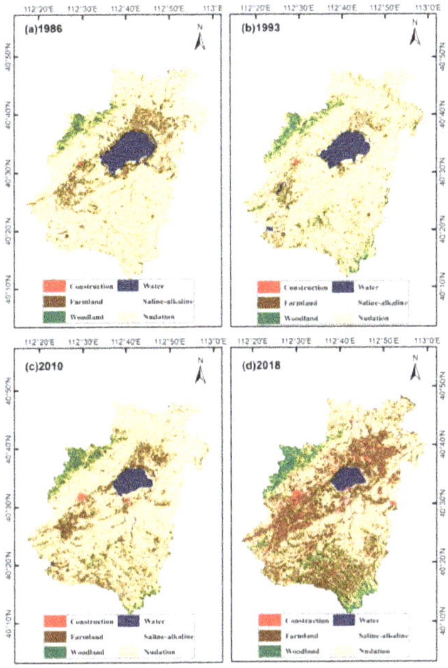

Figure 8. Ground classification of Daihai basin.

Because there are three larger cities in the Huangqihai basin and the economy is more developed than Daihai basin (Figure 9). Thus, the construction area is much larger than that in the Daihai basin. Similarly, the building area also had a huge increase in the Huangqihai basin after 2010. Farmland area fluctuated before 2010. In 1998, there was still a burst of growth, unlike Daihai basin, which lasted until 2001, but the area did not drop sharply until 2006. Next, there has been a steady upward trend. Woodland area has remained at a low level relatively except in 1998, 2001, and 2018. Great fluctuations are evident in excessive human intervention. Nudation reached previous level again in 2006 after a decline in 1999. The water area of the basin has been declining because of the gradual drying up of the Huangqihai Lake. It rebounded in 2014, but fell again in 2018. Saline–alkali land is mainly saline–alkali land formed by the residual salt of bare land after the removal of lake. With the change of lake water area and artificial intervention, the area of saline–alkali land is also changing.

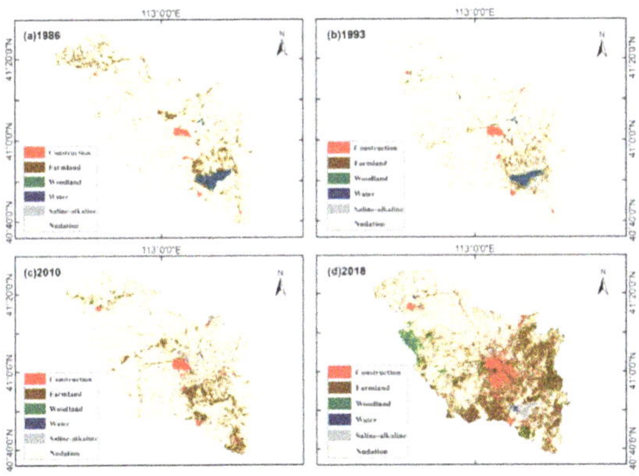

Figure 9. Ground classification of Huangqihai basin.

5. Discussion

5.1. Qualitative Analysis of the Recharge Process

There are two possible sources of groundwater for recharge to lakes. One of them, it comes from local precipitation. The local precipitation assimilates into the groundwater by soil infiltration. The other one, groundwater of recharge lakes comes from external water. Because the lakes are located on the fault zones with transmission channels of water, the exogenous water continuously transports to the lakes and surrounding aquifers through the fault zones.

Generally, the common lake will expand to a maximum in the annual rainy season and shrink to a minimum in the dry season. However, we used remote sensing to make an experiment. Here, the experiment also used the above-mentioned the deep learning method to extract lakes in the target region. Based on the number of remote sensing images in the same year, the randomly selected sample year is 2000. Then the work selected one day of observation month as the representative image of the month, and continued to select the representative image of the next month about one month apart. This work found that the opposite phenomenon has occurred in the Daihai Lake. The lake area is smallest in the rainy season (June to September) and largest in the dry season (January to May and October to December) (Figure 10). In dry season, there was no precipitation to supplement the lake water while the lake water area increased. This indicates that the evaporation of the lake surface is greater than the amount of water entering the lake and the main recharge of the lake is derived from groundwater. The ground water recharge comes from the leakage of the fault zone in this area. However, the above situation did not occur in the experiment of the Huangqihai Lake. During the previous dry season, the lake area continued to decline and the arrival of the rainy season eased the situation. In the next dry season, there was no rainfall and the lake area fell again (Figure 10). This shows that the water supply of the Huangqihai Lake mainly comes from the precipitation and runoff in the basin. The recharge method of the Daihai Lake belongs to the second case and this of Huangqihai Lake is the first case. This is an important evidence to indicate that Daihai Lake receives external water supply to alleviate the lake shrinkage.

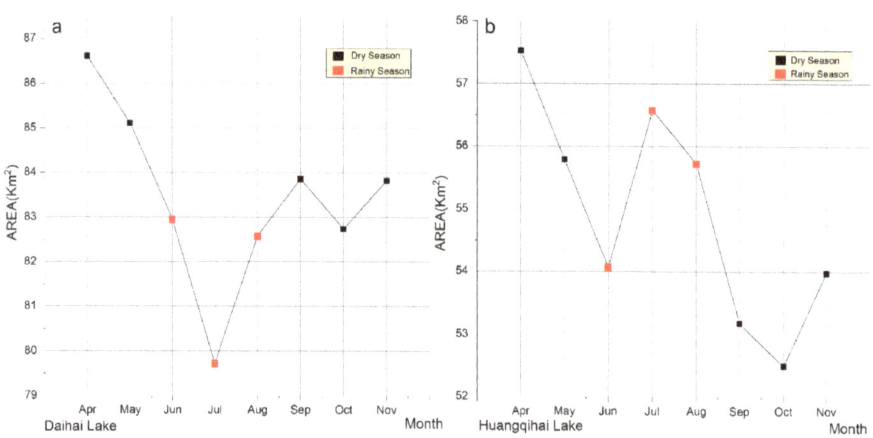

Figure 10. Lake area change in 2000.

In order to further verify our experimental results, a hydrological method was also used. The soil infiltration test showed that the local precipitation cannot enter the groundwater in Daihai basin, because the unsaturated soil water does not reach the maximum water-holding capacity of field (WHCF) and the soil moisture content is in a loss state. In other word, the necessary condition for precipitation infiltration is above the maximum WHCF [14]. We studied the evapotranspiration (ET) of this region is about 395 mm [31] (www.cnern.org.cn), which is basically the equal to the local average annual precipitation of 384 mm (DH) and 374mm (HQH). That is to say, the precipitation of the basin is basically evaporated, and cannot form groundwater of the basin through infiltration. We used remote sensing method to analyze the causes of the lake area change, and we also used soil chemistry and surface evapotranspiration method to verify the results. These experimental results indicated that there is exogenous water supply to Daihai Lake (Figure 11).

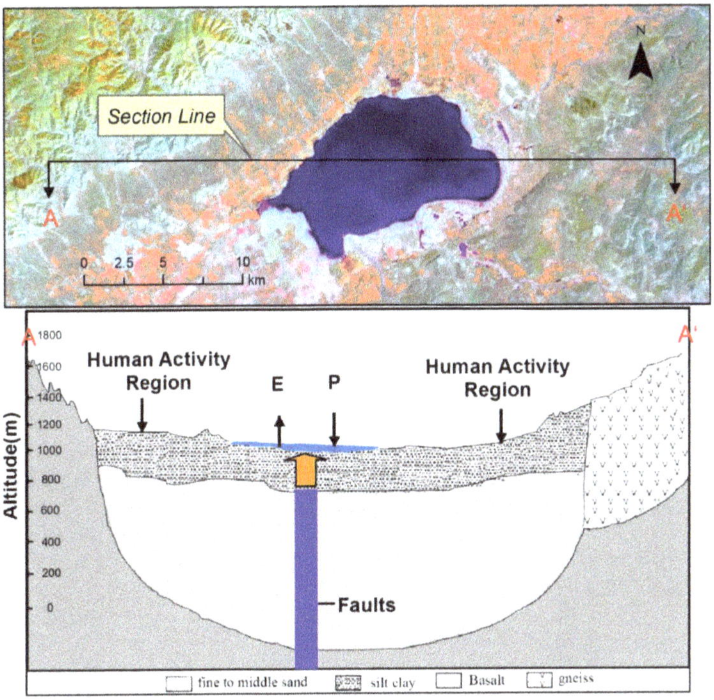

Figure 11. Daihai Lake receives external groundwater recharge.

The above three methods confirm the existence of external groundwater recharge in Daihai Lake, and whether there will be external water recharge in Huangqihai Lake, which is only 64 km away from Daihai Lake. If so, why did the Huangqihai Lake eventually dry up in 2008? Hydrological and isotope experiments showed that the deep underground wells and springs in the Huangqihai basin are the same as those in the Daihai Basin, and the surface springs eventually converge into rivers to supply the Huangqihai Lake. However, because of the artificial construction of river dams to intercept water sources for daily life, economic and agricultural, the lake has lost its recharge.

5.2. Quantitative Analysis of Water Supply and Consumption

The above analysis shows that the Daihai Lake received external groundwater recharge while the Huangqihai Lake received no external groundwater supply. The quantitative value of groundwater recharge was calculated along with the balance relationship between recharge and emission, based on the analysis [32]. If the water content of a basin maintains a dynamic balance, it must be equal to the water flowing in and out. Since the Daihai basin is a closed watershed, the water volume changes of the basin can be expressed through the Daihai Lake.

Firstly, we introduced the related indicators in the next work. The region is special and the crops species are relatively scarce. We use the annual statistical yearbooks of the region to estimate the agricultural water consumption (AWC) for the different crop areas in these years [33–35]. Flood irrigation is the main practice for farmland in Daihai basin, so the water requirement of crops needs to be converted into the total irrigation water consumption. The irrigation efficiency of this area is about 62.7% [36]. We used agricultural irrigation consumption (AIC) as one of the groundwater estimation indicators. Agricultural water consumption in the region accounted for 54.3% of the total water consumption of human activity (TWCHA) in 2003, which makes us to estimate TWCHA by AIC (excluding power consumption of power plants). The local government introduced a power plant that needed to use the lake water for water cooling for economic development after 2005, which increased the evaporation of the lake water directly. Annual water consumption (PPWC) reached 1.206*10^7m^3 according to the estimation. In this experiment, the annual water volume reduction (WVR) was estimated using the annual water level relationship and the area obtained by remote sensing.

For the water flowing into the Daihai Lake, external water, precipitation, and runoff are the main resources. The hypothesis that rainfall infiltrates into groundwater recharge lakes has already been negated through the special phenomenon of change of the lake area in different seasons by remote sensing and has proven that this conclusion is a reliable the work of water chemistry. The runoff volume of the basin is only 6*10^3m^3 yearly [13]. The supply of lake water is less, so it can be ignored. Therefore, water flowing into the lake can be reduced to surface precipitation and external groundwater (DGR). The water flowing out of the Daihai Lake, first of all the direct annual evaporation water consumption (EWC) the lake was calculated by evaporation of the basin, which is need to multiply conversion coefficient (0.55) [37]. Then, the Daihai Power Plant needs to consume lake water every year. Finally, there is about 9.12*10^7m^3 of groundwater in the basin itself and this part of the groundwater has been extracted by people (TWCHA). Although the surface precipitation cannot infiltrate to form groundwater, the lake replenishes lost water of the basin to maintain the water balance in the basin. So, this element of artificial consumption is also part of the lake water consumption. The calculation method of parts of the indicators are more intuitive in Figure 12.

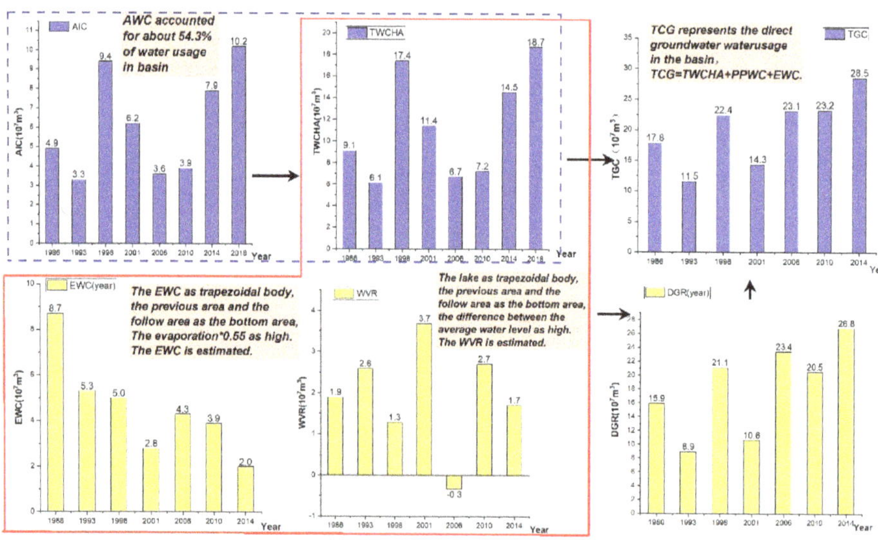

Figure 12. Water consumption and recharge in Daihai Basin.

All of these inflows and outflows are change of lake water volume (WVR). The resulting equation is

$$\text{WVR} = \text{PPWC} + \text{EWC} + \text{TWCHA} - \text{DGR}. \tag{3}$$

$$\text{DGR} = \text{PPWC} + \text{EWC} + \text{TWCHA} - \text{WVR}. \tag{4}$$

The recharge volume of exogenous groundwater is not a fixed value calculated by the water chemistry method Table 6. At the same time, in order to better observe the proportion of water consumption and the variation, we presented the whole calculation results in the form of histogram (Figure 12). The reason of inconsistency is that the calculation method is to use the surface area to calculate EWC in this paper. The area of the Daihai Lake after 2000 year is about 100 km^2 less than that of 1960s, so the calculation of evaporation loss will also be reduced. However, this value is maintained range 0.89*10^8 m^3 to 2.68*10^8 m^3 and the average annual replenishment is 1.81*10^8 m^3, which is close to the 1.8*10 ^{8}m^3 calculated by the water chemistry method. The method just measured the recharge amount for one year while our method measured the recharge amount for many years. The experimental results in this paper are more reliable. Finally, the total annual consumption of groundwater is calculated to be an average of 2.0*10^{8}m^3. This work quantitatively calculated the recharge of groundwater to Daihai Lake, which provided substantial evidence for groundwater to alleviate the shrinkage of the lake. This annual groundwater consumption is slightly larger than the average annual external groundwater recharge. This imbalance state indirectly proves that lakes are shrinking every year.

Table 6. Water consumption and recharge in Daihai Basin (10^{7}m^3)

Date	AWC	AIC	TWCHA	PPWC [1]	WVR [2]	EWC	DGR [3]	TGC [4]
1986.06	3.1	4.9	9.1	0.0	1.9	8.7	15.9	17.8
1993.09	2.1	3.3	6.1	0.0	2.6	5.3	8.9	11.5
1998.05	5.9	9.4	17.4	0.0	1.3	5.0	21.1	22.4
2001.08	3.9	6.2	11.4	0.0	3.7	2.8	10.6	14.3
2006.06	2.3	3.6	6.7	1.2	-0.3	4.3	23.4	23.1
2010.06	2.5	3.9	7.2	1.2	2.7	3.9	20.5	23.2
2014.08	4.9	7.9	14.5	1.2	1.7	2.0	26.8	28.5
2018.05	6.4	10.2	18.7	1.2	/	/	/	/

[1] PPWC: power plant water consumption; [2] WVR: water volume reduction of lake yearly; [3] DGR: direct groundwater recharge; [4] TGC: total groundwater consumption.

Although, we concluded that the area change of the Huangqihai Lake is mainly related to the huge evaporation in the region. Next, this result needs further proof. Because of the irregular lake shape and the lack of the lake water level data, we cannot get the WVR value. So, we can still explain from other aspects. Based on the agricultural planting data and the classification results of this experiment, we obtained AWC value. Then, TWCHA was estimated by the proportion of agricultural water usage in Ulanchab city for many years. Based on the annual runoff data of Jining Hydrological Station of Bawang River (Figure 1), it is concluded that the average annual runoff of the Bawang River has been about 6.3*10^6 m^3 in the past 30 years. Due to many dams in the upper reaches of the river, the inflow of the river into the lake is much lower than this value and other secondary inflow runoff is similar. Therefore, the runoff into the lake can be neglected.

It can be seen from the Table 7 that the water consumption of human activities has been increasing since 1993. In 2001, the unexpected planting events accelerated the consumption of groundwater resources, which prevented the rapid recharge of Huangqihai Lake. In addition, the increase of human activities has also increased the consumption of groundwater resources. Ultimately, the groundwater level will decrease. Because the average depth of the lake was only three m [38], the lake cannot be recharged by groundwater. Besides, saline-alkali land hindered the infiltration of groundwater, which led to the drying up of the lake eventually in 2008. Due to the rapid decrease of lake surface area in the lake, it is obviously different from Daihai Lake and there is no direct groundwater to recharge to it.

Table 7. Water consumption and recharge in Huangqihai Basin ($10^7 m^3$)

Date	AWC	AIC	TWCHA(year)	EWC(year)	TCG
1986.05	1.1	1.7	2.1	11.4	13.5
1993.09	0.7	1.1	1.3	4.1	5.4
1998.05	3.1	5.0	6.2	8.2	14.4
2001.08	6.8	10.8	13.4	4.0	17.4
2006.09	0.7	1.1	1.5	5.3	6.8
2010.07	1.9	3.1	4.3	3.0	7.3
2014.05	2.6	4.2	6.0	2.9	8.9
2018.05	6.2	9.9	14.1	/	/

5.3. Reasons for Lake Degradation

There are two mainstream views on lake shrinkage. The one is that climate change has mainly led to varying degrees of shrinkage of lakes [39–42], and the other one is excessive human activities is main reason [43]. As we know from the above, the reasons for the decline of these two lakes are climate and perceived factors. Using the above experimental data, we calculated the proportion of the annual human water consumption to the two basins.

Evaporation precipitation difference (EPD) is evaporation minus precipitation (Figure 13). Next, we replaced Daihai Basin's EPD and Huangqihai Basin's EPD with DH EPD and HQH EPD solely. Through the investigation of evaporation and rainfall data, we can clearly find that the DH and HQH EPD value are all positive. In other words, the evaporation has always been greater than precipitation in this area, so the water of lakes is in continuous consumption state. Furthermore, we calculated the area of lakes and Pearson correlation coefficients of EPD for 0.83(DH) and 0.77(HQH). Therefore, it is concluded that evaporation is one of the causes of lake water shrinkage. Because the area of Daihai Lake is not consistent with the trend of broken line of DH EPD before 2000, we thought that the reduction of the Daihai Lake water is not only related to the evaporation in the basin, but also may have other factors after 2000. HQH EPD is different from DH EPD. Compared with Daihai Lake, the changes of the Huangqihai Lake area and the trend of the HQH EPD are more undulate. The evaporation loss value is huge in Huangqihai basin. Since 1984, the EPD has increased and the lake area has gradually decreased. In 1990, the EPD value reached the maximum and the lake area presented, correspondingly, the lowest value in this period. After 1990, the EPD decreased and the lake area gradually increased. In 1999, the EPD reached its peak again and the lake area responded very closely to the change. Later, the EPD values remained low, but it seemed that the lake could not stop shrinking and eventually dried up completely in 2008. Excessive evaporation in the basin was the main reason of the drying up of the lake before 2001, but other factors may be lead to the changes after 2001.

The recharge period of groundwater determined is about 30 years by tritium [44]. Since the seasonal variation of recharge flow has been homogenized for 30 years, the recharge flow rate of external water can be regarded as a constant. Therefore, the source of the lake can be simplified as precipitation and relatively stable external water. Furthermore, precipitation and groundwater are stable for recharge source of the lake, and evaporation is stable for consumption source of the lake in the three decades. So, excluding the above two stabilizing factors, the reduction of two lake area should have other fluctuation reasons. It should be related to the large amount of groundwater pumped for agricultural irrigation, industrial production and domestic water.

Before 1998, the climate change in Daihai basin was the main factor leading to the reduction of Daihai Lake area Table 8. After 1998, with the development of economy and the expansion of population, human activities have become the main factor affecting the change of Daihai Lake, and the situation is getting worse and worse. Comparatively speaking, the factors of Huangqihai Lake are more complex. Among them, human activities consume more water than climate before 2001, and climate consumes more water than human activities after 2001 except 2006. This conclusion is consistent with the phenomenon of EPD, and indicates that human activities have more influence on the lake changes since 2001. Therefore, on the whole, the shrinkage of the Huangqihai Lake is a combination of human activities and climate.

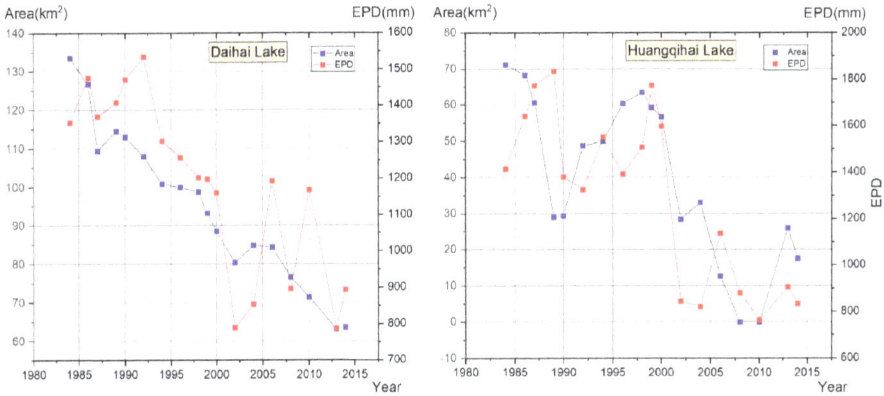

Figure 13. Area, precipitation, and evaporation of the lakes difference in basin.

Table 8. Proportion of water consumption in two basins (%)

			Daihai Basin				
Date	**1986.06**	**1993.09**	**1998.05**	**2001.08**	**2006.06**	**2010.06**	**2014.08**
Human	47.7	40.2	72.7	71.6	77.2	80.8	89.5
Climate	52.3	59.8	27.3	28.4	22.8	19.2	10.5.
			Huangqihai Basin				
Date	**1986.05**	**1993.09**	**1998.05**	**2001.08**	**2006.09**	**2010.07**	**2014.05**
Human	15.5	24.6	43.1	76.9	22.4	58.8	67.3
Climate	84.5	75.4	56.9	23.1	77.6	41.2	32.7

5.4. Environment Effects and Measures

The inland closed lake environment is very sensitive to the feedback of climate change [45], we need to assess the impact of lakes on the surrounding environment. It can be seen Figure 5 that although both lakes have suffered from different degrees of shrinkage, the feedback from the Daihai Lake area to the whole basin is positive because the lake can maintain more water (Figure 14). We calculated these proportions and the proportion of buildings of Daihai basin is 1.19% lower than that of Huangqihai basin average year. In other words, that is the water consumption of agriculture and industry in Daihai basin will be lower than that of Huangqihai basin. The proportion of the total water area of Daihai basin is 2.47% higher than that of Huangqihai basin yearly. That is to say, the Daihai basin has more abundant water resources than the Huangqihai basin. More water resources directly led to an increase 2.64% and 2.70% of the farmland and woodland of Daihai basin compared to the Huangqihai Basin average every year. There is more woodland and farmland, so the proportion of nudation of Daihai basin is 11.80% less than the average of Huangqihai yearly. All in all, more water maintains a better ecological environment in the basins.

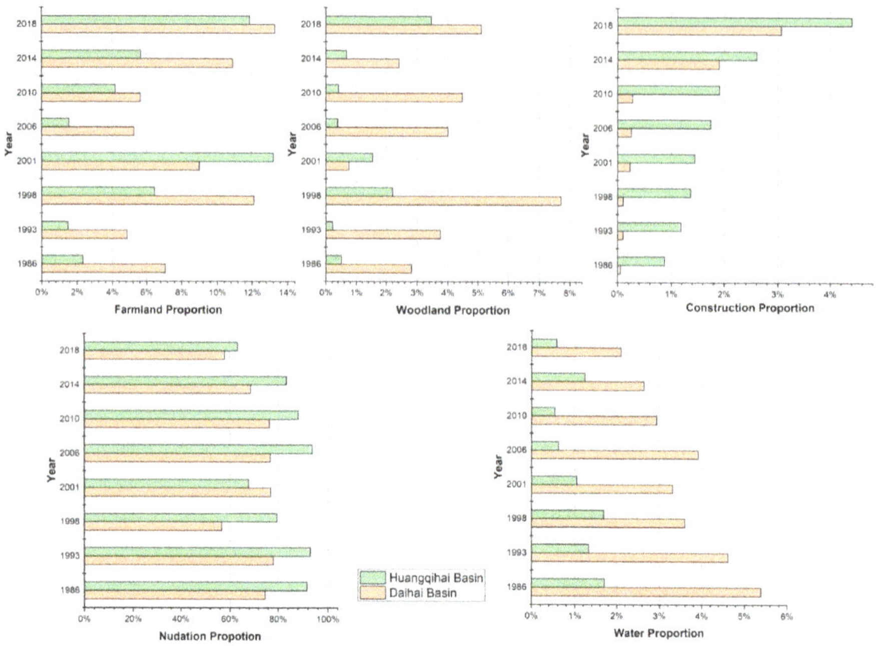

Figure 14. Proportion of various land types in the basin.

Many lakes are shrinking in arid and semi-arid areas [46]. People try to use various methods to alleviate and even change the process. It is not clear whether the effect of the method is effective or ineffective. We found out the reasons for the shrinkage of lakes, in order to implement effective measures rather than carrying out transformation over rules of nature. Therefore, the behavior of returning farmland and saving local water usage can alleviate the shrinkage of the lake. Fortunately, the local government seems to be aware of the crisis of the lake shrinkage and has been carrying out the whole basin activities of returning farmland and woodland since 2016. We still need to observe how it works. Although this plan can alleviate the shrinkage of Daihai Lake, it seems that the surface area of Daihai Lake cannot be restored to former area without more foreign water supply. Hence, the local government proposed to divert the Yellow River to supply Daihai Lake. Whether this project will affect the ecological environment of other places once it is implemented deserves our attention.

The results of the above analysis indicated that excessive groundwater extraction for agricultural irrigation and industrial production is the main reason for the lake decline. Although exogenous groundwater recharge to Daihai Lake, it cannot maintain such a huge consumption of people. Let alone Huangqihai Lake, which has no huge amount of exogenous water supplement. Cognac is normal state for the lake in the future. Therefore, reducing or even stopping the extraction of groundwater is an important measure to alleviate this trend.

6. Conclusions

In this paper, the recharge resources of two lakes were explored based on remote sensing data and site data, and we studied to the reason for degradation of the lakes. We found that Daihai Lake is supplied with external groundwater, which alleviated the decline of the lake in arid areas owing to the existence of a water diversion structure. Huangqihai Lake lacks recharge of external groundwater directly, and humans pumped too much groundwater from the basin, resulting in the lake drying up. Most of the lakes in the Inner Mongolia Plateau are arid or semi-arid climatic conditions, but there are still many lakes. These lakes are basically dependent on direct or indirect recharge of exogenous groundwater. However, these lakes are facing degradation in different degree. Although exogenous water can alleviate the decline of these lakes, the increasing human demand for water cannot prevent these lakes from disappearing eventually. These lakes play an extremely important role in semi-arid fragile ecosystems. People seem to be aware of the urgency of this problem, actions are being taken to curb the lake shrinkage and even to restore the lake to some extent, which we need to keep watch. Due to the limitation of the data quantity of remote sensing images, this paper did not calculate the material around the surface entirely according to the year of the lake area Figure 5 change, but selected the phenomenon of image observation about every five years, which may bring deviation to the experimental results. Besides, this paper cannot estimate the annual direct loss volume of the Huangqihai Lake due to the lack of the lake water level change data, which will make the conclusion insufficient. In the future, we will continue to monitor the changes of these lakes and whether human rescue measures can serve as a model for their protection.

Author Contributions: Conceptualization, J.C. and J.L.; Methodology, J.C. and J.L.; Software, J.L.; Validation, J.C., J.L., and Q.W.; Formal analysis, J.W.; Investigation, J.C. and J.L.; Resources, J.C.; Data curation, J.L.; Writing—original draft preparation, J.L., J.C., and N.L.; Writing—review and editing, N.L. and J.C.; Visualization, Q.W. and J.W.; Supervision, N.L.; Project administration, J.C.; Funding acquisition, J.C. and N.L. All authors have read and agreed to the published version of the manuscript.

Funding: This research was funded in part by the National Key R&D Program of China (no. 2017YFB0502700), in part by the National Natural Science Foundation of China (no. 61771183, 61601437), in part by the Fundamental Research Funds for the Central Universities (no. 2016B07114), in part by the Plan of Science and Technology of Henan Province (no. 192102210082), in part by the Youth Talent Lifting Project of Henan Province (no. 2019HYTP006), and in part by the China Postdoctoral Science Foundation (no. 2013M541035).

Acknowledgments: Thanks for Jiansheng Chen to furnish with professional idea on water hydrochemistry work; thanks for Zhanyang Zhang that provided assistance for Figure 11.

Conflicts of Interest: The authors declare no conflict of interest.

References

1. Crighton, E.J.; Elliott, S.J.; Upshur, R.; van der Meer, J.; Small, I. The Aral Sea disaster and self-rated health. *Health Place* **2003**, *9*, 73–82. [CrossRef]
2. Lioubimtseva, E.; Henebry, G.M. Climate and environmental change in arid Central Asia: Impacts, vulnerability, and adaptations. *J. Arid Environ.* **2009**, *73*, 963–977. [CrossRef]
3. Sarch, M.-T.; Birkett, C. Fishing and Farming at Lake Chad: Responses to Lake-Level Fluctuations. *Geogr. J.* **2000**, *166*, 156–172. [CrossRef]
4. Williams, W.D. Salinisation: A major threat to water resources in the arid and semi-arid regions of the world. *Lakes Reserv. Res. Manag.* **1999**, *4*, 85–91. [CrossRef]
5. Xiao, X.; Ojima, D.S.; Parton, W.J.; Chen, Z.; Chen, D. Sensitivity of Inner Mongolia grasslands to climate change. *J. Biogeogr.* **1995**, *22*, 643–648. [CrossRef]
6. Tao, S.; Fang, J.; Zhao, X.; Zhao, S.; Shen, H.; Hu, H.; Tang, Z.; Wang, Z.; Guo, Q. Rapid loss of lakes on the Mongolian Plateau. *Proc. Natl. Acad. Sci. USA* **2015**, *112*, 2281–2286. [CrossRef]
7. Zhang, P.; Deng, Q.; Zhang, G.; Ma, J.; Gan, W.; Min, W.; Mao, F.; Wang, Q. Active tectonic blocks and strong earthquakes in the continent of China. *Sci. China Ser. D Earth Sci.* **2003**, *46*, 13–24. [CrossRef]
8. Zhang, G.M.; Ma, H.S.; Wang, H.; Wang, X.L. Boundaries between active-tectonic blocks and strong earthquakes in the China mainland. *Chin. J. Geophys. Chin. Ed.* **2005**, *48*, 602–610. [CrossRef]
9. Jie, B.; Xi, C.; Liao, Y.; Hui, F. Monitoring variations of inland lakes in the arid region of Central Asia. *Front. Earth Sci.* **2012**, *6*, 147–156. [CrossRef]
10. Chiew, F.H.S.; McMahon, T.A. Groundwater Recharge from Rainfall and Irrigation in the Campaspe River Basin. *Aust. J. Soil Res.* **1991**, *29*, 651–670. [CrossRef]
11. Jiansheng, C.; Bichen, J.I.; Zhen, L.I.U.; Zhiwei, Z.; Shiyin, Z. Isotopic and hydro-chemical evidence on the origin of groundwater through deep-circulation ways in Lake Daihai region, Inner Mongolia plateau. *J. Lake Sci.* **2013**, *25*, 521–530. [CrossRef]

12. Dong, C.; Wang, N.a.; Chen, J.; Chen, H.; Chen, L.; Li, Z.; Ma, N. New observational and experimental evidence for the recharge mechanism of the lake group in the Alxa Desert, north-central China. *J. Arid Environ.* **2016**, *124*, 48–61. [CrossRef]
13. Qun, H.; Jiahu, J. Analysis of Water Level Descent in Daihai Lake. *J. Lake Sci.* **1999**, *11*, 304–310. [CrossRef]
14. Wang, T.; Chen, J.S.; Xu, Y.; Zhan, L.C.; Huang, D.W. Isotopes and hydrochemistry of Daihai Lake recharging sources, Northern China. *J. Radioanal. Nucl. Chem.* **2017**, *312*, 615–629. [CrossRef]
15. Xiao, J.; Xu, Q.; Nakamura, T.; Yang, X.; Liang, W.; Inouchi, Y. Holocene vegetation variation in the Daihai Lake region of north-central China: A direct indication of the Asian monsoon climatic history. *Quat. Sci. Rev.* **2004**, *23*, 1669–1679. [CrossRef]
16. Jin, D.; Rui, G.; Yong, W.; Shihong, Z.; Peiyi, Y.; Zhenqing, C.; Zhili, Z. Magnetic Fabric Study of Late Holocene Sediments in Huangqihai Lake, Inner Mongolia and its Sedimentary Significance. *Acta Geol. Sin.* **2013**, *87*, 186–196. [CrossRef]
17. Xiaoqiang, L.; Jie, Z.; Ji, S.; Chengyu, W.; Hongli, Z.; Qianli, S. Vegetation history and climatic variations during the last 14 ka BP inferred from a pollen record at Daihai Lake, north-central China. *Rev. Palaeobot. Palynol.* **2004**, *132*, 195–205. [CrossRef]
18. Zhang, J. Timing and possible forcing mechanisms of Huangqihai Lake fluctuations in semi-arid northern China since the late Glacial. *Quat. Int.* **2012**, *279–280*, 559. [CrossRef]
19. Huazhang, L.; Qingsi, L.; Jiaxing, W. Study of evolution of Huangqihai and Daihai Lakes in holocene in inner mongolia plateau. *J. Lake Sci.* **1992**, *4*, 31–39. [CrossRef]
20. Peng, Y.; Xiao, J.; Nakamura, T.; Liu, B.; Inouchi, Y. Holocene East Asian monsoonal precipitation pattern revealed by grain-size distribution of core sediments of Daihai Lake in Inner Mongolia of north-central China. *Earth Planet. Sci. Lett.* **2005**, *233*, 467–479. [CrossRef]
21. Sun, Q.; Wang, S.; Zhou, J.; Chen, Z.; Shen, J.; Xie, X.; Wu, F.; Chen, P. Sediment geochemistry of Lake Daihai, north-central China: Implications for catchment weathering and climate change during the Holocene. *J. Paleolimnol.* **2010**, *43*, 75–87. [CrossRef]
22. Jin, Z.; Li, F.; Cao, J.; Wang, S.; Yu, J. Geochemistry of Daihai Lake sediments, Inner Mongolia, north China: Implications for provenance, sedimentary sorting, and catchment weathering. *Geomorphology* **2006**, *80*, 147–163. [CrossRef]
23. Xu, L.; Liu, Y.; Sun, Q.; Chen, Z.; Chen, J.; Cheng, P. Climate change and human occupations in the Lake Daihai basin, north-central China over the last 4500years: A geo-archeological perspective. *J. Asian Earth Sci.* **2017**, *138*, 367–377. [CrossRef]
24. Zhang, J.; Jia, Y.; Lai, Z.; Long, H.; Yang, L. Holocene evolution of Huangqihai Lake in semi-arid northern China based on sedimentology and luminescence dating. *Holocene* **2011**, *21*, 1261–1268. [CrossRef]
25. Frantz, D.; Röder, A.; Stellmes, M.; Hill, J. An Operational Radiometric Landsat Preprocessing Framework for Large-Area Time Series Applications. *IEEE Trans. Geosci. Remote Sens.* **2016**, *54*, 1–16. [CrossRef]
26. Jia, P.; Zhang, M.; Yu, W.; Shen, F.; Shen, Y. Convolutional neural network based classification for hyperspectral data. In Proceedings of the 2016 IEEE International Geoscience and Remote Sensing Symposium (IGARSS), Beijing, China, 10–15 July 2016; pp. 5075–5078. [CrossRef]
27. Zhao, H.; Shi, J.; Qi, X.; Wang, X.; Jia, J. Pyramid Scene Parsing Network. In Proceedings of the 2017 IEEE Conference on Computer Vision and Pattern Recognition (CVPR), Honolulu, HI, USA, 21–26 July 2017; pp. 6230–6239. [CrossRef]
28. Shelhamer, E.; Long, J.; Darrell, T. Fully Convolutional Networks for Semantic Segmentation. *IEEE Trans. Pattern Anal. Mach. Intell.* **2016**, *39*, 640–651. [CrossRef]
29. Zhang, T.; Tang, H. A Comprehensive Evaluation of Approaches for Built-Up Area Extraction from Landsat OLI Images Using Massive Samples. *Remote Sens.* **2018**, *11*. [CrossRef]
30. Chen, M.; Liu, J. Historical trends of wetland areas in the agriculture and pasture interlaced zone: A case study of the Huangqihai Lake Basin in northern China. *Ecol. Model.* **2015**, *318*, 168–176. [CrossRef]
31. Chen, S.; Chen, J.; Lin, G.; Zhang, W.; Miao, H.; Wei, L.; Huang, J.; Han, X.-G. Energy balance and partition in Inner Mongolia steppe ecosystems with different land use types. *Agric. For. Meteorol.* **2009**, *149*, 1800–1809. [CrossRef]
32. Dyer, J. A GIS-Based Water Balance Approach Using a LiDAR-Derived DEM Captures Fine-Scale Vegetation Patterns. *Remote Sens.* **2019**, *11*, 2385. [CrossRef]

33. Xiao, J.; Liu, Z.; Duan, A.; Liu, Z. Water Production Function during the Whole Growing Stage for Main Crops in China. *Chin. Agric. Sci. Bull.* **2008**, *3*, 430–434.
34. Penman, H. Natural Evaporation From Open Water, Bare Soil and Grass. *Proc. R. Soc. A Math. Phys. Eng. Sci.* **1948**, *193*, 120–145. [CrossRef]
35. Guoyan, P.; Zhu, O.; Qunying, L.; Qiang, Y.; Jishun, W. Water Consumption of Seven Forage Cultivars Under Different Climatic Conditions in the North China Plain. *J. Resour. Ecol.* **2011**, *2*, 74–82.
36. Jiang, L.; Yang, Y.; Shang, S. Evaluation on irrigation efficiency of irrigation district in arid region based on evapotranspiration estimated from Remote Sensing data. *Nongye Gongcheng Xuebao Trans. Chin. Soc. Agric. Eng.* **2013**, *29*, 95–101. [CrossRef]
37. Shunjun, H.; Changyan, T.; Yudong, S.; Bing, C.; Fang, W. Conversion Coefficient of Water Surface Evaporation in Tarim River Basin. *J. Desert Res.* **2005**, *5*, 649–651.
38. Dong, J.; Wang, Y.; Zhang, S.H.; Fu, X.M.; Yao, D.L. Grain size analysis of Holocene lacustrine sediments in the Huangqihai Lake of Inner Mongolia and its sedimentological significance. *Geol. Bull. China* **2014**, *33*, 1514–1522.
39. Fang, L.; Tao, S.; Zhu, J.; Liu, Y. Impacts of climate change and irrigation on lakes in arid northwest China. *J. Arid Environ.* **2018**, *154*, 34–39. [CrossRef]
40. Liang, K.; Yan, G. Application of Landsat Imagery to Investigate Lake Area Variations and Relict Gull Habitat in Hongjian Lake, Ordos Plateau, China. *Remote Sens.* **2017**, *9*, 1019. [CrossRef]
41. Liu, J.; Kang, S.; Gong, T.; Lu, A. Growth of a high-elevation large inland lake, associated with climate change and permafrost degradation in Tibet. *Hydrol. Earth Syst. Sci.* **2010**, *14*, 481–489. [CrossRef]
42. Yao, J.Q.; Chen, Y.N.; Zhao, Y.; Yu, X.J. Hydroclimatic changes of Lake Bosten in Northwest China during the last decades. *Sci. Rep.* **2018**, *8*, 9118. [CrossRef]
43. Gibson, J.J.; Prowse, T.D.; Peters, D.L. Partitioning impacts of climate and regulation on water level variability in Great Slave Lake. *J. Hydrol.* **2006**, *329*, 196–206. [CrossRef]
44. Chen, J.S.; Wang, Q.Q. A discussion of groundwater recharge sources in arid areas of North China. *Water Resour. Prot.* **2012**, *28*, 1–8. [CrossRef]
45. Ma, M.; Wang, X.; Veroustraete, F.; Dong, L. Change in area of Ebinur Lake during the 1998–2005 period. *Int. J. Remote Sens.* **2007**, *28*, 5523–5533. [CrossRef]
46. Zedler, J.B.; Kercher, S. Wetland resources: Status, trends, ecosystem services, and restorability. *Annu. Rev. Environ. Resour.* **2005**, *30*, 39–74. [CrossRef]

MDPI
St. Alban-Anlage 66
4052 Basel
Switzerland
Tel. +41 61 683 77 34
Fax +41 61 302 89 18
www.mdpi.com

Remote Sensing Editorial Office
E-mail: remotesensing@mdpi.com
www.mdpi.com/journal/remotesensing

www.ingramcontent.com/pod-product-compliance
Lightning Source LLC
Chambersburg PA
CBHW040756250926
51850CB00007BA/1012
* 9 7 8 3 0 3 9 2 8 8 0 7 6 *